SAGE was founded in 1965 by Sara Miller McCune to support the dissemination of usable knowledge by publishing innovative and high-quality research and teaching content. Today, we publish over 900 journals, including those of more than 400 learned societies, more than 800 new books per year, and a growing range of library products including archives, data, case studies, reports, and video. SAGE remains majority-owned by our founder, and after Sara's lifetime will become owned by a charitable trust that secures our continued independence.

Los Angeles | London | New Delhi | Singapore | Washington DC | Melbourne

Natural
Hazards
Management
in Asia

Natural Hazards Management in Asia

Indrajit Pal
Tuhin Ghosh

Los Angeles | London | New Delhi
Singapore | Washington DC | Melbourne

First published in 2018 by

SAGE Publications India Pvt Ltd
B1/I-1 Mohan Cooperative Industrial Area
Mathura Road, New Delhi 110 044, India
www.sagepub.in

SAGE Publications Inc
2455 Teller Road
Thousand Oaks, California 91320, USA

SAGE Publications Ltd
1 Oliver's Yard, 55 City Road
London EC1Y 1SP, United Kingdom

SAGE Publications Asia-Pacific Pte Ltd
3 Church Street
#10-04 Samsung Hub
Singapore 049483

Published by Vivek Mehra for SAGE Publications India Pvt Ltd, typeset in 10/12 pts Adobe Garamond by Zaza Eunice, Hosur, Tamil Nadu and printed at Chaman Enterprises, New Delhi.

Library of Congress Cataloging-in-Publication Data

Names: Pal, Indrajit, author. | Ghosh, Tuhin, co-author.
Title: Natural hazards management in Asia / Indrajit Pal and Tuhin Ghosh.
Description: New Delhi, India ; Thousand Oaks, California : SAGE
   Publications, 2018. | Includes bibliographical references and index.
Identifiers: LCCN 2017027756| ISBN 9789386602183 (hardback) |
   ISBN 9789386602206 (ePub 2.0) | ISBN 9789386602190 (ebook)
Subjects: LCSH: Natural disasters—Asia—Management. | Hazardous geographic
   environments—Asia. | Hazard mitigation—Asia. | Emergency management—Asia.
Classification: LCC GB5011.69 .P35 2017 | DDC 363.34095—dc23 LC record available at
https://lccn.loc.gov/2017027756

ISBN: 978-93-866-0218-3 (HB)

SAGE Team: Rajesh Dey, Vandana Gupta, Shobana Paul and Ritu Chopra

Dedicated to all the victims…

Thank you for choosing a SAGE product!
If you have any comment, observation or feedback,
I would like to personally hear from you.

Please write to me at **contactceo@sagepub.in**

Vivek Mehra, Managing Director and CEO, SAGE India.

Bulk Sales

SAGE India offers special discounts
for purchase of books in bulk.
We also make available special imprints
and excerpts from our books on demand.

For orders and enquiries, write to us at

Marketing Department
SAGE Publications India Pvt Ltd
B1/I-1, Mohan Cooperative Industrial Area
Mathura Road, Post Bag 7
New Delhi 110044, India

E-mail us at **marketing@sagepub.in**

Get to know more about SAGE

Be invited to SAGE events, get on our mailing list.
Write today to **marketing@sagepub.in**

This book is also available as an e-book.

Contents

List of Tables

List of Figures

Preface

We are quite familiar with the constant media barrage of natural hazards and their impacts, amplifying the threat of such disasters. The natural hazards can no longer be left solely to nature to determine its fate. Infrastructure, property, economic activities, recreation, and tourism along with all the environmental components are under constant threat. Natural disasters are manifestations of natural hazards and human interventions. Some of the natural hazards could end up causing a disaster or catastrophe due to the unplanned development and lack of risk governance mechanism at various levels. Combating natural hazards is like a war against nature, and at present a widely examined topic.

Hazards are inevitable, but disasters are not. By virtue of the geoclimatic conditions, most of the South Asian countries are exposed to multiple hazards, which sometimes lead to potential disasters and catastrophes. Enhancing disaster risk knowledge through the various levels of education is one of the major tasks to be performed for safer community. The disaster risk information at various levels through education will enhance the community capacity for effective and sustainable risk management. The resilient community-building process will also support us to achieve the seven targets and four priorities under the Sendai Framework for Disaster Risk Reduction (SFDRR 2015–30). This book on natural hazards management examines the key issues that encompass the wise policy formulation for minimizing the impacts of such natural hazards in an Asian context. This documentation provides multiple perspectives and decisions surrounding this emergent issue while also plotting a course for the future of natural hazards management.

The authors have attempted to characterize the probable hazards and the scenario of natural hazards management efforts in Asia, working from basic principles, with necessary scientific approach. The book also provides examples and real-time case studies that illustrate the magnitude of the impact of natural hazards, across the subcontinent. The authors have also discussed at length about the paradigm shift in disaster management with exposure to the use of modern technology, tools, and techniques. This book attempts to provide a holistic introductory outlook to a large cross-section of the academic communities, practitioners, and community-based organizations.

Acknowledgments

This book is the result of a huge amount of work, research, and dedication. This would not have been possible if we did not have the support of many individuals and organizations. Therefore, we would like to extend our sincere gratitude to all of them. We wish to express our sincere thanks to the Asian Institute of Technology, Thailand, and Jadavpur University, India, for providing us with all the necessary facilities and support for the research.

Putting this book together would not have been possible without the immense amount of behind-the-scenes and often unnoticed work done by the production team. Therefore, our sincere word of thanks also goes to the entire team of SAGE Publications for their continued support since the beginning of our publishing endeavors.

Nobody has been more important to us in the pursuit of this project than the members of our family. We take this opportunity to express sincere thanks to our family for the unceasing encouragement and support throughout this venture. We are grateful to the God for the good health and well-being that were necessary to complete this book.

Chapter 1

Introduction

1.1 Introduction

A disaster is a sudden catastrophe, causing extreme disruption of the functioning of a society and leading to widespread human, material, or environmental losses that transcend the ability of the affected society to cope with its own resources. Disasters have the characteristic to occur at random, spontaneously, and at speed. These acute events are either natural or human-influenced, surpass the tolerable magnitude within or beyond certain time limits, pose difficulty in adjustment, result in catastrophic losses of property and income, and paralyze life processes. These events exacerbate natural environmental processes to cause disasters to human society such as sudden tectonic movements leading to earthquakes and volcanic eruptions, continued dry conditions leading to prolonged droughts and also floods, atmospheric disturbances, collision of celestial bodies, etc. (Joshi 2008).

Disasters are common phenomena with the modernization of society. Technological advancement paved the way for development of huge infrastructure and permanent assets, which gradually detached man from nature, resulting in increased vulnerability of the human population. Natural disasters have been a comprehensive part of the advancement of human civilization, manifested in the traditional coping mechanisms, cultural practices, and even myths. However, it is a matter of great concern for increased frequency and intensity of disasters in last few decades, which claimed thousands of lives and caused massive infrastructure loss across the globe.

Classification of disasters is often done according to their origin, whether they are "natural" or "man-made" disasters. For example, disasters caused by floods, droughts, tidal waves, and earth tremors are generally considered "natural disasters." Disasters triggered by chemical, industrial, or technological accidents, environmental pollution, transport accidents, terrorism, war, and political unrest are classified as "human-made" or "human-induced" disasters as they pose prominent involvement of human action.

Modern and social understanding of disasters, however, views this variance as artificial since most disasters are caused from the action or inaction of people along with their social and economic structures. Disasters mostly originate as repercussion of people practicing ways that degrade their environment, growing or over-saturated urban centers, with a dynamic process in altering or perpetuating social and economic systems to fulfill their need. Communities and populations settled in areas that are susceptible to the impact of a raging river or the violent tremors of the earth are always exposed to high risk due to their poor socioeconomic condition. This is intensified by every aspect of nature subjected to seasonal, annual, and sudden fluctuations and also due to the unpredictable timing, frequency, and magnitude of occurrence of the disasters.

1.2 Definition

Almost every day, newspapers, radio, television channels, and social media highlight reports on disaster striking several parts of the world. Disasters are manifested as "Hazards aggravating vulnerable conditions and exceeding individuals' and communities' means to survive and thrive." But what is a disaster? A disaster (from Greek, meaning, "bad star") is a natural or man-made event that negatively affects life, property, livelihood, or economy, often resulting in permanent changes to human society, ecosystem, and environment.

The United Nations (UN) defined disasters as "A serious disruption of the functioning of a community or a society causing widespread human, material, economic and environmental losses which exceed the ability of the affected community/society to cope using its own resources" (UNDP).

Disaster has been defined in the Webster dictionary as "[g]rave occurrence having ruinous results," while the World Health Organisation (WHO) defines disaster as "[o]ccurrence that causes damage, economic destruction, loss of human life and deterioration of health and health services on a large scale, sufficient to warrant an extraordinary response from outside the affected community area."

The University of Delaware's Disaster Research Center differentiates disasters from emergencies and catastrophes.

- *Emergency*: An event that may be managed locally without the need of added response measures or changes to procedure.
- *Disaster*: An event that involves more groups who normally do not need to interact in order to manage emergencies requires the involved parties to relinquish the usual autonomy and freedom to special

response measures and organizations changes the usual performance measures, and requires closer operations between public and private organizations.

- *Catastrophe*: An event that destroys most of a community prevents local officials from performing their duties, causes most community functions to cease, and prevents adjacent communities from providing aid.

A disaster is an event or series of events that gives rise to casualties and damage or loss of properties, infrastructures, environment, essential services, or means of livelihood on such a scale that is beyond the normal capacity of the affected community to cope with. Disaster is also sometimes described as a "[c]atastrophic situation in which the normal pattern of life or eco-system has been disrupted and extra-ordinary emergency interventions are required to save and preserve lives and or the environment." The Disaster Management Act, 2005 (India), defines disaster as

a catastrophe, mishap, calamity or grave occurrence in any area, arising from natural or man-made causes, or by accident or negligence which results in substantial loss of life or human suffering or damage to, and destruction of, property, or damage to, or degradation of, environment, and is of such a nature or magnitude as to be beyond the coping capacity of the community of the affected area.

The UN defines disaster as "the occurrence of sudden or major misfortune which disrupts the basic fabric and normal functioning of the society or community."

A disaster results from the combination of hazard, vulnerability, and insufficient capacity to reduce the potential chances of risk. A disaster occurs when a hazard impacts on the vulnerable population causing damage, casualties, and disruption.

Any hazard—flood, earthquake, or cyclone—which is a triggering event along with greater vulnerability (inadequate access to resources, sick and old people, lack of awareness, etc.) leads to a disaster, causing massive loss of life and property. For example, an earthquake in an unsettled desert cannot be accepted as a disaster, no matter how strong the intensities produced. An earthquake is considered disastrous only when it adversely affects human lives, and their properties and activities. Thus, disaster occurs only when hazards combine with vulnerability. But it is also to be noted that with greater capacity of the individual/community and environment to cope with these disasters, the impact of a hazard reduces. Therefore, we need to understand the three major components, namely, hazard, vulnerability, and capacity, with suitable examples to have a basic concept of disaster management. Natural events in combination with various natural elements may create some major natural disasters; for example, clouds and heavy rain may create floods, and a combination of clouds, heavy rain, and mountains may create flash floods. Likewise, a combination of cloud, heavy rain, and high wind may create a flood, cyclone, typhoon, or hurricane. Earth shakes, submarine landslides, and volcanic eruptions below the ocean can create a tsunami, and less cloud, less rain, with high sunshine may produce drought, heat wave, and water scarcity. Tectonic events and plate movements can also produce the hazards such as earth shakes, earthquakes, and volcanic eruptions.

It is interesting to note that except in earthquake and volcanic eruptions rainfall (excess or shortages) and wind velocities play a very vital role in all other types of natural hazards.

1.3 Disaster Management Cycle

Disaster management aims to reduce, divert, or avoid the potential losses from hazards, assure prompt and appropriate assistance to victims of disaster, and achieve rapid and effective post-disaster recovery (Warfield 2008). The disaster management cycle portrays the ongoing process by which governments, businesses, and civil societies plan for and minimize the impact of disasters, respond during and immediately after a disaster, and intervene to recover the post-disaster situation. Appropriate actions at all points in the disaster management cycle lead to greater preparedness, effective warnings, and reduced vulnerability of disasters during the next repetition of the cycle. The complete disaster management cycle includes the shaping of public policies and plans that modify the causes of disasters and either minimize or mitigate their adverse effects on people, property, infrastructure, and environment.

Disaster risk management (DRM) includes the sum total of all activities, programs and measures taken before, during, and after a disaster with the purpose to avoid a disaster, reduce its impact, and effectively recover from its losses. The three key stages of activities that are taken up within DRM are as follows,

1. *Before a disaster (pre-disaster):* Pre-disaster activities are those that are carried out to reduce human and property losses caused by a

potential hazard. For example, conducting awareness campaigns, strengthening the existing weak structures, preparation of the disaster management plans at household- and community-level, etc. These risk-reduction measures taken under the pre-disaster stage are termed as mitigation and preparedness activities.

2. *During a disaster (disaster occurrence):* During a disaster, activities include initiatives to meet the needs and provisions of victims and minimize their sufferings. Activities taken under this stage are called emergency response activities.

3. *After a disaster (post-disaster):* Post-disaster activities include initiatives taken in response to a disaster to achieve early recovery and rehabilitation of affected communities after a disaster strikes. These are called response and recovery activities. The DRM cycle (DRMC) highlights the range of initiatives that are likely to occur during both the emergency response and recovery stages of a disaster. Some of these include both stages (such things as coordination and the provision of ongoing assistance), while other activities are unique to each stage (e.g., early warning and evacuation during emergency response, and reconstruction and economic and social recovery as part of recovery).

The DRMC also elaborates the role of the media, where there is a strong relationship with funding opportunities. The DRMC model is effective for relatively sudden-onset disasters, such as floods, earthquakes, bushfires, tsunamis, cyclones, etc., but is less reflective for slow-onset disasters, such as drought, where there is no obviously recognizable single event that triggers the movement into the emergency response stage.

Developmental factors play a key role in the mitigation and preparation of a community to effectively cope with a disaster. The mitigation and preparedness phases occur as disaster management improvements are made in anticipation of a disaster event. As a disaster occurs, disaster management actors, in particular humanitarian organizations, become involved in the immediate response and long-term recovery phases. The four disaster management phases illustrated here do not always, or even generally, occur in isolation or in this precise order. Often phases of the cycle overlap and the length of each phase greatly depends on the severity of the disaster.

- *Mitigation*: Minimizing the effects of a disaster.
  Examples: building codes and zoning; vulnerability analyses; public education.

- *Preparedness:* Planning how to respond.
  Examples: preparedness plans; emergency exercises/training; warning systems.
- *Response:* Efforts to minimize the hazards created by a disaster.
  Examples: search and rescue (SAR); emergency relief.
- *Recovery:* Returning the community to normal.
  Examples: temporary housing; grants; medical care.

Disaster management cycle refers to the steps that include pre-disaster, during disaster, and post-disaster elements (Figure 1.1).

Figure 1.1:
Disaster Management Continuum

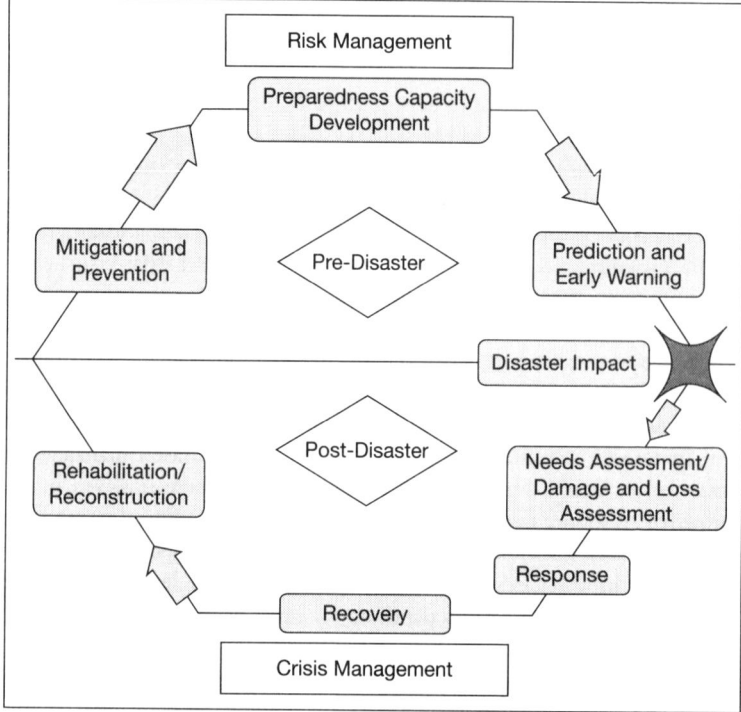

Source: Authors.

1.3.1 Needs Assessment/Damage and Loss Assessment: Impact?

The impacts of all major disasters are basically threefold and they affect all the primary resources such as life and gender, water, land, and biomass; cause damage to forest and greeneries, livestock and wildlife, population and property; and negatively impact the economy, education, infrastructure, health and sanitation, and employment.

Major factors that contribute to the disastrous consequences of a natural hazards are human vulnerability resulting from poverty and social inequality, environmental degradation resulting from improper land and water use, rapid population growth—especially among the poor and uneducated—and lack of preparedness.

1.3.2 Response

The goal of emergency response is to provide instant assistance to upgrade health, maintain life, and support the morale of the affected population. Such assistance may extent from providing specific but limited aid, such as aiding refugees with transport, temporary shelter, and food to establishing semi-permanent settlements in camps and other locations. Likewise, it may involve initial restorations to damaged infrastructure. The focus in the response phase is to meet the basic needs of the people until more stable and sustainable solutions can be found. Humanitarian organizations usually have a strong presence in this phase of the disaster management cycle.

The response of emergency management includes actions focused on limiting injuries, loss of life, and damage to property and the environment that are taken in advance of, during, and immediately after a hazard event. Response processes are undertaken as soon as it becomes apparent that a hazard event is imminent, and it persists until the emergency situation is called off. Response is the most complex of the four functions of emergency management, since it is conducted during periods of very high stress, in a highly time-constrained environment, and with limited information. During response, wavering confidence and unnecessary delay trigger tragedy and destruction.

Disaster response is centered upon information and coordination. The participants, victims, and community needs; the timing and order of events;

and the actions and processes employed are unique to each event. This section will approach the various functions and processes associated with response in a general sense, as they would apply to all hazards and all nations.

1.3.3 Recovery

As the emergency is under control, the affected population is capable of venturing a large number of activities aimed at reinstalling their lives and infrastructural supports. There is no specific point at which immediate relief changes into recovery leading toward long-term sustainable development. There will be many opportunities during the recovery period to enhance prevention and increase preparedness, thus reducing vulnerability. Preferably, there would be a smooth transition from recovery to ongoing development.

Activities of the recovery stage continue until all systems return to normalcy or betterment. Both short- and long-term recovery measures include reviving vital life-support systems to minimum operating standards, temporary housing, public information, health and safety education, reconstruction; counseling programs, and economic impact studies. Information resources and services include data collection related to rebuilding and the documentation of lessons learned.

Recovery constitutes the last step of post-disaster actions, such as rebuilding livelihood, infrastructure, or the retrofitting of damaged structures. Disaster recovery is the emergency management function by which countries, communities, families, and individuals reinstall, reconstruct, or regain what has been lost as result of a disaster and, preferably, reduce the risk of a similar catastrophe in the future. In an integrated emergency management system, that includes pre-disaster planning, mitigation, and preparedness actions, recovery activities may begin as early as during the planning processes and actions, long before a disaster occurs. Once the disaster strikes, planned and unplanned recovery actions are implemented and may continue for weeks, months, or even years.

1.3.4 Rehabilitation and Reconstruction

Rehabilitation includes the provision of temporary public utilities and housing as interim measures to assist longer term recovery. Reconstruction attempts to return communities to improved pre-disaster functioning. It includes the replacement of buildings, and infrastructure and lifeline facilities so that long-term development prospects are enhanced rather than reproducing the same conditions that made an area or population vulnerable in the first place.

- Reconstruction includes long-term measures, for example, houses, livelihoods, infrastructures, etc.
- It is capital intensive and needs careful planning and community participation.
- It also provides a good opportunity to plan developmental activities that are more robust and resilient.

Chapter 8 will discuss rehabilitation and reconstruction in more detail.

1.3.5 Mitigation

Mitigation actions minimize the probability of disaster occurrence or reduce the effects of unavoidable disasters. Mitigation measures incorporate building codes, vulnerability analyses updates, zoning and land-use management, building use regulations and safety codes, preventive healthcare, and public education.

Mitigation will hinge on the inclusion of appropriate measures in national and regional development planning. Its effectiveness will also depend on the accessibility of information on hazards, emergency risks, and the countermeasures to be taken. The mitigation phase, along with the whole disaster management cycle, incorporates the shaping of public policies and plans, which modify the causes of disasters or mitigate their effects on people, property, and infrastructure.

Short-term measures

Short-term measures are the ways to mitigate the impacts of inevitable disasters such as immediate relief, reducing the response time to turn away any losses, and giving the basic needs of the vulnerable and affected people. Those who have lost their properties may be supplied with minimum essential items or even smaller loans at a concessional rate for long-/short-term loans, etc.

Long-term measures

The long-term measures could be planned according to the disaster. In case of frequent fire accidents, fire service stations and periodic checking, training the people, etc. In case of flood, whether big dams and water storage structures could be completed. Environmental degradation could be addressed by watershed management principles. Biomass production will be another long-term strategy.

Mitigation plans

Structural: The disasters could be alleviated by putting structural provisions such as dams, embankments stone walls (sea erosion) dykes, water storage pumps, etc. These are costly and needs one-time heavy investment. For example, barns, etc.

Non-structural: The nonstructural methods of handling disasters have gradually unfolded with the traditional knowledge and strength of the community, which have been time-tested, economic, cost-effective, and user-friendly technology. The nonstructural mitigation plans should be made popular and promoted as people themselves could do it. Financial incentives will motivate people to embrace the method. For example, coastal mangrove plantations, casuarina and cashew nut trees, etc.

1.3.6 Preparedness/Capacity Development

The primary objective of the emergency preparedness is to attain a reasonable level of readiness to respond to any emergency situations through government and nongovernment programs and projects that strengthen the technical and managerial capacity of governments, organizations, and communities. These actions can be described as developing efficient response mechanisms and procedures, rehearsals, developing long-term and short-term strategies, public education, and building early warning systems. Preparedness can also take the form of ensuring that strategic reserves of food, equipment, water, medicines, and other essentials are maintained in cases of national or local catastrophes.

During the preparedness phase, governments, organizations, and individuals develop plans to save lives, reduce disaster damage, and enhance disaster response operations. Preparedness measures include preparedness plans, emergency exercises/training, warning systems, emergency communications systems, evacuation plans and training, resource inventories, emergency personnel/contact lists, mutual aid agreements, and public information/education.

Preparedness involves providing the victims of disaster or people who may help those impacted by disaster with the tools to increase their chance of survival and to reduce their financial and other losses. The primary goals of disaster preparedness are to understand the consequences of disasters and equipped with the right tools to do the task efficiently. This tough process may take years before attaining competent levels, and maintaining such levels is an ongoing effort. Preparedness minimizes hazards' adverse effects through effective precautionary measures that ensure a timely, appropriate, and efficient organization and the delivery of response and relief action. Reciprocating to any disaster, especially a catastrophic event, is guaranteed to be idiosyncratic, complex, and confusing. In recent years, disaster managers have confirmed more effective ways to upgrade public knowledge of disaster preparedness and response actions and to get the public to act upon that knowledge. Until recently, it was thought that the public was not capable of acting rationally in an emergency situation. Response officials feared that the public would panic or would be unable to use preparedness information effectively. However, studies of actual post-disaster scenarios found that the public acts rationally and effectively, even when stressed. These studies highlight the need for governments and other agencies to help the public prepare for such disasters.

1.3.7 Prediction and Early Warning System

A warning is a statement that a high probability of a hazardous event will occur that is based on a prediction or forecast. When a warning is issued, it should be taken as a statement that "normal routines of life should be altered to deal with the danger imposed by the imminent event."

The effectiveness of a warning depends on

- the timeliness of the warning;
- effective communications and public information systems to inform the public of the imminent danger; and
- the credibility of the sources from which the warning came.

If warnings are issued too late, or if there is no means of circulating the information, then there will not be enough time or responsiveness to the warning. If warnings are issued incautiously without reliable data or sources, then they will likely be disregarded. Thus, the people responsible for taking action in the event of a potential disaster will not respond.

1.3.8 Sustainable Development

Developmental factors contribute to all facets of the disaster management cycle. One of the primary aims of disaster management, and its strongest links with development, is the venturing of sustainable livelihoods and their protection and recovery during disasters and emergencies. People have a greater potential to deal with disasters, and their recovery is more prompt and long lasting if the sustainable perspectives goals are achieved. In a development-oriented disaster management approach, the objectives are to reduce hazards, prevent disasters, and prepare for emergencies. Hence, developmental factors are strongly represented in the mitigation and pre-paredness phases of the disaster management cycle. Inappropriate developmental processes can lead to increased vulnerability to disasters and loss of preparedness for emergency situations.

1.3.9 Challenges in the Lifecycle of a Disaster

It is presumed that being prepared in the pre-disaster phase will shepherd in better response operations during and after the disaster. In fact, the UN estimates that for every dollar spent in prepositioning, seven dollars can be saved in disaster response. However, the particular strategies will differ based on the type of the disaster, since the operational difficulty of the entire humanitarian assistance is influenced by the nature of a disaster.

The classification of disasters based on man-made versus natural disasters does not offer substantial inference for research in humanitarian operations. However, a classification based on time and location can be very helpful in assessing the operational difficulty (Apte 2009). As seen in Figure 1.2, disasters can be categorized in terms of location as localized versus dispersed and in terms of time as slow-onset versus sudden-onset. Some examples of recent disasters are provided in the resulting 2×2 matrix in the figure. Effectiveness and efficiency of transportation and distribution of critical supplies and services suffer if the disasters are dispersed and sudden-onset, such as the 2004 tsunami in the Indian Ocean, which affected many

Figure 1.2:
Classification of Disasters

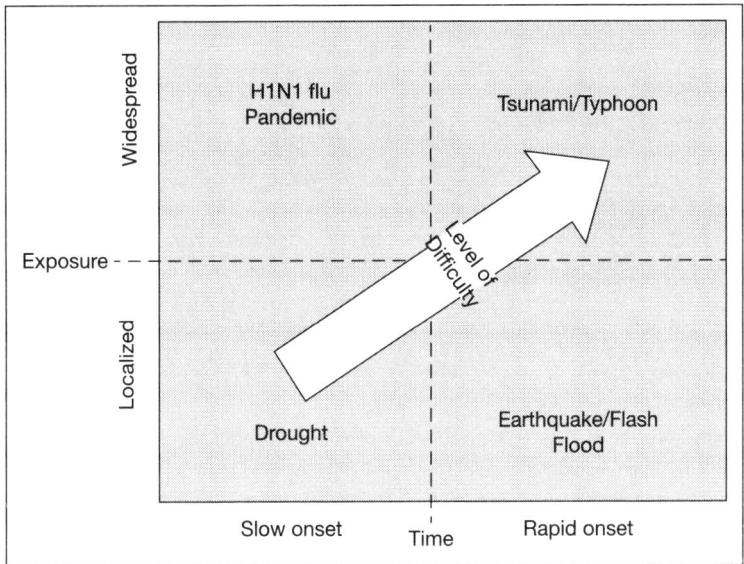

Source: Adapted from Apte (2009).

countries. On one hand, we have localized and slow-onset disasters (such as the 2011 floods of Mississippi), and on the other hand, there are dispersed and sudden-onset disasters. In between are the rest, localized and sudden-onset (such as the 2010 earthquake in Haiti), or dispersed and slow-onset (such as the 2009 H1N1 flu pandemic).

The other concept that is particularly useful for conducting research in humanitarian operations is the lifestyle of a disaster (Apte 2009). The lifecycle of a disaster can be divided into three phases (Figure 1.3): preparedness in the pre-disaster phase, response immediately after the disaster strikes, and recovery in the post-disaster phase.

Being prepared is the first stride in the right direction for alleviating the impact of a disaster or conflict. Therefore, readiness is rated high in any military operation where uncertainty is the name of the game. Some of the principal challenges in being prepared are planning for and establishing sufficient capacity and resources, establishing critical infrastructure, and investing in technology that will enable the flow of information. Military organizations routinely encounter many such challenges, and therefore they have policies and procedures in place that have been tested through the

Figure 1.3:
Lifecycle of a Disaster

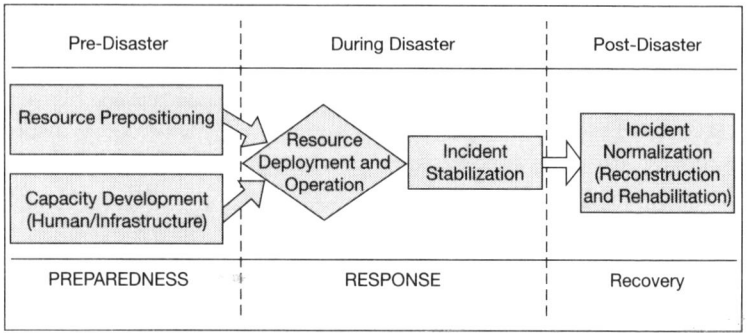

Source: Authors.

events of war or other disasters. Commercial supply chains have also been dealing with similar challenges. Some of the effective policies utilized in the private-sector enterprises such as agility, adaptability, and alignment of the global supply chains (van Wassenhove 2006) can help in meeting the challenges of the response operations in a disaster. In the pre-disaster phase, critical issues such as location of emergency facilities and allocation of resources can often be dealt with by using classic analytical models such as facility location and capacity constrained network-optimization models that have proven to be useful in military and private sectors. However, certain challenges associated with needs assessment and information management require innovative methodologies due to their unique nature and place in a humanitarian response supply chain. Right through the response phase, managing relations and developing trust to facilitate collaboration in the supply chain is the key to reach smooth flow of relief to the affected population. A commonly observed phenomenon is that the humanitarian operations would not succeed, even if the infrastructure exists, if the host community is not willing to collaborate.

The collaboration may be within the local community or among the stakeholders, but it can be improved by building confidence. Confidence may be harder to achieve, especially in a response supply chain of transient suppliers and volunteers. The existing infrastructure may comprise of the actual roads and utility networks, or the distribution and transportation processes of the supply chain. Infrastructure can also be in the form of education of the community and awareness within the community toward disaster, leading to the swift formation of confidence and collaboration.

The post-disaster phase mostly involves recovery actions. After the recovery, the condition of the affected community may or may not attain or exceed the initial condition of the community before the disaster. One of the lead points in the recovery phase is obtaining the knowledge, maintaining it, and learning from the lessons. The information gathered in the post-recovery phase enables preparedness for future disasters. In general, information and knowledge management facilitate all phases of the disaster.

1.3.10 Response and the Role of Scientists, Public Officials, and Average Citizens to Assessments, Predictions, and Warnings

Everyone has a responsibility to understand the consequences of a natural hazard and respond to the assessments, predictions, and warnings. Hence, one of the most important features of disaster management and planning is education. Since everyone is not expected to completely understand everything about a potential natural disaster, one of the most important links between all is the effective communication between various groups of people.

Responsibilities of Scientists and Engineers

Hazard assessment: Scientists have the greatest ability to determine where natural hazards exist, and the effects of such hazards when an event occurs.

Prediction: Scientists and practitioners have access to the monitoring of processes that enable these predictions. They should be able to communicate probabilities to appropriate public officials for dissemination to the general public.

Reduction of risk: Scientists and engineers should make information known to public officials about ways to reduce vulnerability and risk, by suggesting zoning regulations and building codes to public officials.

Early warning: Scientists with access to monitoring and hazard information should help develop early warning systems to effectively communicate such warnings to the public officials responsible for communicating the warning to the general public.

Communication: Scientists need to be able to present the information available in a form that is understandable to all concerned.

Responsibilities of Public Officials

Risk assessment: Public officials need to understand hazard assessments and develop risk assessments to decide where and how resources are to be expended to reduce risk.

Planning and code enforcement: Public officials need to work with scientists and engineers to help minimize vulnerability by making planning decisions (zoning laws) and by building codes that help reduce risk and vulnerability.

Early warning: Public officials have the primary responsibility to inform the public about imminent dangers based on predictions and warnings issued by scientific community.

Response: Public officials have the primary responsibility of maintaining an infrastructure that can deal with the emergencies created by a natural disaster. They need to develop plans for evacuation, emergency response, rescue, and recovery.

Communication: Public officials must be able to communicate effectively with scientific community and the general public to disseminate information.

You (Citizens)

Understanding of hazards: The general public needs to be aware of the effects of natural hazards on their communities, to have some understanding of what might occur in the event of a disaster.

Understanding of early warning systems: The general public must be informed about what their response should be when a warning is issued.

Communication: At least in the developed countries, people can communicate with the public officials to acquire necessary information and effectively carry out their responsibilities for hazard and risk reduction.

1.4 Natural Disaster Global and Asian Perspectives

1.4.1 The Global Scenario on Natural Hazards

Worldwide, from 1994 to 2013, floods were the most frequent natural disasters: 43 percent of the hazards occurring during the period were floods and

2.5 billion people were affected. During the same period, storms/cyclones were the second-most frequent events which killed 244,000 people, and the damage caused was estimated at US$936 billion. Earthquakes and tsunamis killed more people than all other types of disasters put together, with 750,000 deaths. Tsunamis are the deadliest hazard in nature, with 79 deaths per thousand persons affected, against four deaths per thousand affected persons in case of earth movements only. Droughts were only 5 percent of the disasters, but affected over a billion people. Number of people dying due to droughts are not on the same scale as other disasters but the resultant impacts of droughts are evident through malnutrition and other after effects, which linger for a long time, on food production, health and availability of potable water. Disease follows droughts and takes a heavy toll (CRED 2016).

1.4.2 The International Strategy for Disaster Reduction

There has been considerable concern over natural disasters at a global level. Through the scientific interventions and technological advancement has made significant progress, the loss of lives and property due to disasters has not decreased. The economic losses and human toll increased manifold with the increased frequency and fury of disasters. With this background, the UN General Assembly, in 1989, declared 1990–2000 as the International Decade for Natural Disaster Deduction with the objective to reduce loss of lives and property and restrict socioeconomic damage through concerted international action, especially in developing countries.

The partners in the International Strategy for Disaster Reduction (ISDR), in particular the Inter-Agency Task Force on Disaster Reduction and its members, in collaboration with relevant national, regional, international, and UN bodies and supported by the interagency secretariat for the ISDR, are requested to assist in implementing this framework for action as follows, subject to the decisions taken upon completion of the review process of the current mechanism and institutional arrangements:

1. Develop a matrix of roles and initiatives in support of the follow-up to this framework for action involving individual members of the task force and other international partners.

2. Facilitate the coordination of effective and integrated action within the organizations of the UN system and among other relevant

international and regional entities, in accordance with their respective mandates, to support the implementation of this framework for action, identify gaps in implementation, and facilitate consultative processes to develop guidelines and policy tools for each priority area, with relevant national, regional, and international expertise.

3. Consult with relevant UN agencies and organizations, regional and multilateral organizations, and technical and scientific institutions as well as interested States and civil societies, with the view to developing generic, realistic, and measurable indicators, keeping in mind the available resources of individual States. These indicators could assist the States to assess their progress in the implementation of the framework of action. The indicators should be in conformity with the internationally agreed development goals, including those contained in the Millennium Declaration.

Once that first stage has been completed, states are encouraged to develop or refine indicators at the national level reflecting their individual disaster risk reduction (DRR) priorities, drawing upon the generic indicators.

4. Ensure support to national platforms for disaster reduction, including through the clear articulation of their role and value-added as well as regional coordination to support the different advocacy and policy needs and priorities set out in this framework for action, through coordinated regional facilities for disaster reduction and building on regional programs and outreach advisors from relevant partners.

5. Coordinate with the secretariat of the Commission on Sustainable Development to ensure that relevant partnerships contributing to implementation of the framework for action are registered in its sustainable development partnership database.

6. Stimulate the exchange, compilation, analysis, summary, and dissemination of best practices, lessons learned, and available technologies and programs to support DRR in its capacity as an international information clearinghouse; maintain a global information platform on DRR and a web-based register "portfolio" of disaster risk-reduction programs and initiatives implemented by States and through regional and international partnerships.

7. Prepare periodic reviews on the progress toward achieving the objectives and priorities of this framework for action, within the context of the process of integrated and coordinated follow-up and implementation of UN conferences and summits as mandated by

the General Assembly, and provide reports and summaries to the Assembly and other UN bodies, as requested or as appropriate, based on information from national platforms, regional and international organizations, and other stakeholders, including on the follow-up to the implementation of the recommendations from the Second International Conference on Early Warning (2003).

1.4.3 The Yokohama Strategy: Lessons Learned and Gaps Identified

The Yokohama Strategy for a Safer World: Guidelines for Natural Disaster Prevention, Preparedness and Mitigation and its Plan of Action (Yokohama Strategy), adopted in 1994, provides landmark guidance on reducing disaster risk and the impacts of disasters.

The review of progress made in applying the Yokohama Strategy distinguishes major challenges for the coming years in ensuring more systematic action to deal with disaster risks in the context of sustainable development, and in building resilience through enhanced national and local capabilities to manage and minimize risk.

The review stresses the importance of DRR being underpinned by a more proactive approach to informing, motivating, and involving people in all aspects of DRR in their own local communities. It also highlights the scarcity of resources assigned specifically from development budgets for the realization of risk-reduction objectives, either at the national level or the regional level or through international cooperation and financial mechanisms, while taking into account the significant potential to better exploit existing resources and establish practices for more effective DRR.

Specific gaps and challenges are identified in the following five main areas:

1. governance: organizational, legal, and policy frameworks;
2. risk identification, assessment, monitoring, and early warning;
3. knowledge management and education;
4. reducing underlying risk factors; and
5. preparedness for effective response and recovery.

These are the key areas for developing a relevant framework for action for 2005–15.

1.4.4 The Hyogo Framework for Action 2005–15

The *Hyogo Framework for Action 2005–15: Building the Resilience of Nations and Communities to Disasters*, endorsed by the UN General Assembly, aims for substantial reduction of disaster losses, in lives and in the social, economic, and environmental assets of communities and countries. What has become clear over the implementation period of the Hyogo Framework of Action (HFA) is that DRR is a cross-cutting issue that requires a long-term planning perspective, mainstreaming and integration across sectors, and a change in mindset from response, to preparation and prevention.

Since the adoption of the HFA, many global, regional, national, and local efforts have dealt with DRR more systematically; however, much remains to be done. The UN General Assembly has called for the implementation of HFA, reconfirmed the multistakeholder ISDR system and the Global Platform for DRR to support and promote it. The General Assembly has encouraged member states to establish multisectoral national platforms to coordinate DRR in countries. Many regional bodies have formulated strategies at a regional scale for DRR in line with the HFA, in the Andean region, Central America, the Caribbean, Asia, Pacific, Africa, and Europe. More than 100 governments have designated official focal points for the follow-up and the implementation of the HFA (March 2007). Some have taken actions to mobilize political commitment and establish centers to promote regional cooperation in DRR.

The HFA is the key instrument for implementing DRR, adopted by the member states of the UN. Its overarching goal is to build resilience of nations and communities to disasters, by achieving substantive reduction of disaster losses by 2015—in lives, and in the social, economic, and environmental assets of communities and countries. The HFA offers five areas of priorities for action, guiding principles and practical means for achieving disaster resilience for vulnerable communities in the context of sustainable development.

1. **Priority 1: Make DRR a priority**

   Ensure that DRR is a national and a local priority with a strong institutional basis for implementation.

   Strong national and local commitment is required to save lives and livelihoods threatened by natural hazards. Natural hazards must be taken into account in public and private-sector decision-making in the same way that environmental and social impact assessments are currently required. Countries must, therefore, develop or modify policies, laws, and organizational arrangements, as well as plans, programs, and projects, to integrate DRR. They must also allocate sufficient resources to support and maintain them. This includes

- Creating effective, multisector national platforms to provide policy guidance and to coordinate activities;
- Integrating DRR into development policies and planning, such as poverty reduction strategies; and
- Ensuring community participation, so that local needs are met.

2. **Priority 2: Know the risks and take action**

    Identify, assess, and monitor disaster risks, and enhance early warning.

    To reduce their vulnerability to natural hazards, countries and communities must know the risks that they face and take actions based on that knowledge. Understanding risk requires investment in scientific, technical, and institutional capabilities to observe, record, research, analyze, forecast, model, and map natural hazards. Tools need to be developed and disseminated; statistical information about disaster events, risk maps, disaster vulnerability, and risk indicators are essential.

    Most importantly, countries need to use this knowledge to develop effective early warning systems, appropriately adapted to the unique circumstances of the people at risk. Early warning is widely accepted as a crucial component of DRR. When effective early warning systems provide information about a hazard to a vulnerable population, and plans are in place to take action, thousands of lives can be saved.

3. **Priority 3: Build understanding and awareness**

    Use knowledge, innovation, and education to build a culture of safety and resilience at all levels.

    Disasters can be reduced substantially if people are well informed about measures they can take to reduce vulnerability and if they are motivated to act. Key activities to increase awareness of disaster prevention include:

- Providing relevant information on disaster risks and means of protection, especially for citizens in high-risk areas;
- Strengthening networks and promoting dialogue and cooperation among disaster experts, technical and scientific specialists, planners and other stakeholders;
- Including DRR subject matter in formal, non-formal, and informal education and training activities;

- Developing or strengthening community-based DRM programs; and
- Working with the media in DRR awareness activities.

4. **Priority 4: reduce risk**
Reduce the underlying risk factors.
Vulnerability to natural hazards is increased in many ways, for example,

   - locating communities in hazard-prone areas, such as flood plains;
   - destroying forests and wetlands, thereby harming the capacity of the environment to withstand hazards;
   - building public facilities and housing that are unable to withstand the impacts of hazards; and
   - not having social and financial safety mechanisms in place.

   Countries can build resilience to disasters by investing in simple, well-known measures to reduce risk and vulnerability. Disasters can be reduced by applying relevant building standards to protect critical infrastructure, such as schools, hospitals, and homes. Vulnerable buildings can be retrofitted to a higher degree of safety. Protecting precious ecosystems, such as coral reefs and mangrove forests, allow them to act as natural storm barriers. Effective insurance and microfinance initiatives can help to transfer risks and provide additional resources.

5. **Priority 5: Be Prepared and Ready to Act**
Strengthen disaster preparedness for effective response at all levels.

   Being prepared, including conducting risk assessments, before investing in development at all levels of society will enable people to become more resilient to natural hazards. Preparedness involves many types of activities, including

   - the development and regular testing of contingency plans;
   - the establishment of emergency funds to support preparedness, response, and recovery activities;
   - the development of coordinated regional approaches for effective disaster response; and
   - continuous dialogue between response agencies, planners and policy-makers, and development organizations.

   Regular disaster preparedness exercises, including evacuation drills, also are key to ensuring rapid and effective disaster response.

Effective preparedness plans and organization also help to cope with the many small- and medium-sized disasters that repeatedly occur in so many communities. Natural hazards cannot be prevented, but it is possible to reduce their impacts by reducing the vulnerability of people and their livelihoods.

1.4.5 Sendai Framework for Disaster Risk Reduction (2015–30)

The Sendai Framework was adopted in the third UN World Conference on Disaster Risk Reduction (WCDRR) in Sendai in 2015. The document puts forward a framework to deal with Disaster Risk Reduction Management (DRRM) within the context of Sustainable Development. This framework's timeframe is set between 2015 and 2030, and works in synergy with the established sustainable development goals (SDGs). In many ways, it was made to build upon the HFA that was adopted between 2005 and 2015 and develops further on the goals set in that document.

The document incorporates several sectors and actors to create a holistic view of issues regarding climate and disaster risk. It also incorporates local, regional, and global goals for all the respective areas and goals which it sets out.

The Sendai Framework is centered around four distinct priorities:

1. **Priority 1: Understanding disaster risk**

   • Covering the assessment of risk in all its dimensions of vulnerability, capacity, exposure of persons and assets, hazard characteristics, and the environment are discussed.

     o This discussion is then used for outlining pre-disaster risk assessment, prevention, and mitigation and for the development and implementation of appropriate preparedness and effective response to disasters.

2. **Priority 2: Strengthening disaster risk governance to manage disaster risk**

   • The priority emphasizes the importance of an effective and efficient management of disaster risk. Here, a clear vision, plans, competence, guidance, and coordination within and across sectors, as well as participation of relevant stakeholders are created. By doing so, this priority wants to strengthen disaster risk governance for prevention, mitigation, preparedness, response, recovery, and rehabilitation.

  o   Collaboration and partnership across mechanisms and institutions are subjects which are key in making this a reality and are developed in this part of the framework.

3. **Priority 3: Investing in DRR for resilience**

   •   The importance of investing, both with public and private measures, into disaster risk and resilience have great consequences for how well future disasters are resisted. Investments can be drivers of innovation, growth, and job creation as well as protecting previously existing sources of income.

     o   The ways to make this a reality and in turn create stability and better coping and rehabilitation mechanisms in the event of future disasters are discussed in this priority.

4. **Priority 4: Enhancing disaster preparedness for effective response and to "build back better" in recovery, rehabilitation, and reconstruction**

   •   The need to further strengthen disaster preparedness, anticipate events, and integrate DRR has come due to an increment in disaster risk over time. Disasters have shown that the recovery, rehabilitation, and reconstruction phase can contribute not only to rehabilitate but to actually "Build Back Better," making communities more prepared if disasters were to strike again. In this part of the framework, this is discussed further.

The Sendai Framework establishes links between intergovernmental negotiations; the post-2015 development agenda; financing for development, climate change, and DRR; and the calls for enhanced coherence across nations, institutions, and policy tools used.

1.4.6 Paris Agreement

The Paris agreement is an agreement created in the framework of the UN Framework Convention on Climate Change (UNFCCC). According to the UNFCCC briefing to the COP-21 agreement, the new climate changes will have significant impact on development in almost all parts of the world. The Sendai Framework is a comprehensive document showcasing climate change and climate action, and providing measures, guiding principles, and means of implementation.

The briefing has called for greater integration of the two documents, where the Sendai Framework could be partially viewed as an action plan for the implementation of the Paris Agreement that formed after COP-21. Understanding disaster risk requires knowledge of climate change scenarios. It also identifies actions to maintain and strengthen in situ- and remotely sensed earth and climate observations and investments in effective early warning mechanisms. The Sendai Framework gives better disaster preparedness and response. There is also a focus on recovery and reconstruction in the Paris Agreement, which could be integrated into it. The Paris Agreement sets out to create further guidelines and help enforce the UNFCCC. This document sets its focus on the national level, and urges nations to implement the framework. As stated in Article 2 of the agreement, the primary goal is to not raise the global average temperature by more than 2°C while also preparing for the disasters and other consequences that might occur as a consequence of climate change. This makes the document a good supplement in assessing both educational and practical efforts in DRR.

The Paris Agreement strongly emphasizes the disparity between developed and developing countries in implementing the goals set out, and it implores countries with better conditions to assist developing countries in achieving the goals set out by the agreement. The agreement also encourages that countries take action beyond the goals set by the agreement that are progressively ambitious.

1.5 Asian Scenario on Natural Hazards

The Asia-Pacific region, by virtue of its geo-climatic location, is the most disaster-prone region in the world. Nearly two million people across the region were killed between 1970 and 2011, which represents 75 percent of the global disaster fatalities (UN ESCAP and UNISDR 2012). The Asia-Pacific region is more vulnerable to hydro-meteorological events, which occur more frequently, affect more people, and cause greater economic losses compared to other natural hazards. The frequency of occurrence of the natural disasters are also increased manifold due to the climate anomalies and extreme weather events. More than a billion people have been exposed to hydro-meteorological events since 2000, and this number is expected to rise due to the increased frequency and intensity of extreme weather events (UN ESCAP and UNISDR 2012).

In 2014, the number of hydro-meteorological or climate-induced disasters was equivalent to the annual 2004–13 average. Though the numbers of geophysical and hydrological disasters (17 and 65) were the third lowest and the lowest since 2004 respectively, on the other hand, the number of meteorological disasters (57) was the highest since 2004 and showed a 21 percent increase when compared to its annual 2004–13 average (Guha-Sapir, Hoyois, and Below 2015). Many countries in the Asia-Pacific region achieved significant progress in reducing disaster risk at local, national, regional, and global levels through the adaptation of HFA 2005–15. The Sendai Framework for DRR 2015–30 also continued the efforts through its holistic and more development-oriented priorities. Reducing disaster risk is a cost-effective investment in preventing future losses; hence, it contributes to sustainable development. The countries in the Asia-Pacific region have enhanced their capacities in DRM through the international mechanisms for strategic advice, coordination, and partnership development for DRR. The formation of national and local platforms for DRR and a designated national focal point for implementing the Sendai Framework for DRR 2015–30 strengthen government coordination forums composed of relevant stakeholders at national and local levels.

According to the *Annual Disaster Statistical Review 2012*, in the last 10 years, the five countries most hit by natural disasters were China, the United States, the Philippines, India, and Indonesia. In 2012 alone, Asia accounted for nearly 65 percent of global disaster victims, with hydrological events such as floods, storm surge, and landslides accounting for 75 percent of the disasters in Asia during that year (Guha-Sapir et al. 2012). In 2014, the number of hydro-meteorological or climate-induced disasters was equivalent to the annual 2004–13 average. Though the numbers of geophysical and hydrological disasters (17 and 65) were, the third lowest and the lowest since 2004 respectively and showed a decrease of 19.0 percent and 21.7 percent, compared to their annual 2004–13 average, on the other hand, the number of meteorological disasters (57) was the highest since 2004 and showed a 21 percent increase when compared to its annual 2004–13 average. The number of victims in Asia in 2014 (97.8 million) was far below the 2004–13 annual average (160.7 million), and a decrease was observed for all disaster types, except climatological disasters of which the number of victims (31.7 million) increased in 2014 by almost 20 percent, but 87 percent of these victims suffered from one drought in China (Guha-Sapir, Hoyois, and Below 2015).

In Asia, disaster damages only in 2014 (US$64.1 billion) were below their annual average for the years 2004 to 2013 (US$75.3 billion), but this overall figure hides different phenomena. Damages from climatological disasters (US$3.71 billion) were three times their 2004–13 average. Costs from hydrological disasters (US$29.4 billion) were 52 percent higher than their decade's average, and damages from meteorological disasters (US$26.3 billion) were 35 percent above their 2004–13 average. The costliest hydrological disaster was a flood, consecutive to monsoon rains, which made US$16 billion worth of damages in the Jammu region in the north of India. This disaster was also the costliest in all Asia in 2014, and only the floods in Thailand in 2011 (US$42 billion) and in China in 2010 (US$19.5 billion) made more damages. Five other floods—four in China and one in Pakistan—cost more than US$1billion, for a total of US$6.8 billion. The total cost for these floods and the Indian one, accounted for 77 percent of all hydrological damages in 2014 (Guha-Sapir, Hoyois, and Below 2015).

1.6 Indian Scenario on Natural Hazards

1.6.1 India: Disaster Statistics

Data related to human and economic losses from disasters that have occurred between 1980 and 2010 (Table 1.1).

1.6.2 India: Risk Profile

Risk is the combination of the probability of an event and its negative consequences. This risk profile is an analysis of the mortality and the economic

Table 1.1:
Natural Disaster Statistics from 1980 to 2010

No. of events	431
No. of people killed	143,039
Average killed per year	4,614
No. of people affected	1,521,726,127
Average affected per year	49,087,940
Economic damage (US$ × 1,000)	48,063,830
Economic damage per year (US$ × 1,000)	1,550,446

Source: EM-DAT (2016).

Table 1.2:
Average Disaster per Year

Drought	0.23
Earthquake*	0.52
Epidemic	1.81
Extreme temperature	1.23
Flood	5.94
Insect infestation	0.03
Mass movement (dry)	0.03
Mass movement (wet)	1.10
Volcano	—
Storm	2.97
Wildfire	0.06

Source: Authors.
Note: * Includes tsunami.

loss risk for three weather-related hazards: tropical cyclones, floods, and landslides. In addition, new insights have been gained into other hazards such as earthquakes, tsunamis, and droughts (Table 1.2).

1.6.3 Disaster Profile of India: Regional and Seasonal

India is highly vulnerable to all types of natural disasters with the possible exception of volcanic eruptions, about which also we cannot afford to be complacent. The natural disasters take place with various intensities and in different regions due to unique and widely varying climatic, geographical, and geological conditions. Therefore, there is a significant regional and seasonal aspect in the occurrence of natural disasters in India. In this part of the unit, we will concentrate on the pattern of the regional and seasonal occurrence of natural disasters in the country.

However, earthquakes are in a class by themselves. While they do exhibit a regional profile, with the Himalayan belt, Kutch, Sundarbans, and Andaman and Nicobar Islands being most vulnerable and the peninsular part of the country being less vulnerable, earthquakes do not exhibit a seasonal profile for the simple reason that this phenomenon is not climate-related.

North India, comprising the Himalayan mountainous region and the Indo-Gangetic plains, has highly variable topography with some of the tallest mountains and perennial rivers.

Its northern-most boundary also happens to be the zone of collision of two major tectonic plates, namely, the Indian plate and the Asian plate. The area also has many geological faults. North India is also characterized by spells of hot, cold, and rainy weather, and these attributes can vary within wide limits creating unusual situations. Because of these characteristic geographical, climatic, and geological features, The North Indian states (Jammu and Kashmir [J&K], Himachal Pradesh [HP], Punjab, Haryana, Uttaranchal, Delhi, Uttar Pradesh [UP], and Bihar) are visited by natural disasters in the form of earthquakes, landslides, avalanches, floods, droughts, and heat and cold waves.

The location and climate of East and North East India (West Bengal [WB], Sikkim, Assam, Arunachal Pradesh, Nagaland, Manipur, Meghalaya, Tripura, and Mizoram) are such that these states are visited by earthquakes, landslides, floods, and droughts. WB can be affected by cyclones also.

The central parts of the country (Orissa, Chhattisgarh, Jharkhand, Madhya Pradesh [MP], Rajasthan, Gujarat, Maharashtra, and Goa) have a highly variable rainfall regime, both in time and space. Therefore, floods and droughts are major disasters in the area. Orissa and Gujarat suffer heavily from cyclones. Although Goa and Maharashtra escape the fury of cyclones, they suffer from heavy to very heavy rain due to the combined effect of depressions and the topography of the Western Ghats that also creates landslides. In recent years, Orissa has suffered considerably from heat waves.

Peninsular India (Andhra Pradesh, Karnataka, Tamil Nadu, and Kerala) suffers mainly from cyclones, floods, and droughts. While Kerala escapes the fury of cyclones, it suffers from earthquakes and landslides in addition to floods and droughts. The Telangana and Rayalaseema areas of Andhra Pradesh are highly rain-deficient areas and therefore suffer drought conditions often.

Among the island groups, the Andaman and Nicobar Island are vulnerable to earthquakes, heavy rains, and occasionally cyclones. The Andaman Islands also have two sleeping volcanoes, (Narcondam and Barren Island) of which Barren Island Volcano occasionally shows some mild short duration activity that has not proved harmful so far.

The Lakshadweep Islands are coral islands and, therefore, are only a few centimeters above the sea level. In the event of a significant sea level rise as a result of global warming, they could be threatened.

In case there is a significant sea level rise in the coming decades (uncertainties prevail), much of the large coastline of India may be threatened by the rising sea. This could threaten some of India's biggest cities such as Kolkata, Chennai, and Mumbai, which are commercial hubs.

The December 2004 experience of a tsunami has made the entire coastline (especially the eastern coast) of the country vulnerable to this serious disaster, which might be rare in occurrence but is extremely destructive.

1.6.4 Seasonal Disaster Profile of India

From the climatic point of view, India experiences four distinct seasons, which are as follows:

- winter season (December, January, and February);
- pre-monsoon or hot weather season (March, April, and May);
- monsoon season (June to September); and
- post-monsoon season (October and November).

The natural disasters that occur in different parts of India are indicated seasonwise. But it should be noted that earthquakes and tsunamis have no seasonal aspect and can occur any time.

1.6.4.1 WINTER SEASON (DECEMBER, JANUARY, AND FEBRUARY)

During these months, the Himalayan ranges receive copious amounts of rain and snow, and the weather phenomena known as "western disturbances" also bring in strong winds with rain, which at times can be heavy. Hence, the mountainous areas of North India are prone to snow avalanches and landslides. In the aftermath of rainy spells in this cold season, one or two spells of cold waves usually occur. Heavy fog (at times for days together) creates aviation hazard, and hail damages crops and orchards in the plains of North India.

1.6.4.2 PRE-MONSOON OR HOT WEATHER SEASON (MARCH TO MAY)

Cyclones take shape over the Bay of Bengal and the Arabian Sea and move westward or northwestward. Thus, the eastern coast is more vulnerable to cyclones and the accompanying storm surges. The cyclones that generate in the Arabian Sea move west or northwestwards, thus sparing the west coast, but pose serious risk to the oil exploration outfits in the Arabian Sea. If a cyclone recurves, it affects Gujarat adversely and gives considerable rains in Rajasthan as well and sometimes creates floods.

1.6.4.3 Monsoon Season (June to September)

Generally, this is the flood season for the entire country, and floods occur wherever the monsoon becomes more active. Conversely, the areas where the monsoon remains weak suffer from drought in this season. Landslides are a common feature in the hilly areas of the Himalayas, from J&K to the northeastern states. Landslides also occur in the Western Ghats and in the hilly areas of Kerala in this season.

1.6.4.4 Post-Monsoon Season (October and November)

This is again a cyclone season when cyclones generate in the Bay of Bengal and Arabian Sea and move west or northwestwards in the same general fashion as in the pre-monsoon season. But the cyclonic activity is usually more pronounced in this post-monsoon season as compared to that in the pre-monsoon season. This is also the season when the southern states of Andhra Pradesh (coastal areas), Karnataka, Tamil Nadu, and Kerala receive considerable rainfall from the northeast monsoon and are, therefore, vulnerable to the threat of floods.

1.7 Need for Disaster Management

For a comprehensive disaster management effort, the following actions need to be performed:

1. Prevent or stop the disaster to happen—prevention.
2. Reduce its intensity, duration, and frequency—mitigation.
3. Reduce its impact on the individuals, families, and societies—risk transfer.
4. As and when disaster strikes, we should be mentally, physically, and organizationally prepared to face it and to see that minimum damage happens to life, health, properties, and the environment—preparedness.
5. Return to normalcy, as quickly as possible—rehabilitation.
6. Restart the developmental process, which got halted because of the disaster—Reconstruction.
7. Start preparing for the next disaster, which may be round the corner.

Besides these, it should also aim at following aspects:

8. Avoid panic, confusion, tension, and fear while waiting for disaster.
9. Remain organized, streamlined, focused, and cool-headed.
10. Make us more resilient, sturdy, and determined.
11. Sharpen our skills, instincts, and presence of mind.
12. Be mentally, physically, and organizationally ready to face the disaster.
13. Prepare meticulous plans to save life, health, property, and the environment and to minimize the damages during incoming and impending disasters.
14. Mobilize money and resources and to train the concerned manpower to meet the objectives.
15. Remain calm and act according to the plan and the training received before the disaster.

Planning objectives

- Prudent decision-making
- Choices of mitigation
- Sustainable development
- Minimization of impacts
- Sustainability of re-construction
- Green recovery
- Environment and livelihood concerns

1.8 Impacts of Disaster

Disaster impacts various aspects of our society and covers environmental, anthropological, socioeconomic, developmental, infrastructure, and economic losses.

- Environmental aspects
- Anthropological aspects
- Socioeconomic aspects
- Developmental aspects
- Infrastructural and economic losses

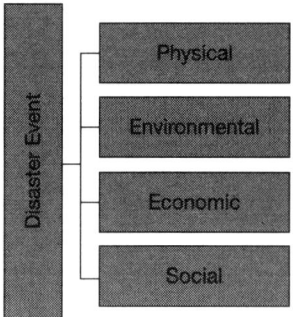

- **Physical**
  - Loss to property and goods
  - Loss of infrastructures (i.e., roads, bridges, dams, industries, etc.)
  - Loss of amenities (i.e., schools, hospitals, hotels, shopping complexes, etc.)

- **Environmental**
  - Loss of environmental reserves (i.e., water, land/soil, land-use, landscape, crops, animals/livestock, wildlife, etc.)
  - Loss of environmental services (i.e., forests, aquaculture, lake/rivers/estuaries, etc.)
  - Loss of environmental functions
  - Loss of resources for livelihood
  - Contamination of resources

- **Social**
  - Loss of lives
  - Loss of health and wellbeing
  - Damage to livelihood
  - Psychological stress and trauma
  - Loss of employment
  - Law and order problems

- **Economic**
  - Loss of services
  - Loss of industry/production
  - Loss of productivity and sustainability
  - Loss of income
  - Insurance losses and relief costs

- Economic costs of physical and environmental losses
- Recovery costs

1.9 Impact of Disasters on the Human Beings and Environment

The impact of all major disasters are basically threefold, and they affect all the primary resources such as life and gender, water, and land and bio-mass, and cause damage to forest and greeneries, livestock and wildlife, and population and property, and negatively impact the economy, education, infrastructure, health and sanitation, and employment (Table 1.3).

The major factors that contribute to the disastrous consequences of a natural hazard are human vulnerability resulting from poverty and social inequality, environmental degradation resulting from improper land and water use, rapid population growth—especially among the poor and uneducated—and lack of preparedness.

How can disasters be measured?

Measurement Units of Disasters

Disaster Type	Measurement Units
Heat and cold wave	Degree centigrade (max temp)
Earthquake	Richter
Flood	Sq. km (area covered)
Drought	Sq. km (area covered)
Extreme temp	Degree centigrade
Insect infestation	Sq. km (area covered)
Epidemic	Number of people vaccinated
Wild fire	Sq. km (area covered)
Tsunami	Richter
Windstorm	Km per hour (wind speed)
Tidal wave	Meters (height)
Radiation	Curies
Chemical spill	Cubic meter

Table 1.3:
Disaster Consequences

	7 Types of Consequences and Their Measures		7 Types of Losses, Both Tangible and Intangible	
Types	*Measures*		*Tangible*	*Intangible*
Deaths	Number of people		Loss of economically active individuals	Social and psychological effects on remaining community
Injuries	Number and injury severity		Medical treatment needs, temporary loss of economic activity by productive individuals	Social and psychological pain and recovery
Physical Damage	Inventory of damaged elements, by number and damage level		Replacement and repair cost	Cultural losses
Emergency Operations	Volume of labor, workdays employed, equipment, and resources		Mobilization costs, investment in preparedness capability	Stress and overwork in relief participants
Disruption to Economy	Number of working days lost and product (ion) lost		Value of lost product (ion)	Opportunities, competitiveness, and reputation
Social Disruption	Number of displaced persons, homeless		Temporary housing, relief, economic production	Psychological, social contacts, cohesion, and community morale
Environmental Impact	Scale and severity		Clean-up costs, repair cost	Consequences of poorer environment, health risks, risk of future disaster

Source: Authors.

Chapter 2
Natural Hazards Classification

2.1 Natural Event Concept

Earth has been experiencing natural events since its origin. By virtue of geological processes within, the Earth is evolving through natural events such as volcanic eruptions, earthquakes, floods, rains, landslides, mudslides, winds, heat waves, thunderstorms, forest fires, snowfalls, etc. The concept and definition of disaster evolved as soon as human civilization faces the impacts of the natural events that disrupt the normal activities of the societies. The abrupt growth of population and its increased density in the vulnerable areas cause more risk to the humans. Whether the natural events and calamities have been increased in severity or our tendency to inhabit the more vulnerable areas increased is yet to be established.

Our earth is constituted of different components of nature and various natural elements. Each of these components has a definite role in our earth system and every element has their own shape, form, and properties. Each element or groups of elements interact with each other and create various natural events and situations.

The components of our nature are sun and sunshine, temperature, cloud, rainfall, rivers, lakes, springs, ocean, tides and wind. The natural elements are of two major types: nonliving and living. The nonliving elements are mountains, valleys, and other landforms; different types of soils; and tectonics and plate movements, etc., while the living elements are human beings, animals and other species, trees, plants, forests, and other botanical species.

Now, the natural events in combinations of various natural elements can create some important natural disasters, such as clouds and heavy rain can create a flood, while clouds, heavy rain and mountains may produce flash floods. Likewise, cloud and heavy rain in combination with high wind can produce floods and cyclones, and typhoons and hurricanes. Earth shakes, submarine landslides, and volcanic eruptions below the ocean can create tsunamis, and less cloud and less rain with high sunshine may produce drought, heat wave, and water scarcity. Tectonic events and plate movements can also produce hazards such as earth shakes or earthquakes and volcanic eruptions.

It is interesting to note that except in earthquake and volcanic eruptions, rainfall in different scales and wind velocities play very crucial roles in all other types of natural hazards.

2.1.1 What Is a Hazard?

A hazard may be defined as "a dangerous condition or event, that threat or have the potential for causing injury to life or damage to property or the environment." Hazards can be grouped into two broad categories, namely, natural and man-made. Natural hazards are hazards that are caused because of natural phenomena (hazards with meteorological, geological, or even biological origin). Examples of natural hazards are cyclones, tsunamis, earthquakes, and volcanic eruptions, which are exclusively of natural origin.

Landslides, floods, drought, and fires are socio-natural hazards since their causes are both natural and manmade. For example, flooding may be caused because of heavy rains, landslide, or blocking of drains with human waste. Manmade hazards are hazards that are due to human negligence. Manmade hazards are associated with industries or energy generation facilities and include explosions, leakage of toxic waste, pollution, dam failure, wars or civil strife, etc. The list of hazards is very long. Many occur frequently, while others take place occasionally. However, on the basis of their genesis, they can be categorized as discussed further (Figure 2.1).

Figure 2.1:
Disaster Classifications

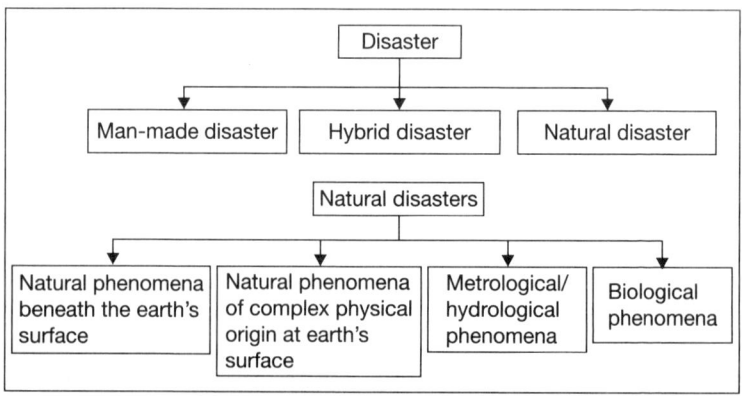

Source: Authors.

2.1.2 Natural Hazards

Natural hazards and the natural disasters that result from them can be divided into several different categories:

Geologic hazards: Earthquake, volcanic eruption, tsunami, landslide, flood, and land subsidence.

Atmospheric hazards: These are also natural hazards, but processes operating in the atmosphere are mainly responsible for them. They will also be considered in this course and include cyclone, typhoon, hurricane, tornado, drought, severe thunderstorm, and lightening.

Other natural hazards: These are hazards that may occur naturally, but don't fall into either of the previously mentioned categories. They will not be considered to any great extent in this course, but include insect infestation, disease, and wildfire/forest fire. Natural hazards can also be divided into *catastrophic hazards*, which have devastating consequences to huge numbers of people or have a worldwide effect, such as impacts with large space objects, huge volcanic eruptions, world-wide disease epidemics, and world-wide droughts. Such catastrophic hazards only have a small chance of occurring, but can have devastating results if they do occur.

Anthropogenic hazards: These are hazards that occur as a result of human interaction with the environment. They include *technological hazards*, which occur due to exposure to hazardous substances such as radon, mercury, asbestos fiber, and coal dust. They also include other hazards that have formed only through human interaction, such as acid rain and contamination of the atmosphere or surface waters with harmful substances, as well as the potential for human destruction of the ozone layer and potential global warming.

2.2 Hazards and Its Characterization

Defining hazard is a difficult task. ISDR has defined hazard as "[a] potential damaging event, whether due to natural physical or human activity may cause life loss, injury, damage of property or infrastructure, social and economic disruption, along with environmental degradation." Interestingly, hazards have both the physical and human drivers for occurrence. For example, a flood situation can arise due to excess rainfall but may be exacerbated for unwise land-use practice within the river basin.

Hazard can be characterized on the basis of a number of criteria. They can be categorized as:

1. Sudden onset
2. Slow onset
3. Industrial/technological
4. Wars/civil strife
5. Epidemics

On the other hand, a hazard may be again categorized as "direct" and "indirect." Or, those can be primary, secondary, and tertiary, depending upon the extent and timing of occurrence. Scientific data analysis can depict the processes toward a disaster may arise out of a usual event.

2.2.1 Natural Hazards

The common and most damaging group of events is classed as "natural hazards." The definition of natural hazard by UN International Strategy for Disaster Reduction (UNISDR 2009) is as follows:

> [A]ny natural process or phenomenon that may cause loss of life, injury or other health impacts, property damage, loss of livelihoods and services, social and economic disruption or environmental damage.

The Earth is being changed due to rigorous anthropogenic intervention. Though all the natural hazards are triggered by physical forces, certain incidents and their resultant factors are sometimes influenced by those human factors also, due to either ignorance or oversimplification. As an example, the conversion of land cover within a river basin for settlements may create more intense river floods.

2.2.2 Technological Hazards

We are not safe from the risks posed by hazardous technologies, and any choice of technology carries with it possible worst case scenarios that must take into account in any implementation decision. The public has the right to know precisely what these worst case scenarios are and participate in all decisions that directly or indirectly affect their future health and well-being. The disaster threats from the built environment are defined by UN as:

[H]azards originating from technological or industrial conditions, including accidents, dangerous procedures, infrastructure failures or specific human activities, that may cause loss of life, injury, illness or health impacts, property damage, loss of livelihoods and services, social and economic disruption or environmental damage.

Technologic hazards are the negative consequences of human innovation that can result in the harm or destruction of life, property, or the environment. They range from chemical spills to power failures, and from computer programming bugs to mass transportation accidents. Typically, by nature, they are new hazards in terms of the wide spectrum of threats that humans have faced; also, relatively little is known about their consequence. They can be very difficult to predict, and a wide range of triggers tends to initiate them, including many natural disasters previously discussed. Depending on the circumstances, seemingly equal technological hazards can affect geographic areas from as small as a single city block to as large as an entire continent. As technology advances, the catalogue of technological disasters only expands.

Technological hazards differ from natural hazards in that societies have chosen to assume technology associated risks (known and unknown) in exchange for some realized benefits. Perhaps the best illustration of this cost/benefit gamble, as well as one of the single greatest technological hazards, is

Table 2.1:
Different Categories of Hazard

Natural Hazards
Geological: earthquakes, volcanic eruptions, landslides, avalanches
Atmospheric: tropical cyclones, tornados, hail, ice and snow
Hydrological: river floods, coastal floods, drought
Biological: epidemic diseases, wildfires

Technological Hazards
Transport accidents: air accidents, train crashes, ship wrecks
Industrial failures: explosions and fire outbreak, release of toxic or radioactive materials
Unsafe public buildings and facilities: structural collapse, fire
Hazardous materials: storage, transport, misuse of materials

Source: Authors.

the automobile. On an average, 1.2 million people worldwide die each year in traffic accidents, yet society has collectively decided to accept that risk for the benefit of rapid transit (Table 2.1).

Some of the hazards have their own hybrid character and they may be classed as natural-technological (na-tec) hazards. Radioactive pollution consequent to the Tohuku earthquake and tsunami (Japan) in March 2011 is a unique example of such hazard (Figure 2.2).

Figure 2.2:
Examples of Natural Disasters: (a) Earthquake, (b) Landslide (c) Flash Flood, (d) Tsunami, (e) Urban Flood, and (f) Cyclone

Sources: https://pixabay.com (a, d, e and f) and Dr Indrajit Pal (b and c).

2.3 Natural Events, Calamities, and Disasters

Natural events such as flood, cyclone, tsunami, avalanches, landslides, volcanic eruptions, earth shakes (earthquakes), droughts, desertification, locust invasion, etc., are normally referred to as natural calamities or natural hazards, though in reality they are all natural events. Human beings are hardly responsible or having any role in creating these natural hazards or calamities. However, some of our actions may accelerate, trigger, or advance some of these natural events to a certain extent. The impact of these natural events on the human beings is largely functions of our alertness to respond, preparedness to face, and precautions to safeguard. Natural events/hazards/ calamities often become natural disasters in the absence of these. Therefore, an empirical analogy could be as follows:

Natural disasters=Natural events or natural calamity (bigger events are known as calamities) +Absence of our preparedness

Natural hazards can also be defined as *rapid onset hazards*, such as volcanic eruptions, earthquakes, floods, landslides, severe thunderstorms, lightening, and wildfires, which develop with little warning and strike rapidly, and also *slow-onset hazards*, such as droughts, insect infestations, and disease epidemics that take a considerably long time to develop. In addition, there are man-made hazards that may generate from bad governance and policy, hoarding, disparity, absence of equity, etc.

2.4 Classification of Disasters

Disasters are not new to mankind. They have been the constant, though inconvenient, companions of human beings since time immemorial. Disasters can be natural or man-made. Earthquake, cyclone, hailstorm, cloud-burst, landslide, soil erosion, snow avalanche, flood, etc. are examples of natural disasters, while fire, epidemics, road, air, rail accidents, and leakages of chemicals/nuclear installations, etc. fall under the category of human-made disasters. The High Power Committee on Disaster Management, constituted in 1999, has identified 31 various disasters categorized into five major subgroups (Table 2.2).

Table 2.2:
List of Various Disasters

Water- and climate-related disasters	• Flood and drainage management
	• Cyclones
	• Tornadoes and hurricanes
	• Hailstorms
	• Cloudburst
	• Heat wave and cold wave
	• Snow avalanches
	• Droughts
	• Coastal or river bank erosion
	• Thunder and lightening
	• Tsunami
Geological-related disasters	• Landslides and mudflows
	• Earthquakes
	• Dam failures/dam bursts
	• Mine fires
Chemical-, industrial- and nuclear-related disaster	• Chemical and industrial
	• Nuclear disaster
Accident-related disasters	• Forest fires
	• Urban fires
	• Mine fires
	• Oil spills
	• Major building collapse
	• Serial bomb blasts
	• Festival-related disasters and fires
Air, road, and rail accidents	• Boat capsizing
	• Village fire
	• Biological disasters
	• Biological disasters and epidemics
	• Pest attacks
	• Cattle epidemics
	• Food poisoning

Source: India High Power Committee Report (1999).

2.4.1 Disaster Category Classification and Peril Terminology: Methodology

The classification of different types of disasters into main categories was primarily based on a matrix including the existing disaster categories from the main database owners: Asian Disaster Reduction Centre (ADRC) GLobal IDEntifier Number (GLIDE), CRED (EM-DAT), La Red (DesInventar), Munich RE (NatCatSERVICE) and Swiss Re (Sigma). The same exercise was done for the definitions of the disaster categories. Taking into account each database specificities, a first working group meeting was set up in order to reach a common disaster category classification and terminology to fit all databases.

2.4.2 Hierarchy of Disaster Categories

The new classification distinguishes two generic disaster groups: natural and technological disasters. The natural disaster category being divided into six disaster groups: biological, geophysical, meteorological, hydrological, climatological, and extraterrestrial (Table 2.3). Each group covers different disaster main types, each having different disaster subtypes. Tables 2.4 to 2.9 give an overview of the grouping of natural disasters.

2.4.3 Disasters Based on Origin and Nature of Onset

Natural and man-made disasters could be divided into two broad categories: (a) based on origin and (b) based on nature of onset (Figure 2.3). Geological, water, and climate-related disasters are categories into disasters that are natural in origin. Accident-related and industrial disasters are categorized as man-made disasters.

Table 2.3:
Disaster Subgroup Definition and Classification

Disaster Subgroup	Definition	Disaster Main Type
Geophysical	Events originating from solid earth	Earthquake, volcano, mass movement (dry)
Meteorological	Events caused by short-lived/small to mesoscale atmospheric processes (in the spectrum from minutes to days)	Storm
Hydrological	Events caused by deviations in the normal water cycle and/or overflow of bodies of water caused by wind setup	Flood, mass movement (wet)
Climatological	Events caused by long-lived/meso to macroscale processes (in the spectrum from intra-seasonal to multi-decadal climate variability)	Extreme temperature, drought, wildfire
Biological	Disaster caused by the exposure of living organisms to germs and toxic substances	Epidemic, insect infestation, animal stampede

Source: India High Power Committee Report (1999).

Table 2.4:
Grouping of Geophysical Disasters

Disaster Generic Group	Disaster Group	Disaster Main-Type	Disaster Subtype	Disaster Sub-subtype
Natural disaster	Geophysical	Earthquake	Ground shaking	
			Tsunami	
		Volcano	Volcanic eruption	
		Mass movement (dry)	Rockfall	
			Avalanche	Snow avalanche
				Debris avalanche
			Landslide	Mudslide, lahar Debris flow
			Subsidence	Sudden subsidence
				Long-lasting subsidence

Source: Authors.

Table 2.5:
Grouping of Meteorological Disasters

Disaster Generic Group	*Disaster Group*	*Disaster Main-type*	*Disaster Subtype*	*Disaster Sub-subtype*
Natural Disaster	Meteorological	Storm	Tropical storm	
			Extra-tropical cyclone (winter storm)	
			Local/Convective storm	Thunderstorm/ lightning
				Snowstorm/ blizzard
				Sandstorm/dust storm
				Generic(severe) storm
				Tornado
				Orographic storm(strong winds)

Source: Authors.

Table 2.6:
Grouping of Hydrological Disasters

Disaster Generic Group	*Disaster Group*	*Disaster Main-type*	*Disaster Subtype*	*Disaster Sub-subtype*
Natural disaster	Hydrological	Flood	General (river) flood	
			Flash flood	
			Storm surge/ coastal flood	
		Mass movement (wet)	Rock fall	
			Landslide	Debris flow
			Avalanche	Snow avalanche
				Debris avalanche
			Subsidence	Sudden subsidence
				Long-lasting subsidence

Source: Authors.

Table 2.7:
Grouping of Climatological Disasters

Disaster Generic Group	Disaster Group	Disaster Main-type	Disaster Subtype	Disaster Sub-subtype
Natural disaster	Climatological	Extreme temperature	Heat wave	
			Cold wave	Frost
			Extreme winter conditions	Snow pressure
				Icing
				Freezing rain
				Debris avalanche
		Drought	Drought	
		Wildfire	Forest fire	
			Landfires (grass, scrub, bush, etc.)	

Source: Authors.

Table 2.8:
Grouping of Biological Disasters

Disaster Generic Group	Disaster Group	Disaster Main-type	Disaster Subtype	Disaster Sub-subtype
Natural disaster	Biological	Epidemic	Viral infectious diseases	
			Bacterial infectious diseases	
			Parasitic infectious diseases	
			Fungal infectious diseases	
			Prion infectious diseases	
		Insect infestation	Grasshoppers/ locusts/worms	
		Animal stampede		

Source: Authors.

Table 2.9:
Grouping of Extra-terrestrial Disasters

Disaster Generic Group	Disaster Group	Disaster Main-type	Disaster Subtype	Disaster Sub-subtype
Natural disaster	Extraterrestrial	Meteorite/Asteroid		

Source: Authors.

Figure 2.3:
(a) Schematic Diagram of Disasters Based on Nature; (b) Schematic Diagram of Disasters Based on Nature of Onset

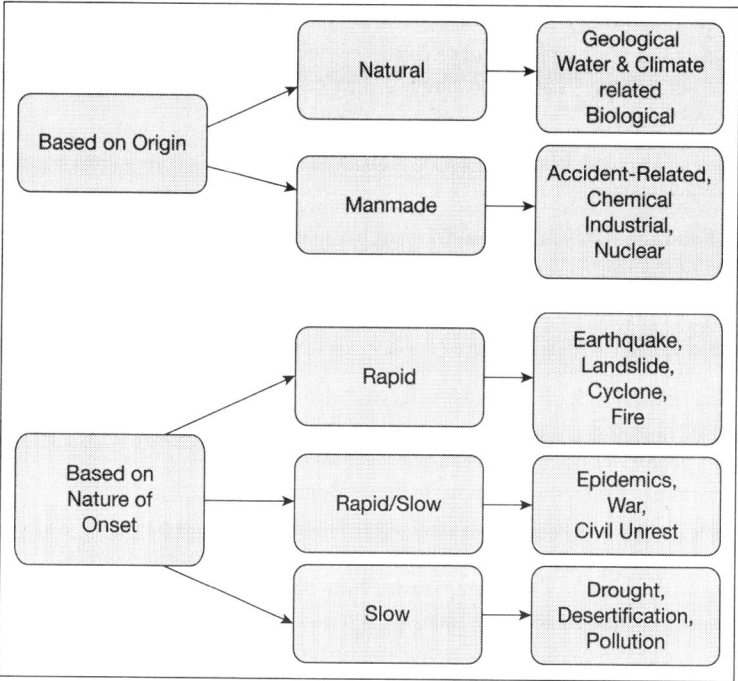

Source: Authors.

2.5 Natural Disasters

2.5.1 Flood

India is one among the most flood-prone countries in the world with a 12.5 percent flood-prone area of its total land area. India accounts for one-fifth of the global death count due to floods. Over 30 million people are displaced annually. The most flood-prone states are Uttar Pradesh, Bihar, West Bengal, Andhra Pradesh, and Gujarat. Floods are a social disaster that affects the poorer section of the society more than the rich. One disaster makes the poor more vulnerable to the next and converts a disaster into disaster process. The same flooded area can also be affected by draught immediately.

2.5.1.1 FLOOD TYPES

A flood is a hazard that can occur in many countries in Europe and in the rest of the world. You have probably seen many floods on the news happening somewhere in the world, and maybe close to where you live.

Floods bring misery to those that live in the area. They can cause loss of life and often cause a great disruption of daily life: Water can enter into peoples' houses; lack of drinking water and also some other consequences may occur such as the electricity supplies may break down, roads can be blocked, and people cannot go to work or to school. Floods all over the world cause enormous damage every year such as economic damages, damage to the natural environment, and damage to national heritage sites.

A flood is a situation in which water temporarily covers land where it normally doesn't. This water comes from the sea, lakes, rivers, canals, or sewers. It can also be rainwater.

Floods can be described according to speed (flash flood), geography, or cause of flooding. Several types of flooding will be described in this chapter together with some aspects of hydrology.

- Flash floods
- Coastal floods
- Urban floods
- River (or fluvial) floods
- Ponding (or pluvial flooding)

Hydrology

- Water cycle
- Water storage
- A little or lots of water (water quantity)
- Rainfall intensity
- Speed (water velocity)
- Catchment area

Flash Floods

In areas with steep slopes, heavy rain can cause a riverbed that held very little or no water at first to suddenly brim with fast-flowing water. The rain water is collected on the slopes, then flows downhill, gathering speed, and all the water comes together in the river bed. The water level rises fast. The water flows over the river banks and floods the area. Speed is the keyword. It all happens fast. It rains heavily. The water flows at high speed. Because of this speed, it has the strength to carry away heavy objects. The flood stops as suddenly as it starts. A flash flood is a very direct response to rainfall with a very high intensity or sudden massive melting of snow.

The area covered by water in a flash flood is relatively small compared to other types of floods. The amount of water that covers the land is usually not very large, but is so concentrated on a small area that it can rise very high.

Because of the sudden onset and the high travelling speed of the water, flash floods can be very dangerous. The water can transport large objects such as rocks, trees, and cars. Never drive through a flash flood, even if it doesn't seem to be very deep: the car may be swept away by the sheer speed of the water.

When a dyke breaches along the sea or a river, the sudden inflow of water with high speed can also be compared with a flash flood.

Coastal Flooding

A simplified explanation of a coastal flood is when the coast is inundated by the sea. The cause of such a coastal surge is a severe storm. The storm wind pushes the water up and creates high waves, which inundate the coastal region.

A storm is formed in a low pressure area, as you may know. An interesting fact is that beneath a low pressure area, the sea level is higher. Think of a balloon. Press it hard and it flattens. Release the pressure, and it bounces back. The rule of thumb is that with each millibar of pressure that lessens, the sea level rises a centimeter. This contributes to the high sea level, but the wind can also have a larger effect.

A flood starts when waves move inland on an undefended coast or over-top, or breach the coastal defense works such as dunes and dikes. The waves attack the shore time and again. When it is a sandy coast, each wave in a storm will take sand away. Eventually, a dune may collapse that way. A vegetated coastal sand dune is comparatively more stable than a bare sand dune.

A unique characteristic of a coastal flood is that the water level drops and rises with the tide. At high tide, the water may flow in, and at low tide, it may recede again. When a sea defense is breached, low tide is the time to repair the breach. In the animation, you see the build-up of force by the sea and how the sea floods the coast. Once it overtops and breaches the defenses, the sea enters fast, but it slows down when it spreads over a larger area.

Urban Flooding

Flooding in urban areas can be caused by flash floods, coastal floods, or river floods, but there is also a specific flood type that is called urban flooding.

Urban flooding is specific in the fact that the cause is a lack of drainage in an urban area. As there is little open soil that can be used for water storage, nearly all the precipitation needs to be transported to surface water or the sewage system. High intensity rainfall can cause flooding when the city sewage system and draining canals do not have the necessary capacity to drain away the amount of rain that is falling. Water may even enter the sewage system in one place and then get deposited somewhere else in the city on the streets.

Urban floods are of a great concern as it disrupts daily life in the city. Roads can be blocked, and people can't go to work or to schools. The economic damages are high, but the number of casualties is usually very limited because of the nature of the flood. The water slowly rises on the city streets. When the city is on flat terrain, the flow speed is less and you can still see people driving through it. The water rises relatively slowly, and the water level usually does not reach life-endangering heights.

River Flooding (Flooding from Rivers Overtopping Their Banks or Breaking through Dikes)

Rainfall over an extended period and an extended area can cause major rivers to overflow their banks. The water can cover enormous areas. Downstream areas may be affected, even when they didn't receive much rain themselves.

With large rivers the process is relatively slow. The rain water enters the river in many ways. Some rain will fall into the river directly, but that alone doesn't make the river rise high. A lot of rain water will run off the surface when the soil is saturated or hard. It will flow to small rivers that flow to larger rivers, and these rivers flow into even larger rivers. In this way, all the rain that fell in a large area (catchment area) comes together in this one very large river. When there is a lot of rain over a long period, you see the river rise gradually as it is fed with water form smaller rivers or rivulets. It takes time for all the rainwater to reach the river, but once it is in the river, it has to flow downstream to the sea.

While the water level slowly rises, officials can decide to evacuate people before the river overflows. The area that is flooded can be huge. Villages are surrounded by large stretches of water were cattle normally graze. As a result, the entire community can become isolated from the outer world as roads are blocked and communications are down.

When a dike or a dam breaks and a lot of water is released suddenly, the speed of the water at the breach can be compared with the speed of a flash flood. As a larger area gets covered, the speed will be reduced. The water spreads out as much as possible, flowing to the lower lying areas before slowly rising. A breach is very dangerous for the people living close to it. The strength of the water may carry cars, trees, and even houses away and cause loss of live.

Ponding (or Pluvial Floods)
Ponding is a type of flooding that can happen in relatively flat areas. On reaching the ground, rain water usually gets infiltrated through the soil pores or poured into ponds, canals, or lakes through surface run-off, later on drained away, or pumped out. When excess rainwater enters a water system rather than being stored, or leaving the system, flooding occurs. In this case, rain is the source of the flood: not water coming from a river; rather water on its way to the river. That's why it is also called "pluvial flood."

Puddles and ponds develop on the land and canals are filled to brim and spill over; gradually a layer of water covers the land. It is like urban flooding, but without the sewage systems and in more rural areas.

Because of the gradual character, people have time to go indoors or leave the area. The layer of water is no more than centimeters or perhaps decimeters high and causes no immediate threat to people's life. Depending on the economic activity and size of the area that is covered, it may cause immense economic damage.

Water Cycle
The surface of the Earth is for a large part covered in water: about 70 percent. Of all the water on Earth, only about 3 percent is fresh water. Most of that fresh water is ice in pole caps and glaciers. Only 0.014percent of all water can be used by people.

The sun warms the water in the oceans and seas and turns the water into water vapor (evaporation). High up in the atmosphere, the air cools down and this vapor turns into tiny water droplets (condensation), which together form clouds. Part of these clouds rain out over the sea. This is called the short water cycle.

In the long water cycle the droplets will go on a longer journey. The clouds float on air currents. Droplets fall down as snowflakes, icicles, or raindrops (precipitation). They may fall on land or on a river or lake. Water that falls on land may stay there as snow or ice, or it may seep into the earth and become soil moisture or ground water (infiltration). Soil water is bound to the soil and is available for plants. Ground water can fill the empty spaces in the ground and can form underground rivers or lakes. Water may also flow over land and eventually enter a river. In the end, it can flow all the way back to sea (transportation) or evaporate again somewhere along the way. The Sun is an integral part of the water cycle as the source of energy for all these processes.

Water Storage in Surface Water

Rain (or precipitation in general) can be stored in surface water such as rivers, lakes, and ponds. The water level will rise. Later, the water will either be transported downstream, seep into the soil (infiltration) or evaporate. Then the level will be normal again.

The amount of water that can be stored depends on the volume/extent of the basin and the height that it can rise before it overflows.

When there is more rain than can be stored, infiltrated, or transported, the area will flood. The surface water is very important for the temporary storage of water (Figure 2.4). Drainage or discharge of water out of the area is always a slower process than storage close to where the water falls. In areas where there's little room for surface-water storage, creative solutions are sought. In this way, less water flows from the roofs into to sewage system.

Water Storage in Soil

Rain water can also be stored in the ground. Soils consist of particles and pores. Those pores can be filled not just with air but also with water. The amount of pores is a soil is different for different types of soil. The pores in a clay soil account for 40 percent to 60 percent of the volume. In fine sand this can be 20 percent to 45 percent (Figure 2.5).

The soil particles have small pores in them where water can enter (soil water), and between the particles are larger pores that can be filled. The soil

Figure 2.4:
The Process of Water Storage in Surface Water

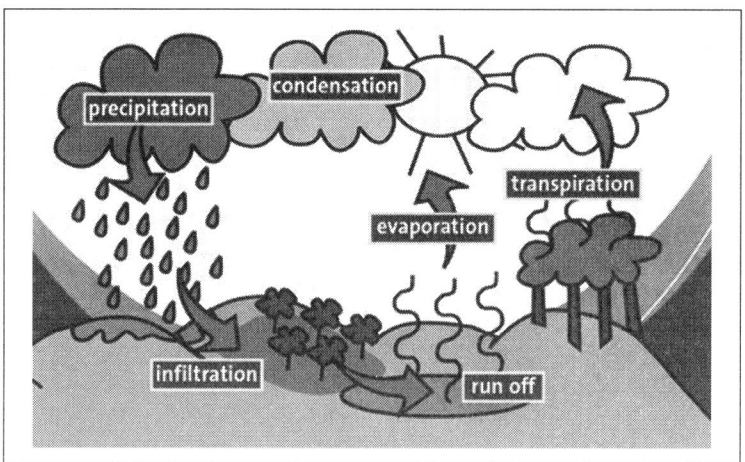

Source: Authors.

Figure 2.5:
The Process of Water Storage in Soil

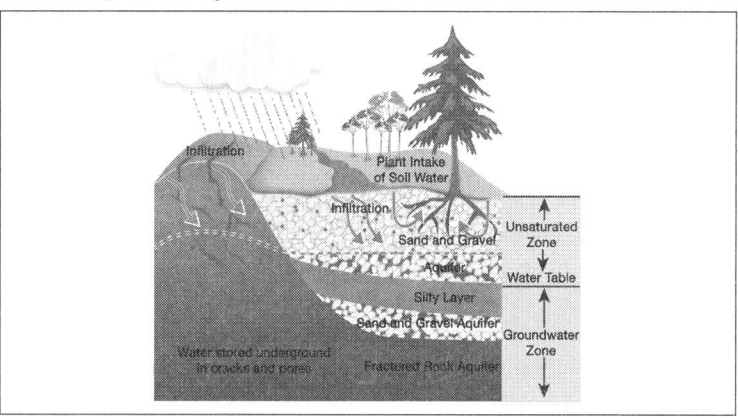

Source: Authors.

is filled with water up a certain level. This level goes up and down with the changing weather conditions. This water level is the ground water level.

The process of water entering the soil is called *infiltration*. When the soil has taken up all the water it can, we say that it is *saturated*. If you walk over a saturated soil, you feel that it is wet and soggy, like biscuits dipped in tea.

Part of the water that infiltrates, will move on. It will go to underground storage reservoirs or to underground rivers, and it may, through ground water flows, eventually reach a river or a lake. Another part will be used by plants or will evaporate.

Water Storage in Sewage Systems
In areas with a lot of hard surfaces, such as cities with streets and houses, the water runs off immediately to sewers or surface water. As there is hardly any water storage in the soil in areas with hard surfaces, you see a quick rising of the surface water level when it rains. When the sewers are used to capacity, the streets will flood.

- An area is 2 hectare (ha). The soil is clay with 40 percent pores, 1 m deep over a hard rock surface. The pores are half full with water.
  - How many liters of water can be stored? So how many millimeters of rain can be stored here?
  - What if the amount of water you calculated fell on a hard surface and it couldn't be drained?

Intensity of Rainfall
The intensity of rainfall is a measure of the amount of rain that falls over time. The intensity of rain is measured in the height of the water layer covering the ground in a period of time. It means that if the rain stays where it falls, it would form a layer of a certain height. We say things like 30 millimeter (mm) of rain fell today or it rained 20 mm in 2 hours. Sometimes people speak of the volume of water that falls on a square meter in a period of time: 10 l per square meter per day for instance. A millimeter of water equals a liter of water on a square meter.

It is hard to say what a high or low intensity is. It depends on the local circumstances. Generally speaking, a relatively low intensity is, for instance, 2 mm of rain a day, and relatively high may be 30 mm an hour. High intensity of rainfall on steep slopes may lead to flash floods. On flat areas, it may lead to ponding or urban floods when the drainage capacity is insufficient for the intensity of the falling rain.

Speed
The flow speed or velocity of the water is mainly determined by the slope of the terrain that the rain falls on. The steeper the slope, the faster the water flows. Try rolling marbles over a plank: the steeper the plank, the faster they will roll.

Next to flow velocity, you also have rise velocity: the speed with which the water level rises. The unit to describe the rising speed is usually centimeter per hour or centimeter per day or something similar.
The rise in speed is determined by a combination of factors.

*   *Rainfall intensity:* The higher the intensity, the higher the rising speed.
*   *The infiltration capacity of the surface:* Rocks, but also streets and buildings, have a very low infiltration capacity. Nearly all of the precipitation (rain or snow) will then run off or will stay and form puddles and ponds (ponding).
*   *Slope of the terrain:* Water will not form a layer on slopes. Water will always flow down and will gather at the lowest point or points, and there it will rise.

Catchment Area

A catchment area is a hydrological unit. Each drop of precipitation that falls into a catchment area eventually ends up in the same river going to the sea if it doesn't evaporate. However, it can take a very long time. Catchment areas are separated from each other by watersheds. A watershed is a natural division line along the highest points in an area. Catchments are divided into subcatchments, also along the lines of elevation (Figure 2.6).

Figure 2.6:
A Theoretical Catchment Area with Flow Paths of Precipitation

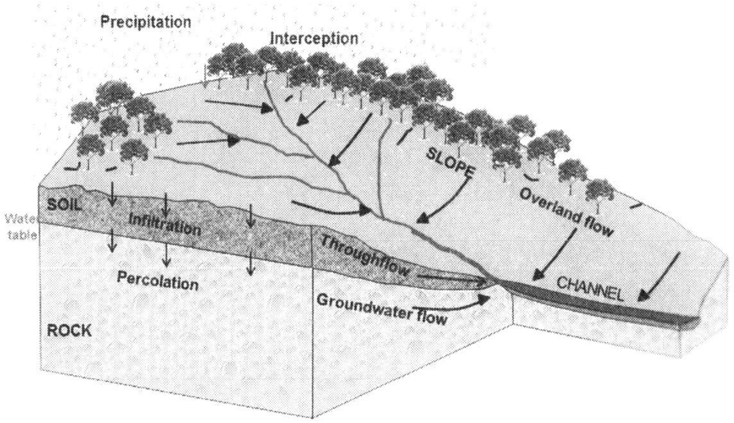

Source: Authors.

2.5.2 Earthquake

Tremors and vibrations in the crust of the earth are called earthquakes. An earthquake (also known as a quake, tremor, or temblor) is the result of a sudden release of energy in the Earth's crust that creates seismic waves (Figure 2.7). Earthquakes are measured with a seismometer, a device that also records is known as a seismograph. The moment of the magnitude (or the related and mostly obsolete Richter magnitude) of an earthquake is conventionally reported, with magnitude 3 or lower earthquakes being mostly imperceptible and magnitude 7 causing serious damage over large areas. Intensity of shaking is measured on the modified Mercalli scale.

Causes of Earthquakes

1. *Crustal instability:* The tectonic forces are generally the main cause of earthquakes. They lead to sudden movements of the crustal blocks.

Figure 2.7:
Schematic Diagram of Focus or Hypocenter, That Is, Earthquake's Point of Initial Rupture and Epicenter, That Is, the Point at Ground Level Directly Above the Hypocenter

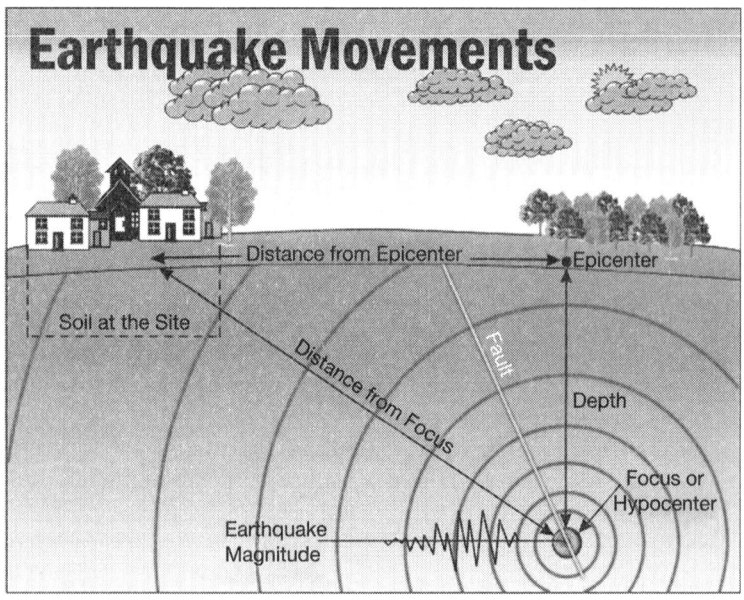

Source: USGS.

Thus, a majority of earthquakes are associated with areas of crustal instability and such earthquakes are called "tectonic earthquakes."

2. *Volcanic eruptions:* Volcanic eruptions also cause many earthquakes. They accompany most of the explosive eruptions. Such earthquakes are said to be of "volcanic origin."

Impacts of Earthquakes

1. Changes in the earth's crust may lead to a number of indirect effects such as landslides, avalanches, tsunamis (even as in the Indian Ocean in 2004), blocking of course of rivers, and subsequent flooding when the blockage is removed by accumulated water.
2. Means of transport are interrupted, due to the damage to roads and twisting of railway lines.
3. It may damage large dams, power installations, and even nuclear power plants.
4. The earthquakes may also damage underground wires, pipelines, and water systems.
5. It interrupts the socioeconomic conditions of the affected area and also hinders its development.
6. Moreover, it causes huge loss of life and property and also the environment of the place.
7. Direct effects are seen as the changes in the earth's surface.
8. The daily life and routine of the people get affected for a long period of time.

The Indian subcontinent is highly prone to natural disasters. Floods, droughts, cyclones, and earthquakes are a recurrent phenomenon. As per the latest seismic zoning map brought out by the Bureau of Indian Standards (BIS), over 65 percent of the country is prone to earthquakes of intensity MSK VII or more. Some of the most intense earthquakes of the world have occurred in India, but fortunately, none of these have occurred in any of the major cities. India has highly populous cities, including the national capital of New Delhi, located in zones of high seismic risk. Typically, the majority of the constructions in these cities are not earthquake resistant. Thus, any earthquake striking in one of these cities would turn into a major disaster.

Six major earthquakes have struck different parts of India over a span of the last 15 years. The damage caused by these earthquakes reiterate the scale of vulnerability. However, if any of these earthquakes had struck populous urban centers, the damages in terms of human lives and property would have been colossal.

Frequent disasters lead to erosion of development gains and restricted options for the disaster victims. Physical safety—especially that of the vulnerable groups—is routinely threatened by hazards. Disasters such as the Gujarat earthquake have very clearly illustrated that we need mitigation, preparedness, and response plans so that the threat to human life and property is minimized. The vulnerability atlas of India published by Building Materials & Technology Promotion Council (BMTPC) of the Government of India, and the Code of Practice (IS 1893:2002, Part 1) for earthquake-resistant design has divided India into four zones depending on the earthquake vulnerability of the area, that is, Zones II, III, IV, and V. Zone II is of low damage risk, Zone III is of moderate damage risk, Zone IV is of high damage risk, and Zone V is of very high damage risk. Earthquakes, which are highly unpredictable and inevitable, have rocked several parts of the country during the last few decades.

An earthquake in Bhuj on January 26, 2001, has exposed the impact of human intervention with environment without the necessary precautionary measures. Rapid industrialization and urbanization combined with the population explosion have added to our vulnerability to earthquakes. All these factors have urged us to have a comprehensive earthquake mitigation strategy.

The concept of earthquake and a brief account of terms associated with the earthquake is placed below:

Earthquake: Earthquake is a series of ground vibrations, caused due to the release of strain energy in the form of seismic waves.

Focus: The originating earthquake source of waves inside the earth. It is also called the hypocenter.

Epicenter: The geographical point on the surface of the earth vertically above the focus of the earthquake.

Magnitude: The magnitude of an earthquake is a number that is a measure of energy released in an earthquake. The magnitude of an earthquake is always constant.

Intensity: The intensity of an earthquake at a place is a measure of strength of shaking during the earthquake. Intensity is determined from effects on people, structures, and the natural environment. The intensity of an earthquake is not constant.

Richter scale, and MSK and MMI scale

A Richter scale is used to measure the magnitude where as MSK (Medvedev S. Karmik) and Modified Mercali Intensity (MMI) scale is used to measure the intensity.

Seismograph and seismogram

A seismograph is the instrument, which records the earthquake.

A seismogram is the record of the earthquake, which is produced by the seismograph.

As of now, prediction of earthquake is scientifically not possible apart from the Chinese method of watching for clues in unusual animal behavior. Apart from awareness and preparedness, consult a structural engineer/architect to make your house earthquake resistant.

2.5.2.1 DIFFERENCE BETWEEN MAGNITUDE AND INTENSITY

- Magnitude is the measure of the earthquake size, and it remains unchanged with distance from the earthquake, whereas intensity describes the degree of shaking caused by an earthquake at a given place and which decreases with distance from the earthquake epicenter.
- Magnitude measurement requires instrumental monitoring for its calculation; however, assigning an intensity requires a sample of the felt responses of the population and damaged structure.

Table 2.10 depicts the correlation between magnitude and intensity.

Table 2.10:
Magnitude and Intensity of Earthquake

Magnitude	Intensity	Description
1.0–3.0	I	I. Not felt except by a very few under especially favorable conditions.
3.0–3.9	II–III	II. Felt only by a few persons at rest, especially on upper floors of buildings. Many people do not recognize it as an earthquake. Vibrations similar to the passing-by truck.
4.0–4.9	IV–V	IV. Felt indoors by many and outdoors by few during the day. Some awakened at night. Dishes, windows, and doors disturbed; walls make cracking sound. Sensation like heavy truck striking building. Standing motor cars rocked noticeably. V. Felt by nearly everyone; many awakened. Some dishes, windows broken. Unstable objects overturned. Pendulum clocks may stop.
5.0–5.9	VI–VII	VI. Felt by all, many frightened. Some heavy furniture moved; a few instances of fallen plaster. Damage is slight. VII. Damage negligible in buildings of good design and construction; slight to moderate in well-built ordinary structures; considerable damage in poorly built or badly designed structures; some chimneys broken.

(Continued)

Table 2.10:
(Continued)

Magnitude	Intensity	Description
6.0	VIII	VIII. Damage slight in specially designed structures; considerable damage in ordinary substantial buildings with partial collapse. Damage great in poorly built structures. Fall of chimneys, factory stacks, columns, monuments, and walls. Heavy furniture overturned. IX. Damage considerable in specially designed structures; well-designed frame structures thrown out of plumb. Damage great in substantial buildings, with partial collapse. Buildings shifted off foundations. X. Some well-built wooden structures destroyed; most masonry and frame structures destroyed with foundations. Rails bent. XI. Few, if any (masonry) structures remain standing. Bridges destroyed. Rails bent greatly. XII. Damage total. Lines of sight and level are distorted. Objects thrown into the air.

Source: USGS.

2.5.3 Landslide

A landslide is a downslope movement of rock or soil, or both, occurring on the surface of rupture—either a curved (rotational slide) or planar (translational slide) rupture—in which much of the material often moves as a coherent or semi-coherent mass with little internal deformation. Although the action of gravity is the primary driving force for a landslide to occur, there are other contributing factors affecting the original slope stability. Typically, pre-conditional factors build up specific subsurface conditions that make the area/slope prone to failure, whereas the actual landslide often requires a trigger before being released. Figure 2.8 is a graphic illustration of a landslide, with the commonly accepted terminology describing its features.

2.5.3.1 LANDSLIDE CLASSIFICATION

Landslides can be classified into different types on the basis of the type of movement and the type of material involved. In brief, the material in a landslide mass is either rock or soil (or both); the latter is described as earth if mainly composed of sand-sized or finer particles and debris if composed of coarser fragments. A more comprehensive and widely accepted

Figure 2.8:
Components of Landslide

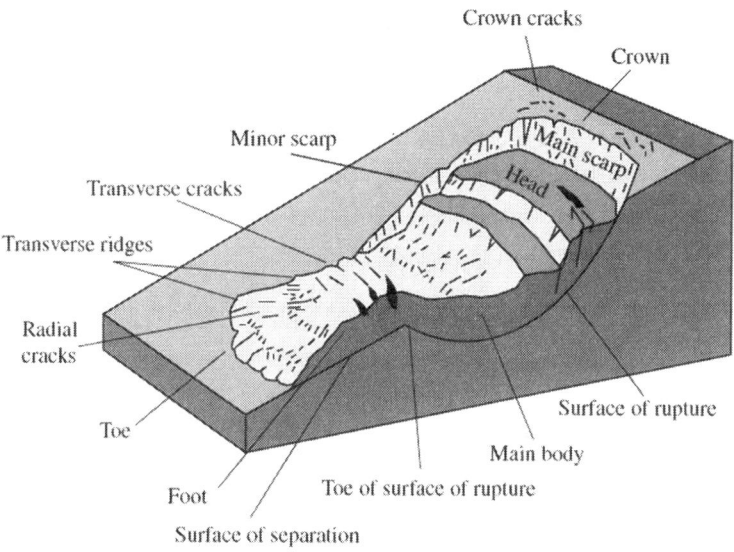

Source: USGS.

classification of landslides, based on the types of movements and material involved, is depicted in Table 2.11. Movement types are divided into five main groups: falls, topples, slides, spreads, and flows. A sixth group, complex slope movements, includes combinations of two or more of the other types of movement (Figure 2.9). Similarly, materials are divided into two classes: rocks and engineering soil. Soil is further subdivided into debris and earth, based on the grain size. A short description of the various landslide types is given in Table 2.11.

It has been frequently postulated that one of the underlying causes of the increase in landslides in mountain areas in less developed countries is road building. In a cloudburst event, which occurs occasionally during the monsoon, landslides have the potential to turn into very devastating debris flows. In such conditions, road users and the inhabitants of houses would be at high risk.

Road building in the mountains is beneficial for the connectivity of the mountain people, but equally, we should take care of the advanced technology for prevention and mitigation. Preventing these types of landslide problems is comparatively simple and inexpensive, and it has the potential

Table 2.11:
Type of Landslides and Classification of Slope

Type of Movement		Bedrock	Type of Material	
			Engineering Soils	
			Predominantly Coarse	Predominantly Fine
Slide	Falls	Rock fall	Debris fall	Earth fall
	Topples	Rock topple	Debris slide	Earth slide
	Rotational	Rock slide	Debris slide	Earth slide
	Transitional			
	Lateral spreads	Rock spread	Debris spread	Earth spread
	Flows	Rock flow (deep creep)	Debris flow	Earth flow
			(Soil creep)	
	Complex	Combination of two or more principal types of movement		

Source: USGS.

Figure 2.9
Landslide Classification

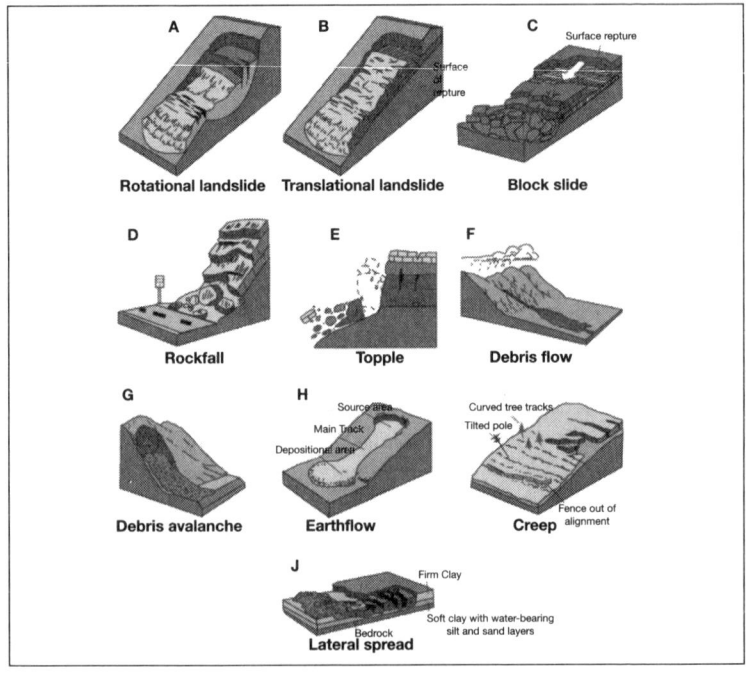

Source: USGS.

to improve both livelihoods and safety. If we are to reduce losses from land-slides, then it is this sort of problem that needs to be tackled systematically.

2.5.4 Cyclone

Cyclones are violent storms, often of vast extent, which are associated with turbulent weather conditions with high velocity winds, cloudiness, and rainfall. Tropical cyclones are known by different names in different regions. They are called depressions in the Bay of Bengal, hurricanes in the Caribbean Sea, willy-willy in Australia, typhoons in China, and tornadoes in the USA and West Africa. Most damage from cyclones is caused by the strong winds, torrential rains, and high storm tides.

A cyclone is a relatively small and intense low latitude warm-core pressure area having a wind circulation in clockwise direction in the Southern Hemisphere and counter-clockwise in the Northern Hemisphere.

The prerequisite for the cyclones are

1. warm ocean temperature, above 27°C;
2. absence of strong vertical wind shear;
3. presence of low pressure;
4. presence of Coriolis force; and
5. cyclone get spinning motion from the rotation of Earth at latitudes greater than 5.

Tropical cyclones form only over warm ocean waters near the equator. To form a cyclone, warm, moist air over the ocean rises upward from near the surface. As this air moves up and away from the ocean surface, it leaves less air near the surface. So basically, as the warm air rises, it causes an area of lower air pressure below. Air from surrounding areas with higher air pressure pushes in to the low pressure area. Then this new "cool" air becomes warm and moist and rises, and the cycle continues. As the warmed, moist air rises and cools, the water in the air forms clouds. The whole system of clouds and wind spins and grows, fed by the ocean's heat and the water evaporating from the ocean surface. As the storm system rotates faster and faster, an eye forms in the center. It is very calm and clear in the eye, with very low air pressure. Higher pressure air from above flows down into the eye (Figure 2.10). When the winds in the rotating storm reach 39 mph (63 kmph), the storm is called a "tropical storm." And when the wind speeds

Figure 2.10:
Schematic Diagram of Cyclone Formation. The white arrows show where warm air is rising. The dark arrows indicate where cool air is sinking

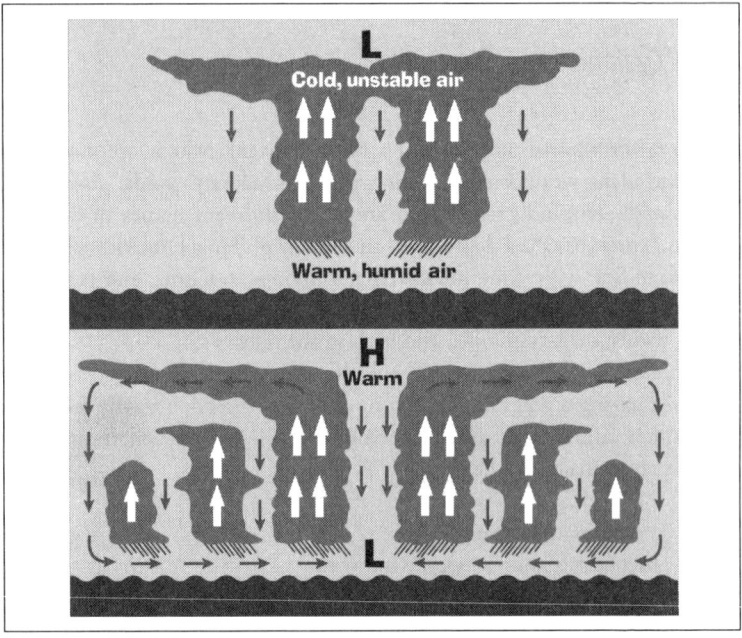

Source: www.pixabay.com

reach 74 mph (119 kmph), the storm is officially a "tropical cyclone" or hurricane. Tropical cyclones usually weaken when they hit land, because they are no longer being "fed" by the energy from the warm ocean waters. However, they often move far inland, dumping many centimeters of rain and causing lots of wind damage before they die out completely.

Tropical storms can and do produce a lot of damage through high wind speed and the "storm surge." A storm surge is caused by strong winds pushing on the ocean's surface. The wind causes the water to pile up higher than the ordinary sea level. Storm surges are particularly damaging when they occur at the time of a high tide, which increases the risk of flooding. Storm surges are extremely destructive to property, and they can cause flooding and large amounts of erosion. Storm surges account for 90 percent of all fatalities associated with cyclones.

Cyclones are divided into categories depending on the strength of the winds produced. There are many different classification scales, but one you may be familiar with is the Saffir-Simpson Hurricane Scale. This scale is used to describe storms. The classifications (1–5) are intended primarily for use in measuring the potential damage and flooding (storm surge) that a cyclone will cause upon landfall (Table 2.12).

India, with its long coastline, is vulnerable to the impacts of tropical cyclones that develop in the North Indian Ocean (the Bay of Bengal and the Arabian Sea). These systems are classified as depressions, deep depressions, cyclonic storms, severe cyclones, and severe cyclones with cores of hurricane winds depending on the surface winds associated with them as indicated in Table 2.13.

2.5.4.1 CYCLONE SEASON

There are two cyclone seasons for India. The first in the pre-monsoon season (April and May) and the second is the post-monsoon season (October to

Table 2.12:
Saffir–Simpson Hurricane Scale for Cyclone

Category	Wind Speed (mph)	Damage at Landfall	Storm Surge (feet)
1	74–95	Minimal	4–5
2	96–110	Moderate	6–8
3	111–30	Extensive	9–12
4	131–55	Extreme	13–18
5	>155	Catastrophic	19+

Source: Modified after National Hurricane Center, National Oceanic and Atmospheric Administration (NOAA).

Table 2.13:
Classification of Tropical Windstorms

Type	Surface Winds
Depression	18 to 27 Kts (36 to 54 km per hour)
Deep depression	28 to 33 Kts (56 to 66 km per hour)
Cyclonic storm	34 to 47 Kts (68 to 94 km per hour)
Severe cyclonic storm	48 to 63 Kts (96 to 117 km per hour)
Severe cyclonic storm with core of hurricane winds	>64 Kts (>117 km per hour)
Kts: Nautical Miles	

Source: Modified after National Hurricane Center, National Oceanic and Atmospheric Administration (NOAA).

December). The cyclones of the post-monsoon season are generally more intense than those of the pre-monsoon season. On an average, six cyclones form in Indian seas out of which the distribution in the Bay of Bengal and the Arabian Sea is in the ratio 4:1.

2.5.5 Tsunami

Tsunami is a Japanese term for sea waves generated by undersea earthquakes. These waves may originate from undersea or coastal seismic activity, or volcanic eruption. Whatever maybe the cause, water is displaced into a violent and sudden motion ultimately breaking over land, even at very long distances, with great destructive power. It is to be noted that while the coasts are hit by very high waves of water, there is hardly any appreciable wave on the high seas. Therefore, ships on the high seas are not affected.

In most cases, a tsunami could be the after-effect of an undersea earthquake due to which the abrupt vertical movement of ocean floor generates waves that travel at high speed in the ocean. As they approach land, their speed decreases while their height increases. It can be highly destructive to coastal areas, as was witnessed during the catastrophic tsunami event in December 2004.

2.5.6 Drought

Drought is a temporary reduction in water availability in an area for an unusually long period. Depending on the resulting water scarcity, a drought has disastrous and long-term socioeconomic impacts, which may last for months and in some cases years. It is a slow-onset phenomenon.

Three types of droughts are recognized:

- *Meteorological drought:* When the monthly or seasonal rainfall over an area is appreciably below normal.
- *Hydrological drought:* When the water scarcity over an area results in reduction in the available water in surface water bodies, and the water table also recedes. Needless to state, prolonged meteorological droughts leads to hydrological droughts.
- *Agricultural drought:* When the water scarcity results in partial or total loss of crops and affects agricultural activity adversely.

A drought is generally caused by adverse water balance or scarcity of water to satisfy the normal needs of agriculture, livestock, or human population. It can also occur in areas that normally enjoy adequate rainfall and moisture levels. Drought may be caused due to excessive evapotranspiration losses, high temperature, and low soil-holding capacity.

Prolonged drought can result in aridity or even desertification when the exceptionally dry soil can no longer sustain any biological activity, whether organisms or vegetation.

A drought is a long period of very dry weather. It is an insidious natural hazard. It is a climatic anomaly characterized by deficient supply of moisture resulting either from subnormal rainfall, erratic rainfall distribution, higher water need, or a combination of all the factors. Most of the droughts are generally associated with arid or semi-arid climates, but it can also occur in areas of adequate rainfall, late arrival, or early departure of monsoon.

Odisha, Chhattisgarh, Jharkhand, internal parts of Karnataka and Maharashtra, Gujarat, Rajasthan, parts of Punjab, Haryana, UP and Tamil Nadu are the most drought-prone areas of India.

2.5.6.1 COPING WITH DROUGHTS

Droughts lead to failure of crops. This gives rise to poverty, unemployment, and shortage of food. It also adversely affects the agro-based industry. Therefore, it is an urgency to reduce or minimize the impact of drought. Some common ways are helpful in reducing the impact of drought are:

• Water shortage areas should be identified.
• Rainwater harvesting should be encouraged.
• Afforestation should be encouraged.
• In the urban areas, misuse and wastage of water should be stopped.

2.5.7 Volcanoes

A volcano is an opening in the crust of the earth through which lava comes out and spreads over settlements, roads, and cultivated areas, destroying houses and making land unsuitable for cultivation. Steam from volcanic eruption may lead to heavy rainfall, causing landslides, mudflows, and floods. Many poisonous gases come out of volcanic eruption and cause environmental pollution.

2.5.8 Avalanches

An avalanche is defined as the event in which a large mass of snow, ice, rock, or other material moves swiftly down a mountain side or over a precipice and crushes everything in its path. An avalanche starts when the large mass of snow, ice, and rock overcomes the frictional resistance of the sloping surface either due to rain, melting of ice base, or vibrations of any kind.

It will be seen that landslides and avalanches are events of mountain regions and are rather similar in nature and impact. The basic difference is that a landslide involves the movement of rock, soil, and mud whereas an avalanche comprises snow, ice, and rock. Landslides can occur in smaller hills or rocky slopes, but avalanches occur in high mountains with snow in abundance.

An avalanche is a mass of snow that comes loose from steep mountain slope and hurtles down to the valley below. It can be huge and frightening, sweeping away trees and burying houses. Avalanches are a danger in any mountainous area that has slopes and heavy snow.

They are worst on bare slopes, with no trees to hold back the snow. In some countries, new forests are being planted to reduce the danger. Snow bridges are built over roads and railways to protect them.

2.5.9 Heat Wave and Cold Wave

As the name implies, these are spells of extreme surface air temperatures over a region for a rather prolonged period of several days or few weeks. When the maximum temperature in the day over an area overshoots in the hot weather months (March to June), it leads to a heat wave. Similarly, when the minimum temperature falls appreciably below normal in the winter months, it is called a cold wave. Both the phenomena are extreme weather events and lead to very considerable discomfort, illness, epidemics, and deaths due to exposure.

Chapter 3

Risk

3.1 Introduction

Risk is the potential where the possibility of any chosen action, inaction or activity will lead to any disruption of any system with certain loss. Those losses can also be termed as "risks." Every individual is under potential risk, but it varies with the time and space. Risk can also be defined as a probability or threat of damage, injury, liability, loss, or any other negative occurrence that is caused by external or internal vulnerabilities and that may be avoided through preemptive action. The components of risk are generally in the domain of finance, food security, insurance, trading, and also with the home and workplace.

The related terms of risk are threat and hazard, and they are sometimes confusing with risk, as they usually indicate toward possible harm.

A "threat" is a situation that indicates the act to elicit a negative impact. It is generally shown as the intent to create harm or loss to any other person. This may include crime, natural calamity, health problem, political or social agitation, insurgency and even attack from wild animals.

A "hazard" is a kind of situation that may pose certain extent of threat to life, property, and/or the environment. Usually, hazards are in a dormant or submissive state with their potential to damage. When a hazard becomes active, it may create an emergency situation. This is called an "incident." Risk is the resultant factor from the interaction between hazard and vulnerability.

Natural disasters disrupt the usual life and livelihood, including the degradation of environment. Moreover, the development process after such natural disasters is severely impacted when the funds earmarked are siphoned off for rescue, relief, and rehabilitation processes. Also, the contingency plans usually do not refer much to the necessary disaster prevention. Thus, a comprehensive and proactive risk management effort is necessary for covering risk assessment and risk evaluation.

A community/locality is said to be at "risk" when it is exposed to hazards and is likely to be adversely affected by its impact. Whenever we discuss "disaster management," it is basically "disaster risk management." DRM includes all measures that reduce disaster-related losses of life, property, or assets by reducing either the hazard or the vulnerability of the elements-at-risk.

3.1.1 What Is Risk?

Risk is a "measure of the expected losses due to a hazard event occurring in a given area over a specific time period. Risk is a function of the probability of particular hazardous event and the losses it would cause." The level of risk depends upon

- Nature of the hazard.
- Vulnerability of the elements that are affected.
- Economic value of those elements.
- Risk includes two elements—the likelihood of something happening, and the consequences if it happens (Beer and Ziolkowski 1995).
- Risk occurs where factors and processes are sufficiently measurable for believable probability distributions to be assigned to the range of possible outcomes (Dovers 1995).
- Risk is the perceived likelihood of given levels of harm (EMA 1995).

Uttarakhand (INDIA) Flash Floods 2013

A major portion of the state of Uttarakhand is located in the Himalayan terrain and has altitudinal range of 200 to 7,784 m above sea level (asl). The state shares its border with Nepal in the east and Tibet (China) in the north. The state has 2 administrative divisions, Garhwal and Kumaun, and 13 districts. Of these, five northern districts, namely, Bageshwar, Pithoragarh, Uttarkashi, Chamoli, and Rudraprayag, were the worst affected by the disaster of 2013.

Geologically, the disaster-affected area falls in the Lesser Himalaya, Central Crystallines, and Higher Himalaya. The Main Central Thrust that is a major tectonic discontinuity of the Himalayas and along which the Central Crystallines is juxtaposed against the Lesser Himalaya along a north-northeast (NNE) dipping thrust traverses through it. The area has particularly high relative relief that promotes mass wastage and erosion. Except for Uttarkashi, some portion of which falls in Zone IV, all the disaster-affected districts fall in Zone V of the *Seismic Zonation Map of India* (IS 1893, 2002). Geological history, ongoing tectonic activities, and high relative relief coupled with peculiar meteorological characteristics make the area vulnerable to a number of hazards of which earthquake, landslide, and flash flood are common.

The disaster-affected area of the state is source to major glacier-fed Himalayan rivers that include the Alaknanda, Bhagirathi, Mandakini, Yamuna, Kali, Dhauli, and Pinder. Alaknanda and Bhagirathi confluence at Devprayag, and thereafter the river is known as the Ganga.

The Mandakini valley of Rudraprayag district that was hit the hardest by the disaster of June 2013, houses the sacred shrine of Kedarnath that is dedicated to lord Shiva, the God of death and destruction. The temple township is located on glacial outwash deposits at an altitude of 3,581 m asl. The main shrine is located on raised middle portion of the deposit that is 20–25 m above the level of Mandakini (3,562 m asl). For reaching Kedarnath, one had to trek upstream along the course of Mandakini from Gaurikund for a distance of 14 km.

Originating from the Chorabari glacier, Mandakini forms the western boundary of the temple township while the abandoned channel of Saraswati, that had confluence with Mandakini to the south of the temple, forms the eastern boundary. Dudh Ganga meets Mandakini to the south of Kedarnath and thereafter, till Gaurikund, Mandakini maintains a tectonically controlled north-northeast–south-southwest (NNE–SSW) course.

A moraine dammed lake, Chorabari Tal, was present a little downstream of the snout of the Chorabari glacier. This lake was located in the depression formed in the glacial material to the west of the right lateral moraine and was fed by the seepage of the glacial melt. The lake did not have a well-defined outlet and its water used to seep out along the moraine slope to the north-northwest (NNW) of Kedarnath. Even though the depression was around 200 m long, 100 m wide, and 15–20 m deep, not more than 2–3 m water used to be there in the lake (Figure 3.1).

Figure 3.1:
View of the Road Disruption in June 2013 Due to Bank Erosion (left) and Debris Slide in Dharali, Uttarkashi (right)

Source: Authors.

The Disaster Impacts of June 16–17, 2013

There was heavy rainfall in the entire state with the onset of monsoon that arrived early in 2013. This is attributed to the clash of the southwest (SW) monsoon front with the westerlies. Prolonged and unprecedented heavy rainfall

(Continued)

(Continued)

for consecutive days between June 14 and 18, 2013, over a large area, resulted in flash floods and landslides at many locations, which eventually turned into a massive disaster. The rainfall in the state between June 15 and 18, 2013 is measured to be 385.1 mm against the normal rainfall of 71.3 mm, which is in excess by 440 percent. In the period of 5 days between June 14 and 18, the state received approximately 2,000 mm of rainfall, which is more than what it receives during the entire monsoon period.

Percent deviation in rainfall clearly shows that the rainfall during the week ending on June 12 (June 6–12, 2013) was more than 100 percent in all the districts except Pithoragarh. The rainfall, however, increased enormously in the subsequent week when it was measured to be 997 percent higher than normal over the state. Except for Pithoragarh and Rudraprayag, deviation from normal in the other three districts was more than 1,000 percent in the week ending on June 19.

A clearer concept has been referred to as the "risk management cycle," or better "spiral," in which learning from a disaster can stimulate adaptation and modification in development planning, rather than as implore construction of pre-existing social and physical conditions. Chapter 1 illustrated the disaster cycle and its various components (relief, recovery, reconstruction, prevention, and preparedness), and how these changed through time. Initially, most of the emphasis was given to disaster relief, recovery, and reconstruction, thereby getting into a cycle where the next disaster was going to cause the same effects or worse. Later on, more attention was given to disaster preparedness by developing warning systems and disaster awareness programs. Now, the efforts are focusing on disaster prevention and preparedness, thus, increasing the time gap between individual disasters, and reducing their effects, requiring less emphasis in relief, recovery, and reconstruction. The eventual aim of DRM is to enlarge this cycle and to only reach the response phase for extreme events with very low frequency. Disaster prevention is achieved through risk management. Figure 3.2 presents the general risk-management framework, which is composed of a risk assessment block and a block in which risk-reduction strategies are defined. A summary of the terminology used in risk management is given in Table 3.1. Central to the procedure is risk analysis, in which the available information is used to estimate the risk to individuals or populations, and property or the environment from various hazards.

Risk analysis generally contains the following steps:

1. hazard identification;
2. hazard assessment;

Figure 3.2:
Risk Management Framework

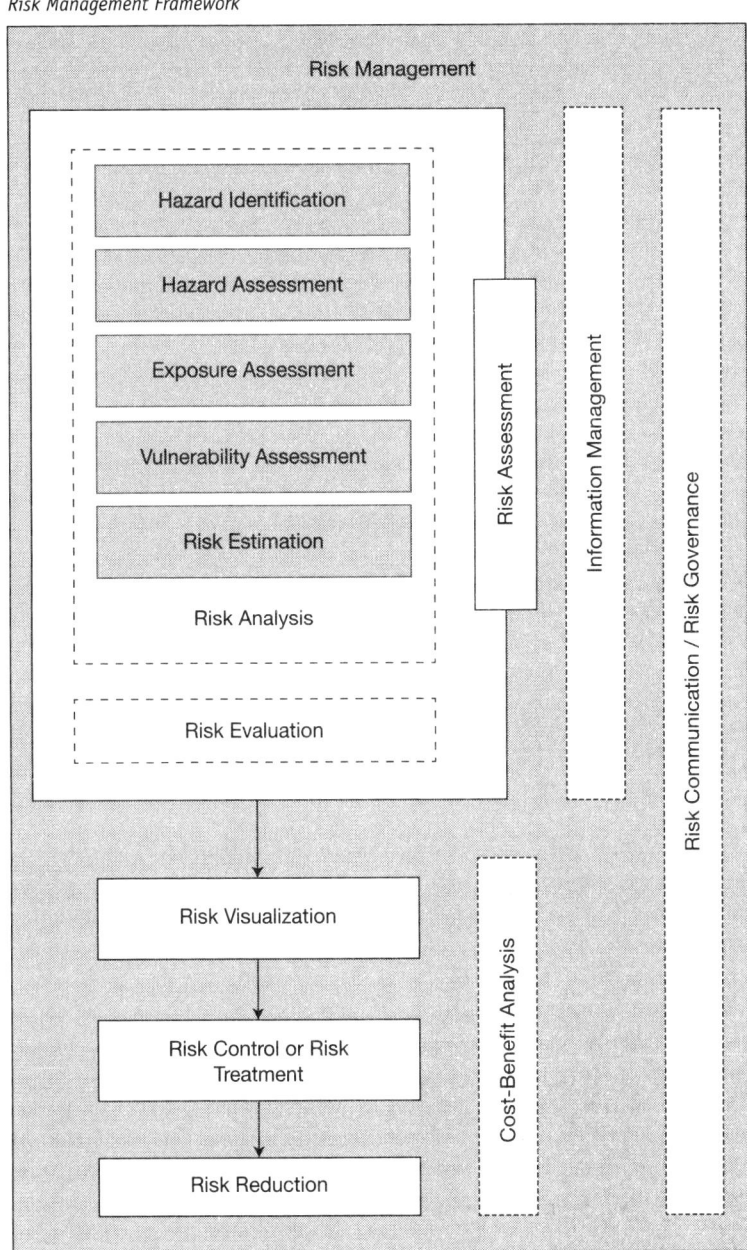

Source: Authors.

Table 3.1:

Summary of Definitions Related to Risk Management Term Definition Risk Analysis

Term	Definition
Risk analysis	The use of available information to estimate the risk to an individual or a population, property, or the environment from hazards. Risk analysis generally contains the following steps: hazard assessment, elements-at-risk/exposure analysis, vulnerability assessment, and risk estimation.
Risk evaluation	The stage at which values and judgements enter the decision process, explicitly or implicitly, by including consideration of the importance of the estimated risks and the associated social, economic, and environmental consequences, to identify a range of alternatives for managing the risks.
Risk assessment	The process of risk analysis and risk evaluation.
Risk control or risk treatment	The process of decision-making for managing risks, and the implementation or enforcement of risk-mitigation measures and the re-evaluation of its effectiveness from time to time using results of risk assessment as one input.
Risk management	The complete process of risk assessment and risk control (or risk treatment).

Source: Based on UNISDR (2004). Terminology of disaster risk reduction. United Nations International Strategy for Disaster Reduction. Geneva, Switzerland. http://www.unisdr.org/eng/library/lib-terminology-eng%20home.htm

3. elements-at-risk/exposure analysis;
4. vulnerability assessment; and
5. risk estimation.

Risk evaluation is the stage at which values and judgments enter the decision process, explicitly or implicitly, by including consideration of the importance of the estimated risks and the associated social, environmental, and economic consequences, in order to identify arrange of alternatives for reducing the risks (UN-ISDR 2004). Risk assessment is the combination of risk analysis and risk evaluation. It is more than a purely scientific enterprise and should be seen as a collaborative activity that brings professionals, authorized disaster managers, local authorities, and the people living in the exposed areas together (Montague 2004; O'Brien 2000). Risk governance is, therefore, an integral component. The final goal, reduction of disaster risk, should be achieved by combining structural and nonstructural measures that focuses on emergency preparedness (e.g., awareness raising, early warning systems, etc.), inclusion of risk information in long-term land-use planning, and evaluation of the most cost-effective risk-reduction measures (Figure 3.3). Spatial information plays a crucial role in the entire risk-management framework, as the hazards, as well as the vulnerable elements-at-risk, are spatially distributed. According to the risk analysis framework, as illustrated in Figure 3.4, there are three important components in risk analysis: (a) hazards; (b) vulnerability; and (c) elements-at-risk (van Westen, Castellanos, and Kuriakose 2008). They are characterized by both spatial and nonspatial attributes. Hazards are characterized by their temporal probability and intensity, which is derived from frequency-magnitude analysis. Intensity expresses the severity of the hazard, for example, water depth, flow velocity, and duration in the case of flooding. The hazard component in the equation actually refers to the probability of occurrence of a hazardous phenomenon with a given intensity within a specified period of time (e.g., annual probability). Hazards also have an important spatial component, both related to the initiation of the hazard (e.g., a volcano)

Figure 3.3:
Risk Analysis and Its Components

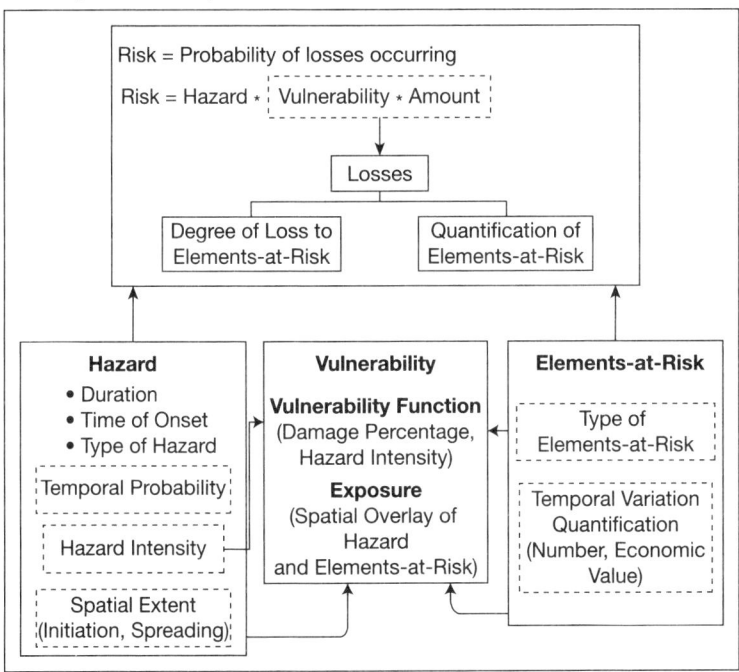

Source: Authors.

Figure 3.4:
The Probable Impacts of Natural Hazards on Different Sectors

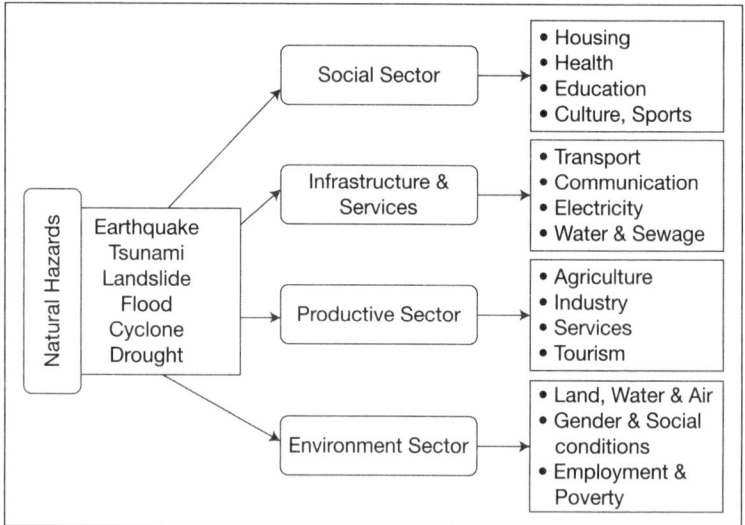

Source: Authors.

and the spreading of the hazardous phenomena (e.g., the areas affected by volcanic products such as lava flows; van Westen 2009). Elements-at-risk or "assets" are the population, properties, economic activities, including public services, or any other defined values exposed to the hazards in a given area (UN-ISDR 2004). Elements-at-risk also have spatial and non-spatial characteristics. There are many different types of elements-at-risk and they can be classified in various ways. The way in which the amount of the elements-at-risk is characterized (e.g., as number of buildings, number of people, economic value, or qualitative rating according to their importance) also defines the way in which the risk is presented. The interaction of the elements-at-risk and the hazard defines the exposure and the vulnerability of the elements-at-risk. Exposure indicates the degree to which the elements-at-risk are actually located in an area affected by a particular hazard. The spatial interaction between the elements-at-risk and the hazard footprints are depicted in a geographic information system (GIS) by spatial overlay of the hazard map with the elements-at-risk map (Van Westen 2009). Vulnerability refers to the conditions determined by physical, social, economic, and environmental factors or processes, which increase the susceptibility of a community to the impact of hazards (UN-ISDR 2004). The vulnerability of communities and households can be analyzed in a holistic, qualitative manner using a large number of criteria that characterize the physical, social, economic, and environmental vulnerability. The importance of each of these indicators is evaluated by assigning weights and combining them using spatial, multicriteria evaluation. Physical vulnerability is evaluated as the interaction between the intensity of the hazard and the type of element at risk, making use of the so-called vulnerability curves. For further explanations on hazard and risk assessment, see Alexander (1993), Okuyama and Chang (2004), Smith and Petley (2008), and Alcantara-Ayala and Goudie (2010).

3.2 Hazard Identification and Hazard Profiling

The first step that must be taken in any effective disaster management effort is the identification and profiling of hazards. It is only logical that a disaster manager concerned with treating a community's or nation's risk must first know what hazards exist and where they exist.

The actual number of possible hazards throughout the world is staggering, and the list is by no means limited to what is found in this, or any other, text. However, disaster managers must be able to identify those hazards that are most likely to occur and those that are most devastating should they occur. Understandably, it is impossible to plan for or prevent every possible contingency, so most governments and other organized emergency-management entities will focus their efforts upon those hazards that would be likely to result in the greatest undesirable consequences.

Disaster managers must attempt to identify every scenario that could possibly occur within a given community or country as result of its geologic, meteorological, hydrologic, biological, economic, technological, political, and social factors. This hazard assessment, as it is often called, must include not only the actual physical hazards that exist but also the expected secondary hazards, including social reactions and conditions.

A range of sectors are also significantly affected by the natural hazards as long- and short-term consequences of disasters. Social, infrastructure and services, productive sector and environment services related sectors are affected by single or multiple hazards flood, cyclone, earthquake, tsunami, landslide, flash floods, drought, etc.

3.3 Sequential Effects of Hazards

Hazardous process of all types can have primary, secondary, and tertiary effects.

> *Primary effects* occur as a result of the process itself. For example, water damage due to a flood and collapse of buildings due to an earthquake, landslide, cyclone, etc.
>
> *Secondary effects* occur only because a primary effect has caused them. For example, fires ignited by earthquakes, disruption of electrical power and water service as a result of an earthquake or flood, and flooding caused by a landslide or mudslide.
>
> *Tertiary effects* are long-term effects that are set off as a result of a primary event. These include things such as loss of habitat caused by a flood, permanent changes in the position of river channel caused by a flood, and crop failure caused by a floods or droughts.

3.4 Classifying Hazard and Risk

A hazard can create potential damage due to the release of stored energy in the form of chemical, mechanical, thermal, radioactive, electrical, or in any other form. Another possibility is a responsible hazardous situation, such as an infrastructural disorder, policy change, etc.

In the methodology of classifying a hazard, the important factors considered are most "likelihood" of the hazard turning into an incident and the "intensity" or "severity" of the incident.

Statistical analysis about risk is toward formulating the expected value of an undesirable incident. This generally comes out from the combination of the probabilities of possible events and the gross assessment of the consequent harm into a single value. In a simple way, there is a binary possibility of "incident" or "no incident." Then, the formula for calculating risk is

Risk = (probability of the incident) × (expected loss due to that incident)

As an example, if the potential activity "X" has a probability of 0.05 of suffering out of an incident A, with a loss of 2,000, then total risk is a loss of 1,000, the product of 0.05 and 2,000. In reality, the situations are often very complex and do not behave in a simple binary mode. In that case, a situation will have multivariate possible incidents, and then the total risk will be the sum of the overall risk considering each different incident, with a comparable outcome:

Risk = All incidents $\sum$ (probability of the incident) × (expected loss due to that incident)

As an example, if the potential activity "X" has a probability of 0.01 of suffering out of an incident A, with a loss of 1,000, and a probability of 0.000001 of suffering out of another incident B, with a loss of 2,000,000, then total loss expectancy is 12, which is equal to a loss of 10 from an accident of type A and 2 from an accident of type B.

Similarly, the common practice is to consider both the likelihood of occurrence and severity on a numerical scale and formulate as below:

Risk = Hazard × Vulnerability (−) Capacity to cope

This score is generally used to identify the hazards that are necessarily to be mitigated. A lower score on the likelihood of occurrence indicates a dormant hazard, while a high score is toward an active hazard. The severity of the incident occurred is very much relative, as the character of the different social groups are variable in nature, as are their capacity to cope with the incident.

So risk is the likelihood of an accident plus the severity of the potential consequences. The factors influencing the risk are new or unforeseen hazards, increasing population and complexity, exposure, energy, and policy and technological change in any system. So, there is a strong interrelation between hazard, risk and vulnerability, and all their components (Figure 3.5). In the process of accumulating risk and hazard, the exposure to natural hazards is increasing.

Risk assessment involves not only the assessment of hazards from a scientific point of view but also the socioeconomic impacts of a hazardous event. Risk is a statement of probability that an event will cause X amount of damage, or a statement of the economic impact in monetary terms that an event will cause. Risk assessment involves

- hazard assessment;
- location of buildings, highways, and other infrastructure in the areas subject to hazards;
- potential exposure to the physical effects of a hazardous situation; and
- the vulnerability of the community when subjected to the physical effects of the event.

Risk assessment aids decision-makers and scientists to compare and evaluate potential hazards, and set priorities on what kinds of mitigation are possible where to focus resources and further study. Vulnerability is the level of

Figure 3.5:
The Linkage between Hazard, Risk, and Vulnerability

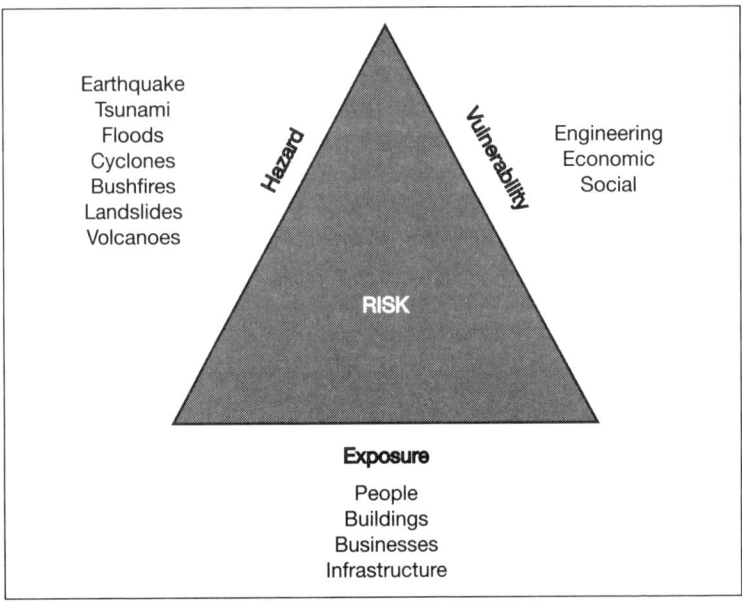

Earthquake
Tsunami
Floods
Cyclones
Bushfires
Landslides
Volcanoes

Hazard

Vulnerability

Engineering
Economic
Social

RISK

Exposure
People
Buildings
Businesses
Infrastructure

Source: Authors.

relative damage that can occur to a community/individual/structure during a disaster. Resilience is the capacity to recover after a disaster.

- Identification
- Vulnerability index and profile
- Implication of vulnerability with hazard and risk of the area
- Poverty and vulnerability

Disaster = Vulnerability + Hazard

can be redefined as

Natural Disaster = Natural Events + Absence of Preparedness

3.5 Risk Assessment

Recipients of warning messages respond better when the warnings include the hazard's potential impacts to people's safety, livelihood systems, infrastructure,

Figure 3.6:

Interaction of Climate-related Hazards with Vulnerability and Exposure of Human and Natural Systems

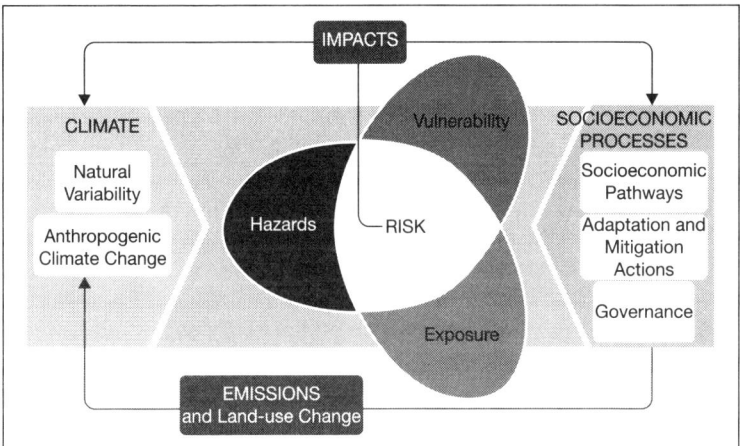

Source: IPCC (2014).

etc. Potential impacts or risk results from the interaction of vulnerability, exposure, and hazard (Figure 3.5). Vulnerability refers to the predisposition to be adversely affected, while exposure is the presence of people, livelihoods, infrastructure, ecosystems, and other assets in the places and settings that could be adversely affected. Hazard, in the context of this toolkit, is a weather event that could trigger a transboundary flood (Figure 3.6).

3.6 Assessing Hazard and Risk

Hazard assessment and risk assessment are not synonymous!

Hazard Assessment consists of determining the following:

- when and where hazardous processes have occurred in the past;
- the severity of the physical effects of past hazardous processes (magnitude);
- the frequency of occurrence of hazardous processes;
- the likely effects of a process of a given magnitude, if it were to occur now; and

- making all this information available in a form that us useful to the planners and public officials responsible for making decisions in event of a disaster.

A *vulnerability index* is the measure of exposure of any population to the identified hazard. Eventually, this index is the composite of multiple quantitative indicators considering established formulas to deliver a single numerical output. This index has the capacity to combine diverse issues into a common framework, with possible comparison. In a more simple form, the physical indicators can be combined with the social, health, and often with the psychological variables too to estimate potential complexity in DRR.

What are the ways, we can avoid disasters?

- Do nothing.
- Avoid or remove the hazard.
- Weaken the hazards.
- Send away the hazards somewhere else.
- Do not allow the vulnerable people and society to live in the hazardous areas. Also do not allow the environment and economy to become vulnerable in hazardous areas.
- Convert vulnerable people/society/environment and vulnerable economy in the hazardous areas into nonvulnerable ones.
- Reduce the vulnerability of people, society, environment, and economy.
- Prepare very well to face the disasters, and take all steps much before disaster strikes.

3.7 Risks and Development

3.7.1 Disaster and Development: Key Concepts and Issues

Disaster impacts considerably all the sectors of development and thus results in a serious social and economic setback to the development. On the other hand, the process of development, and the kind of development choices made in many countries, sometimes creates disaster risks. The intricate relationship between disaster and development is outlined in Table 3.2.

Further, mainstreaming is a cross-cutting issue that requires political commitment, public understanding, scientific knowledge and know-how, responsible risk-sensitive development planning and practice, a

Table 3.2:
Three Dimensions of Development and Disaster Linkage

	Economic Development	*Social Development*
Disaster limits development	• Destruction of fixed assets. Loss of production capacity, market access, or material inputs. • Damage to transport, Communications or energy infrastructure. Erosion of livelihoods, savings, and physical capital.	Destruction of health or education infrastructure and personnel. Death, disablement or migration of key social actors leading to an erosion of social capital.
Development causes disaster risk	Unsustainable development practices that create wealth for some at the expense of unsafe working or living conditions for others or degrade the environment.	Development paths generating cultural norms that promote social isolation or political exclusion.
Development reduces disaster risk	Access to adequate drinking water, food, waste management, and a secured welling increases people's resiliency. Trade and technology can reduce poverty. Investing in financial mechanisms and social security can cushion against vulnerability.	Building community cohesion, recognizing excluded individuals or social groups (such as women), and providing opportunities for greater involvement in decision-making, enhanced educational and health capacity increases resiliency.

Source: Reducing Disaster Risk: A Challenge for Development, UNDP.

people-centered early warning system and disaster response mechanisms. In addition, safeguarding human rights and integrating gender concerns are central to achieving mainstreaming concepts at the local and national levels. Because disaster risks impact multisectoral development activities (such as education, health, environment, governance, employment, and livelihoods), they influence development gains. Therefore, an assessment of the extent to which these social domains consider natural or human-induced factors of risks (existing and prospective) in the conceptualization and implementation of programs is crucial. This also means that development programs need to assess whether a development project could cause/ increase the risk of any kind of disaster in future, and if it is necessary to identify/introduce counter-measures for risk control.

There is an emerging consensus that the key to achieving sustained reductions in disaster losses lies in factoring risk considerations into both development and post-disaster recovery activities. Managing risks could become a means of reducing future disaster risks through "corrective" development planning that ensures, through measures such as land-use planning, building controls and others, that development activities do not generate new risks.

On the emerging contours of development and disasters, the Second Asian Ministerial Conference on Disaster Risk Reduction (AMCDRR) in November 2007 at New Delhi had organized a special panel discussion covering three aspects: (a) disaster impending development, (b) developing creating disasters, and (c) development without disasters. It was recognized that most of the countries of Asia have very high risks of disasters and that they are also on a high trajectory of economic growth. Therefore, the challenge of the Asian countries in the coming years and decades would be to develop in a manner that would reduce the risks of disasters. It was felt that the challenges are more formidable for the high-risk and fast-developing South Asian region.

It was also recognized that the economic development that has a spinoff effect on housing, education, nutrition, health, etc., does help in vulnerability reduction; however, there is always the danger that unplanned growth of human settlements and unhindered exploitation of natural resources, especially in low-income, high-growth South Asian economies, would create new risks in the long run. Therefore, mainstreaming DRR in development would be one of the most challenging tasks of development planning in the coming years. Innovative tools and methodologies have to be developed to ensure that development does not create new disasters and that the risks of disasters created by unplanned developments in the past are reduced in the future. These tools have to be tested; further adapted according to the local needs, capacities, and resources; and applied in a systematic and sustainable manner through a participatory process. Yet another challenge would be incorporating DRR in climate change adaptation and coping with high-density urban growth. Innovative solutions are required to address these challenges. Capacity development of various stakeholders at national, regional, and local levels, exchange of information and good practices, and regional cooperation would be the key components in any innovative solutions.

3.7.2 Risk and Development

With 23 percent of the global population but only 1.3 percent of the global income, South Asia remains one of the most underdeveloped regions of the world. This is reflected in the poor physical, social, economic, and human development index (HDI) of almost all the countries of the region. The region as a whole is home to more than 40 percent of the world's poor, malnourished, and illiterate. The geoclimatic conditions of the region are such that make the countries of the region highly vulnerable to natural hazards of every description. Unplanned human settlements, unsafe building practices, and high density of population, particularly in the growing urban areas, have further compounded the complex matrix of hazards, risks, and vulnerabilities of the region. The end results are disasters of every type and magnitude that visit the region at regular intervals, consuming life, property, and livelihood of hundreds of thousands of people every year. In the year 2007, for example, out of 10 worst natural disasters of the globe, seven were in South Asia. This is not an isolated example of a year, but the general trend of years and decades.

The disasters have been eroding, over minutes, hours, or days, hard earned gains of development of years and decades. It is estimated that the countries of the region have been losing between 2 percent to 20 percent of their GDP and 12 percent to 66 percent of the revenues on account of disasters every year. These do not include the losses in some of the informal sectors of the economy that generally go unaccounted or the long-term damage and loss of environment and ecology that cannot be measured immediately: Some of the countries are not been able to spend as much on social sectors like public health or education because of the loss due to disasters. On top of it, almost all the countries of the region are forced to divert scarce resources to disaster relief, rehabilitation, and reconstruction, which create further setbacks to development. It is now quite evident that natural and manmade disasters in South Asia are one of the important barriers to the realization of the Millennium Development Goals of the United Nations.

The link between disaster and development has been appreciated by the countries of South Asia only recently. All the countries of the region have endorsed the *HFA 2005–2015: Building the Resilience of Nations and Communities to Disasters* that envisages "integrating risk reduction into development policies and plans at all levels of Government including poverty reduction strategies and multisectoral policies and plans." The countries in the South Asian Association for Regional Cooperation (SAARC) have adopted the SAARC Comprehensive Framework on Disaster Management that identifies "mainstreaming disaster risk reduction into the development policies and practices of the government at all levels" as one of the key priority areas for developing resilience to disasters in SAARC region. Each country has developed its disaster management framework that commits mainstreaming these commitments into practices—words into actions.

The South Asian countries are poorly placed in almost all the selected world development indicators published annually by the World Bank. In terms of HDI of the UNDP for the year 2008, the global ranking of the eight South Asian countries out of the 177 countries of the world are Afghanistan (166), Bangladesh (140), Bhutan (133), India (128), Maldives (100), Nepal (142), Pakistan (136), and Sri Lanka (99).

3.7.3 Disaster Proofing Development

Development is definitely linked with the processes involved in the disaster management efforts and strategies. When the development process is positive, it turns the people or society more resilient or less vulnerable. Thus, the impact will also be minimal. It contributes to the effort of disaster mitigation, as proper development can minimize the cost of the recovery processes after any disaster (Figure 3.5).

3.8 DRR Mainstreaming

3.8.1 Mainstreaming DRR in Development in South Asia

South Asian countries have just made a beginning toward mainstreaming DRR in development. Pursing the HFA in the respective countries has led to some "foundation"-level initiatives that would facilitate more specific national and local-level activities on mainstreaming DRR. These include:

- Development of a legislative framework and the institutional capacity to prevent, mitigate, prepare, and manage hazards and disasters;
- Undertaking hazard, risk, and vulnerability assessments;
- Developing education, training, and information exchange programs;
- Raising awareness of the community;
- Development of partnerships with the stakeholders at each level;
- The utilization of cooperative and information-sharing mechanisms and institutions across the region.

While there are efforts in South Asian countries to implement DRR in development through National Adaptation Plans of Action (NAPAs), their integration to DRR needs specific priority. In order to address adaptation concerns as part of their national development plans, the explicit focus on disaster risk is seen only in few cases. For example, the Safe Island program of the Maldives is an integrated effort on addressing vulnerability through strategic planning for climate change adaptation. Similarly, Bhutan has initiated plan of action in this direction through NAPA. It is expected that all member countries develop respective NAPAs with an aim to mainstream DRR in development.

Specific entry point activities for mainstreaming DRR in development have been taken up in the multihazard-prone regions.

3.8.2 Entry Points for Mainstreaming Disaster Risk Reduction in Development

Education: The building of appropriate school structures, which not only adhere to safety measures but may also be useful as disaster shelters, and the development of curricula and institutionalization of safety drills that provide information on DRR, particularly targeting women and children.

Health: Ensuring the suitability of health infrastructure, compliance with building codes, availability of and accessibility to goods and services (especially in times of emergency), and increased capacity to prepare for disaster events and the outbreak of infectious diseases.

Environment: Integrating disaster risk concerns into existing environmental assessment tools and planning mechanisms (environmental impact assessments and strategic environmental assessments), promoting greater compliance to existing environmental and risk management regulations, promoting integrated approaches to spatial planning, strengthening capacities to protect ecosystem services that reduce disaster risk (wetlands, coastal forests, watersheds, coral reefs, etc.), identifying potential sources of hazardous materials that can trigger acute environmental emergencies, and strategically assessing the environmental impacts of proposed post-disaster recovery plans.

Governance: Efficiency and accountability of governance structures at Central and local levels should be strengthened, encouraging more inclusive and participatory decision-making processes. Local and national governments designing and applying regulatory frameworks that ensure a safer environment, reduce structural vulnerabilities, and guide social behavior and economic decisions toward risk reduction and disaster prevention.

Employment and livelihoods (including informal sector): Considering the possible impacts of disasters on livelihoods and jobs, particularly those affecting the informal sector and youth. Promoting innovative mechanisms to reduce underlying risk such as microfinance and risk-transfer schemes, targeting especially women. Promoting greater compliance to existing workplace safety regulations and environmental standards and raising awareness of DRR measures in relevant sectors (e.g., engineers/construction sector, chemical industry, etc.).

Agriculture: Increasing agricultural productivity through investments in soil health, water management, extension services, and research increases food availability for subsistence farmers. However, special focus is needed to mitigate the impact of hydrometeorological fluctuations through multiple cropping, water conservation, and biological control measures, with contingency cropping strategies linked to weather monitoring and early warning systems.

Gender: Improving women's participation in decision-making processes. Productive activities should specifically include awareness of disaster risks, preparedness, and preventive measures that reinforce traditional coping measures undertaken by women and increase disaster resilience of communities. Research on the degree to which women suffer the negative impact of disasters could be undertaken to better understand and address their specific vulnerabilities and needs.

Information and communication technologies: Taking steps to strengthen science advisory mechanisms, invest in higher education and research, promote private-sector development, and improve access to communication technologies can also be linked to better hydrometeorological monitoring, seismic risks monitoring, the and possibility of feeding into better early warning systems to save both lives and livelihoods (UNDP Disaster Risk Reduction Module, March 2007).

3.9 Tools and Techniques for DRR Mainstreaming

Realizing the regional imperatives, national needs and prospects of mainstreaming, the following mechanisms and instruments need to be developed and further strengthened while formulating plan of action for mainstreaming DRR in development.

1. *Identify development-induced disasters:* It is a well-known fact that inappropriate development processes are contributing to risk accumulation. There are many examples demonstrating how economic growth and social improvement lead to increase in disaster risk. Rapid urbanization is an example. The growth of informal settlements and inner city slums, whether fueled by international migration or internal migration from smaller urban settlements or the

countryside, has led to the growth of unstable living environments. These settlements are often located in ravines, on steep slopes, along flood plains, or adjacent to noxious or dangerous industrial or transport facilities. One such development has led to increase in risk due to landslides in the urban areas of Chittagong in Bangladesh. This is true in other megacities as well and in rapidly expanding small- and medium-sized urban centers. When the population expands faster than the capacity of urban authorities or the private sector to supply housing or basic infrastructure, risk in informal settlements can accumulate quickly. In cities with transient or migrant populations, social and economic networks tend to be loose. Many people, especially minority or groups of low social status, can become socially excluded and politically marginalized, leading to a lack of access to resources and increased vulnerability.

2. *Develop guidelines on mainstreaming:* All development projects should have a mandatory guideline to address how exactly it is going to implement DRR in terms of social and physical vulnerability. Risk can be reduced by making efforts wherein either the vulnerability or the exposure is reduced. Risk can also be reduced by reducing the hazard probability, for example, while undertaking a road construction in hilly area, the slope stability measures can be built in such a way that the hazard probability can decrease, thus reducing the overall risk. Similarly, the poverty alleviation or education program can also reduce the social vulnerability, thus reducing overall disaster risk. Similarly, limiting development in high-risk areas could reduce exposure and thus overall risk will be reduced.

3. *Develop sector-specific guidelines on mainstreaming:* It is necessary that appropriate strategy is developed to mainstream DRR into following specific sectors with clear-cut guidelines and objectives. Some of the suggestive sectoral guidelines could be as under:

Infrastructure: public works, roads, and construction

- Promote use of hazard risk information in land-use planning and zoning regulations.
- Conduct disaster risk impact assessments as part of the planning process before the construction of new roads or bridges.

4. *Carry out cross-sectoral risk analysis:* Cross-sectoral risk analysis needs to be carried out at national, local, and regional levels. Ongoing schemes across the sectors should be critically revisited and, wherever possible, the development aspects of these schemes should be

integrated for a better result. This should be done in a futuristic mode with immediate medium- and long-terms planning. For example, if a hydroelectric project is being implemented, attempts must be made to assess the change in the hydrological regime and its impact on soil erosion and landsliding. This would require a multidisciplinary approach across sectors.

5. *Develop area-specific guidelines on mainstreaming:* Area-specific guidelines for mainstreaming DRR in development should be formulated with particular reference to coastal and hilly areas that are prone to disasters.

   • *Coastal zone management:* Coastal zone management would be critical for environment, natural resources, climate change adaptation, and DRR as well. It would then lead to a holistic development of the coastal zones in the region that caters to a significant population of South Asia, the majority of whom are poor and vulnerable to any type of disaster. Therefore, in any coastal zone management effort, DRR with respect to multiple hazards must be considered.

   • *Hilly area development:* As South Asia encompasses large tracts of hilly area, it is important to use all developmental initiatives specific to hilly area to implement a DRR strategy that is very critical for environmental protection and sustainable development. It would then lead to a holistic development of the hilly area and its population, the majority of whom is poor and vulnerable to disaster and is often isolated from the mainland of development.

6. *Create techno-legal regime for mainstreaming:* It is necessary that an appropriate techno-legal mechanism is developed to implement the regulations made with respect to the DRR strategy. There may be a statutory organization responsible for the undertaking of the assessment on compliance and the implementation on the ground. For example, the hydroprojects have a mandatory provision of afforestation and it is imperative that it is implemented on ground and that proper assessment is done with respect to its positive impact.

7. *Conduct disaster impact assessment:* The assessment of the potential risks to any place (village, city, nation, etc.) or elements (infrastructure/land use, etc.) is the major part of the disaster impact assessment (DIA) related to any developmental activity. Therefore, it is necessary to consider all possible impacts of various hazards that may arise due to the implementation of a project. This entire exercise could be very complex and may require comprehensive assessment

of data related to natural as well as social sectors. Some elements of the Disaster Impact Assessment (DIA) are similar to well known practice of Environmental Impact Assessment (EIA) and therefore, it must be pursued under similar guidelines.

8. *Private–public partnership:* In the present scenario, it is visualized that more and more unorganized and organized private sectors would play a major role in developmental activities. It is important to foster collaboration with the private sector in a public-private partnership to address the implementation of DRR in a development initiative. This partnership could play a key role in communication, infrastructure, market, health, and many others areas. Recently, a leading software industry in Hyderabad, India, has demonstrated a disaster response system for the citizens of the city, which is operational 24/7 and is fully endorsed by the government.

9. *Research and development:* It is one of the major elements of mainstreaming disaster mitigation/reduction into development. Research and development (R&D) capacity in earthquake, flood, drought, climate change, industrial and nuclear disaster, and many other fields must identify areas and strategies on how to identify risk at an early stage in a holistic manner and minimize it by suitably integrating mitigation measures into the development model. Various professional scientific organizations must re-orient their program to support the safe developmental needs. For example, the road development agencies must take into account the present requirement of mass transport and suggest suitable infrastructure that is viable and environmentally sustainable.

10. *Awareness generation, training, and capacity building:* It is important to make all stakeholders aware about the coupling of disaster and development. It must be understood and communicated that there exist a mechanism by which development can be implemented with DRR provisions. This awareness will lead to public demand for disaster audit and in turn will ensure sustainable development. It is important to note that awareness development must be initiated at all levels, starting from school curricula to basic training in safe construction to advance project management. Capacity-building through education, training and mid-career intervention using on-campus as well as off-campus models must be implemented for quickly covering a large manpower base. Building on capacities that deal with existing disaster risk is an effective way to generate capacity to deal with future risks arising out of new context, which is often not visualized.

11. *Recognition of best efforts:* Recognition of efforts is one of the best incentives that promotes and attracts many to emulate the good practice in implementing DRR in development. It also acts as a stimulant for the recipients to carry on the good work and innovate ways in which the efforts will have far reaching results across the society. Numerous such examples can be cited from the drought management and poverty alleviation programs that are being implemented in the western part of India and have received international accolades.

Figure 3.7 presents the series of steps necessary for successful mainstreaming of disaster risk reduction in hazard-prone countries. Not all the steps are necessarily sequential in order as presented, but rather overlap with each other in practice.

Figure 3.7:
Steps to Successful DRR Mainstreaming

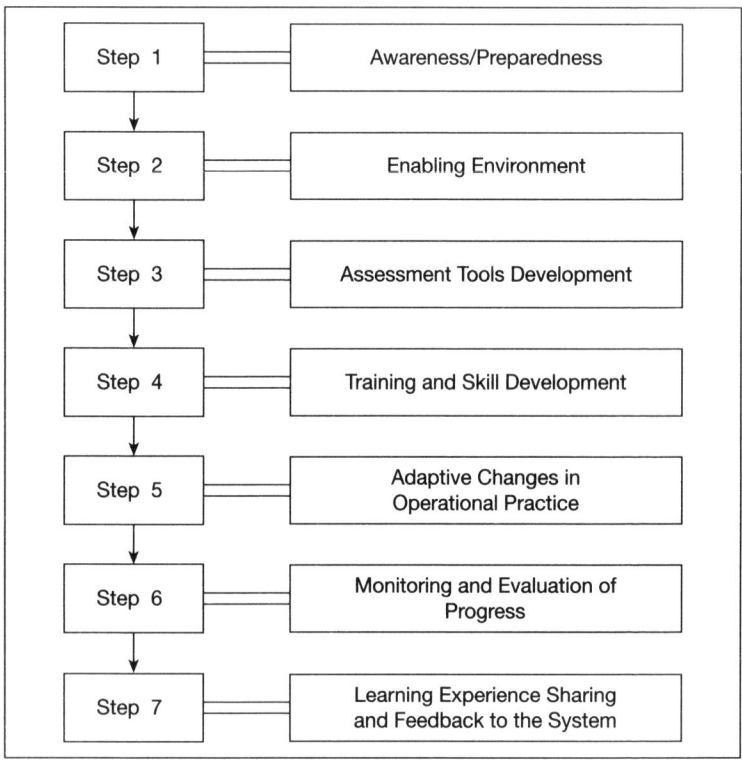

Source: Modified after Benson, Charlotte, and Twigg (2007).

Chapter 4

Vulnerability

4.1 What Is Vulnerability?

Vulnerability may be defined as "[t]he extent to which a community, structure, services or geographic area is likely to be damaged or disrupted by the impact of particular hazard, on account of their nature, construction and proximity to hazardous terrains or a disaster prone area." Vulnerabilities can be categorized into physical and socioeconomic vulnerability. Physical vulnerability includes notions of whom and what may be damaged or destroyed by natural hazards such as earthquakes or floods. It is based on the physical condition of the people and elements-at-risk, such as buildings, infrastructure, etc., and their proximity, location, and the nature of the hazard. It also relates to the technical capability of the building and structures to resist the forces acting upon them during a hazard event.

Vulnerability refers to not only the possible physical effects of a natural hazard but also the way it affects human life and property, and also the capacity to recover from the impact. Vulnerability to a given hazard depends on the following elements or characters (see Figure 3.4):

- proximity to a possible hazardous event;
- population density in the area proximal to the event;
- scientific understanding of the hazard;
- public education and awareness of the hazard;
- existence or nonexistence of early-warning systems and lines of communication;
- availability and readiness of emergency infrastructure;
- construction styles and building codes; and
- cultural factors that influence public response to warnings.

A number of parameters from the existing conditions are generally considered to identify the adverse effects on the community's ability to withstand, recover, mitigate, and prepare for a hazard. The absence of preparedness is a major component of vulnerability and has to be incorporated in the vulnerability assessment. The geographic proximity to the source and origin of the disasters is a prime factor for estimation of physical vulnerability. Physical vulnerability also includes the difficulty in access to water resources, health services, law enforcement departments, fire services, means of communications, roads, bridges, etc., in case of disasters. The socioeconomic vulnerability of a community is usually indicated by a weak family structure and financial capabilities, lack of understanding for decision making and conflict resolution, etc., and if discriminated on caste, ethnicity, linguistic, or religious basis. Varied sources of income, the access and control over means of production (e.g., farmland, livestock, irrigation, capital, etc.), and the availability of natural resources in the area are common criteria to assess the economic vulnerability of any community. A vulnerable environmental system cannot cater the people with its resources and services to overcome their physical and socioeconomic vulnerability.

In developing countries, the poorest person are socially vulnerable because of the lack of information and access to resources. Within this community, children, women, and the elderly are considered to be the most vulnerable. Knowledge and understanding of the local conditions provided by the local actors must be considered to reduce the socioeconomic vulnerability.

Sometimes, a community may have negative attitude toward change and lack of initiative and may become increasingly dependent on external support. Their livelihood options may not have variety, lack entrepreneurship, and never possess the concept of collectivism. This produces a brittle society and individualism in the society. In this process, that society becomes victim of conflicts, unproductive, and pessimistic, which reduces their capacity of coping with a disaster.

4.1.1 Physical Vulnerability

4.1.1.1 HUMAN VULNERABILITY

Humans are vulnerable to environmental extremes of temperature, pressure, and chemical exposures that can cause death, injury, and illness. For any hazard agent—water, wind, ionizing radiation, toxic chemicals, and infectious agents—there often is variability in the physiological response of the affected population. That is, given the same level of exposure, some people will die, others will be severely injured, still others slightly injured, and the rest will survive unscathed. Typically, the most susceptible to any environmental stressor will be the very young, the very old, and those with weakened immune systems.

4.1.1.2 Agricultural Vulnerability

Like humans, agricultural plants and animals are also vulnerable to environmental extremes of temperature, pressure, chemicals, radiation, and infectious agents. Like humans, there are differences among individuals within each plant and animal population. However, agricultural vulnerability is more complex than human vulnerability because there is a greater number of species to be assessed, each of which has its own characteristic response to each environmental stressor.

4.1.1.3 Structural Vulnerability

Structural vulnerability arises when buildings are constructed using designs and materials that are incapable of resisting extreme stresses (e.g., high wind, hydraulic pressures of water, and seismic shaking) or that allow hazardous materials to infiltrate into the building. The construction of most buildings is governed by building codes intended to protect the life safety of the building occupants from structural collapse—primarily from the dead load of the building material themselves and the live load of the occupants and furnishings—but do not necessarily provide protection from extreme wind, seismic, or hydraulic loads. Nor do they provide an impermeable barrier to the infiltration of toxic air pollutants.

4.1.2 Social Vulnerability

The social vulnerability perspective (e.g., Cannon, Twigg, and Rowell 2003; Cutter, Boruff, and Shirley 2003) represents an important extension of previous theories of hazard vulnerability (Burton et al. 1978). As a concept, social vulnerability has been defined in terms of people's "capacity to anticipate, cope with, resist and recover from the impacts of a natural hazard" (Wisner et al. 2004). Whereas people's physical vulnerability refers to their susceptibility to biological changes (i.e., impacts on anatomical structures and physiological functioning), their social vulnerability refers to their susceptibility to behavioral changes. As discussed in greater detail further, these consist of psychological, demographic, economic, and political impacts.

The central point of the social vulnerability perspective is that, just as people's occupancy of hazard-prone areas and the physical vulnerability of the structures in which they live and work are not randomly distributed, social vulnerability is also not randomly distributed—either geographically or demographically. Thus, just as variations in structural vulnerability can increase or decrease the effect of hazard exposure on physical impacts (property damage and casualties), so too can variations in social vulnerability. Social vulnerability varies across communities and also across households within communities. It is the variability in vulnerability that is likely to be of the greatest concern to local emergency managers because it requires that they identify the areas within their communities having population segments with the highest levels of social vulnerability.

4.1.2.1 Socio-Economic Vulnerability

The degree to which a population is affected by a hazard will lie not merely in the physical components of vulnerability but also in the socioeconomic conditions (Figure 4.1). The socioeconomic conditions of the people also

Figure 4.1:
Components of Vulnerability

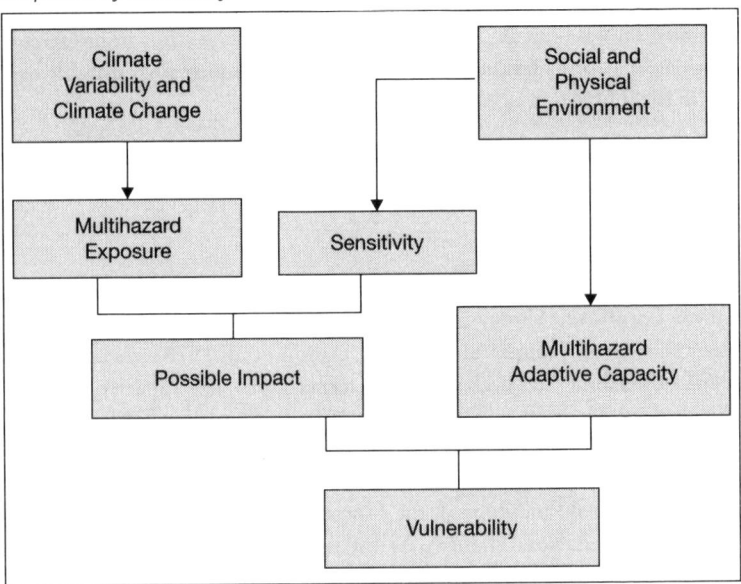

Source: Modified after Adelphi/EURAC 2014, p. 58.

determine the intensity of the impact. For example, people who are poor and living on the sea coast don't have the money to construct strong concrete houses. They are generally at risk and lose their shelters whenever there is strong wind or cyclone. Because of their poverty, they are not able even to rebuild their houses.

4.1.2.2 WHAT IS CAPACITY?

Capacity can be defined as "[the] resources, means and strengths which exist in households and communities and which enable them to cope with, withstand, prepare for, prevent, mitigate or quickly recover from a disaster." People's capacity can also be taken into account. Capacities could be classified into physical and socioeconomic capacities.

Physical capacity: People whose houses have been destroyed by the cyclone or crops have been destroyed by the flood can salvage things from their homes and from their farms. Some family members have skills that enable them to find employment if they migrate, either temporarily or permanently.

Socioeconomic capacity: In most of the disasters, people suffer their greatest losses in the physical and material realm. Rich people have the capacity to recover soon because of their wealth. In fact, they are seldom hit by disasters because they live in safe areas and their houses are built with stronger materials. However, even when everything is destroyed, they have the capacity to cope with it. Hazards are always prevalent, but the hazard becomes a disaster only when the frequency or likelihood of a hazard and the vulnerability of the community increases the risk of being severely affected.

4.1.3 Environmental Vulnerability

Any natural environment is unequivocally the life support system for the society dependent on that, with its entire human well-being. Successful environment management is increasingly becoming the weapon for the success or failure of a robust economic and social system, and the lucrative lifestyle in developed countries can really "afford" it. Environmental management is being treated as a high priority in countries for development projects and also at a global scale through international agreements. The approaches being used are largely concerned about the state of the environment and the anthropogenic interventions on the ecosystem, its resources, and wastes generated. The concern has culminated into developing guidelines for action, wise practice, way of conservation, and limiting exploitation, degradation, and pollution. This is a critical approach toward effective environmental management, which is necessary to ensure a sustainable future. They need not to always focus on optimization or the cumulative outcome of all the actions and management approaches over varied scales of time or space. Even a country with a decent current state of the environment may be converted to a highly vulnerable one, in due course of time, for several environmental changes.

4.2 Poverty and Disaster Vulnerability

Poverty and risk to disasters are inextricably linked and mutually reinforcing. The poor section of the society is worst affected in case of disaster. The situation further aggravates due to the compulsion of the poor to exploit environmental resources for their survival, increasing the risk and exposure of the society to disasters, in particular those triggered by flood, drought, and landslides. Poverty also compels the poor to migrate and live at physically more vulnerable locations, often on unsafe land and in unsafe shelters. These poor inhabit such locations either due to the fact that there is no other land available at reasonable cost or it is close to the employment opportunities. The poor people who inhabit on marginal land are prone to all types of disasters. The type of construction of these houses further deteriorates the condition. These dwellings made up of low-cost material, without giving much consideration to technical aspects, are easy targets of various hazards.

We need to remember that we cannot stop the geologic process (you can't stop a volcano from erupting or a cyclone from forming) and we also cannot stop the population from growing. So, we can only attempt to reduce the hazard to life and property.

There are number of vulnerability indicators used to assess the actual scenario. Those indicators are variable in weightage and character, depending on the time and space, and also on the nature of hazard that may occur. Vulnerability indicators (Table 4.1) generally account for the (a) identification of mitigation targets; (b) identification of vulnerable people, communities, regions, etc.; (c) raising awareness; (d) allocation of adaptation funds; (e) monitoring of adaptation policy; and (f) conducting scientific research.

Table 4.1:
Indicators of Vulnerability

Vulnerability Indicator	Proxy for Vulnerable Conditions	Mechanism for Conversion to Vulnerability
Poverty	Marginalization	- Narrowing of coping strategies - Less diversified and restricted entitlement - Lack of empowerment
Inequality	Degree of collective responsibilities, formal and informal insurance coverage, along with underlying social welfare function	Direct: Concentration of available resources in smaller population affecting collective entitlement. Indirect: Inequality to poverty links as a cause of entitlement concentration
Institutional adaptation	- Architecture of entitlements determines exposure - Institutions as conduct for collective perceptions of vulnerability - Endogenous political institutional constrain or enable adaptation	Responsiveness, evolution and adaptability of all institutional structures

Source: Authors.

In that way, it indicates the possible identification at the local scale, with more primary data.

In general, less developed countries are more vulnerable to natural hazards than industrialized countries due to lack of understanding, education, infrastructure, building codes, etc. Poverty also plays an important role, since poverty leads to poor building structure, increased population density, and lack of communication and infrastructure. Human intervention in natural processes can also increase vulnerability (Table 4.2 and Figure 4.2). Development of lands and habitation are many cases exposed to multiple hazards, for example, building on floodplains subject to floods, sea-cliffs subject to landslides, coastlines subject to hurricanes and floods. The severity or frequency of natural hazard is increasing (Figure 4.3). For example: overgrazing or deforestation leading to more severe erosion (floods and landslides), mining groundwater leading to subsidence, and construction of roads on unstable slopes leading to landslides or even contributing to global warming leading to more severe storms.

Table 4.2:
An Example of Assessing Housing Vulnerability

Vulnerability Index	Vulnerability Class	Criteria
	I. House vulnerability	
1	Less vulnerable	The houses having more than 2 feet floor height, usually not flooded or low risk of flooding.
2	Moderately vulnerable	The houses having 1–2 feet floor height, subject to flooding.
3	Highly vulnerable	The houses having less than 1 foot floor height, very often flooded.
	II. Distance/proximity vulnerability	
1	Less vulnerable	The households that are positioned beyond 100 m from the embankment.
2	Moderately vulnerable	The distance of the houses is between 21–100 m from the embankment.
3	Highly vulnerable	The households that are positioned within 20 m from the embankment.
	III. Economic vulnerability	
1	Not/Less Vulnerable	The families belonging to upper level of APL-possessing pump set, fishing boat, T.V., radio, mobile, agricultural land, insurance, source of family income either service or business or large-scale agriculture, educated families
2	Moderately vulnerable	i) The families belonging to lower level of APL-without pump set, fishing boat, T.V., radio, mobile, medium landholding, middle income. ii) The upper income BPL families are included in this type-possessing pump set, land, radio, boat, mobile.
3	Highly vulnerable	Mainly BPL families are included in this type who do not possess pump set, boat, TV, radio, mobile. Often they do not have agricultural land or possess very small agricultural land and family income is significantly low.
	IV. Social vulnerability	
1	Less vulnerable	No physically challenged members/well educated members/having <50% women/ <30% old age/ child.
2	Moderately vulnerable	The families that do not fit into the criteria of Vulnerability Index-1 and 3
3	Highly vulnerable	The family members are illiterate/having >50% women members/ >30% aged or child members/ physically handicapped members.

Figure 4.2:
Vulnerability, Its Components, and the Relation with Disaster

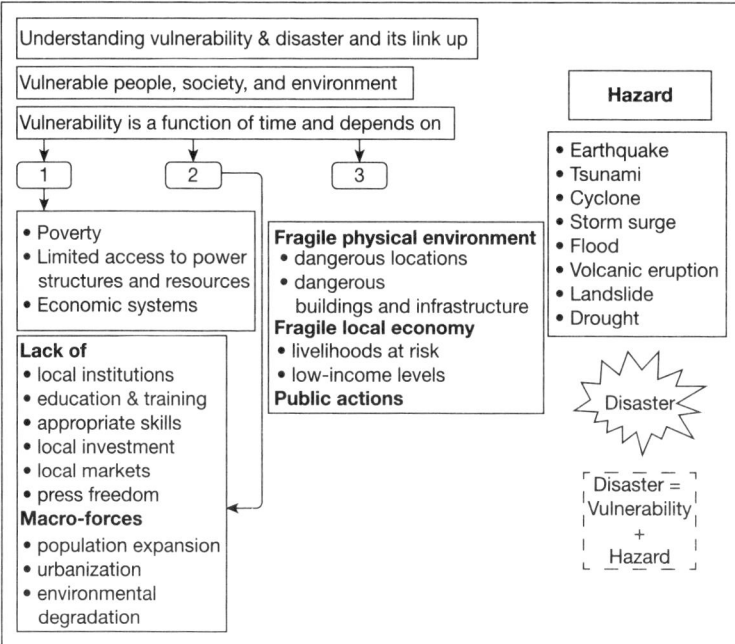

Source: Authors.

- Vulnerability is the coping capacity of those at risk (Handmer 1995).
- Vulnerability is the degree of susceptibility and resilience of the community and environment to hazards (EMA 1995).
- Vulnerability depends on the characteristics of a person or group in terms of capacity to anticipate, cope with, resist and recover from the impact of a hazard (Blaikie et al. 1994).
- Risk can only be managed if those who are vulnerable are identified. As Salter (1995) comments, risk and vulnerability are inextricably linked and, therefore, vulnerability must be understood if risk is to be managed.

Figure 4.3:
Natural Disaster Impact Model

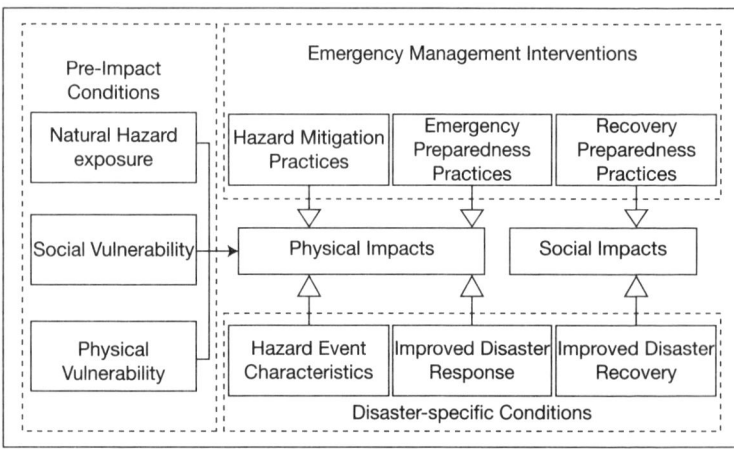

Source: Modified after Lindell and Prater (2003).

4.3 Contemporary Approaches to Risk and Vulnerability Assessment

In general, earlier approaches to risk assessment, which were primarily quantitative and favored technical solutions, have been replaced by more holistic approaches, which recognize not only the need for technical assessment of risk but also the interlinkages of technical elements with socioeconomic and political factors. Social scientific approaches, including the recognition that risk is interpreted as a social construct, now play a vital role in the assessment of risk and vulnerability (Salter 1995). This fundamental change parallels changes in related philosophies and processes such as:

- *Development:* People-centered and sustainable rather than solely economic.
- *Impact assessment:* Social as well as environmental, and sociopolitical as well as technical or quantitative.
- *Planning:* An adaptive process (people-oriented) in which the experience gained in taking part (the process itself) is the goal rather than a prescriptive process in which the plan itself is the goal.

4.4 Social Constructs of Risk and Some Implications for Vulnerability

The following important human attributes (some quantifiable and some nonquantifiable) affect how different people assess risk, and how vulnerable they are.

- *Socioeconomic characteristics (e.g., age, gender, ethnicity, income, education, employment, and health):* Older people and children may be much more vulnerable than active adults. Poorer people, with fewer capital resources, are likely to suffer far more from the effects of hazards such as a flood invasion of their homes. Some specific ethnic groups (e.g., Aborigines, people for whom English is a second language) may be much less able to take advantage of the assistance offered because of communication problems and cultural differences.
- *People's knowledge of the environment and of the hazards that environment poses to them (e.g., traditional ecological knowledge [TEK]):* TEK maybe effectively used to cope with a situation that outsiders perceive to be threatening, and generally provides a much more detailed understanding of local environments. It can be valuable in predicting the threats posed by hazards (e.g., when significant floods are actually likely) and can also provide people with alternative food supplies.
- *Their ignorance,* where even the direction of change or broad nature of outcomes are unclear, and where there is hold effects, surprise, and even chaotic processes may operate (Dovers 1995). For example, people who have newly moved into a vulnerable area often lack knowledge of the actual threats posed by hazards, such as severe bushfires, and fail to take suggested precautions seriously.
- *Their ability to cope with those hazards (risks)* through technology, financial attributes, education, political power, and having a voice. Knowledge, high levels of education, and high incomes generally give people more confidence in articulating their feelings and needs and hence they may be able to cope better with adversity.
- *Their ability to access help from outside.* Having confidence (as previously mentioned) makes asking for assistance much easier.

Attributes such as these highlight some key points:

- Other things being equal, it is generally those living at the margins who are most vulnerable to risk and uncertainty. Marginality in this context is defined in two main ways:
  - o *Socioeconomic:* people who, for social or economic reasons, are outside the mainstream.
  - o *Geographic:* people who live in very isolated locations and who as a result are often socioeconomically marginalized.
  - o The better the understanding of these factors, the more effective are the coping mechanisms.

4.5 Coping with Risk

The key to coping with risk is being sensitive to differences in people's perceptions of the problem and, hence, understanding their levels of vulnerability. Working with communities at risk is essential if these ideas are to be incorporated into risk management. In this approach, technical knowledge, essential for dealing with many elements of disaster mitigation, is combined with local knowledge, and the administration of risk management plans becomes a shared responsibility, integrated with local institutional structures and tapping into external forms of support.

4.5.1 What Is a Vulnerability Map?

A vulnerability map gives the precise location of sites where people, the natural environment, or property are at risk due to a potentially catastrophic event that could result in death, injury, pollution, or other destruction. Such maps are made in conjunction with information about different types of risks. A vulnerability map can show the housing areas that are vulnerable to a chemical spill at a nearly factory. But it just as likely could delineate the commercial, tourist, and residential zones that would be damaged in case of a 100-year flood or, more devastation, a tsunami. Vulnerability maps are most often created with the assistance of computer technology called GIS and digital land survey equipment designed for use in the field. However, vulnerability maps can also be created manually using background maps such as satellite imagery, property boundaries, road maps, or topographic maps. In such cases, the municipality's planning office should be involved in order to take advantage of the base maps that have already been made for other purposes.

4.5.2 Benefits of Vulnerability Mapping

Vulnerability mapping can allow for improved communication about risks. It allows for better visual presentations and understanding of the risks and vulnerabilities so that decision-makers can visualize where resources are needed for protection of these areas. The vulnerability maps will allow them to decide on mitigating measures to prevent or reduce loss of life, injury and environmental consequences before a disaster occurs. An interdisciplinary risk group considers where mitigating measures should be taken before, for example, a flood occurs. Those preparing the maps can overlap flood inundation and slope stability zones with property maps in order to determine which properties and buildings are at risk. They can then notify the landowners and inform them of government subsidies or other support available for undertaking a measure that would protect their homes from potential damage by, for example, water inundation or slope failure. Vulnerability maps can be useful in all phases of disaster management.

Prevention, mitigation, preparedness, operations, relief, recovery and lessons-learned. In the prevention stage planners can use vulnerability maps to avoid high risk zones when developing areas for housing, commercial, or industrial use. Technical experts can be alerted about places where the infrastructure can be affected in case of a disaster. Fire departments can plan for rescues before a potentially dangerous event is at hand. During an exercise where a predetermined scenario takes place, the rescue crews may use the map to determine where to respond first to save human lives, the environment, or property. They can also be used to evacuation routes to test the effectiveness of these routes for saving large numbers of residents and tourists and moving special groups such as senior citizens, children, and those with handicaps. The operations officer can be updated about the disaster situation and the need for and the location of sensitive areas. The vulnerability map can also include evacuation routes to test their effectiveness for saving lives. After the disaster the vulnerability map and a new map showing the extent of the damage, can assist in assessing how well the emergency was managed. During a post-disaster review, the consequences of the disaster can be easily assessed with the help of field data. The evaluators can see if an accurate assessment of vulnerable areas was made and if they were adequately protected. It will also be apparent how effective the mitigation measures were.

4.5.3 Planning the Vulnerability Map

Collecting information for a vulnerability map is one of the major steps for preparing the vulnerability maps. As with any risk management endeavor, a group of experts that are familiar with the risks is an asset for creating a vulnerability map. Working together in an interdisciplinary risk group will provide comprehensive information about risks and vulnerable sites. The members of the group will increase their knowledge about hazards and the type and extent of disasters that can be expected. Risk groups are most effective when they include an assortment of experts at the municipal level. The group preparing the vulnerability map needs to select those risks that are found in the area to be mapped and decide which risks will be addressed.

Natural risks that can cause disasters include:

- seismic activity;
- landslide/slope failure;
- avalanche;
- volcanic eruption;
- windstorm, cyclone, tornado, typhoon;
- snowstorm;
- flood;
- excessive precipitation, flash floods;
- tsunami;
- extreme frost, blizzard;
- extreme drought; and
- wildfire.

Man-made caused hazards or threats include:

- industrial activities;
- commercial activities;
- traffic and transport on land, sea, or air;
- sabotage;

- terrorist attack; and
- arson.

Once the risks have been selected, the group then discusses types of scenarios where one of these risks becomes an actual event. The scenarios will describe the date, the day of the week, the time of day, the intensity of the event, the weather conditions, season, etc., in order to determine an adequate picture of how the natural or human-caused event will take shape and what will be affected by its impact at the particular time and place that it occurs. A cyclone with a certain speed can hit a particular area with specified impact. A domino-effect scenario can also be written for an event when one of these risks, a natural event or human-caused, can trigger another making the disaster even more devastating. A major flood might be severe enough to encroach upon developed areas. An industrial storage tank containing dangerous goods could be damaged by the rushing waters, causing chemicals to spill into the environment. Whatever the scenario will be, it needs to be written with enough detailed to provide information about the boundaries of the risk zones. Risk zones should be estimated with the best available knowledge and techniques. With expert help, more accurate risk zones can be calculated for high tides, floods, tsunamis, landslides, etc.

The next stage is to determine the objects within the risk zones that will be considered vulnerable and therefore, will be mapped (Table 4.3). Vulnerable sites are those where people live, work and visit. They can also be areas where farming, forestry, grazing or industry prevail. Or they can be sites that are difficult to replace or rebuild or which possess historical or cultural values.

The baseline data collected for the vulnerability map should be the basis for decisions taken about where prevention and mitigation work needs to be accomplished. Here is an example checklist of objects that could be threatened by a risk. The user can fill in the municipal or regional office that could take responsibility for locating the threatened objects to be mapped.

The office responsible for determining what is vulnerable may vary depending on how the municipal offices are organized. Such offices can be an environmental protection office, technical office, planning office, health-care services, education department, culture and recreation department, social services department, agricultural department, forestry department, fire and rescue service department or police station.

Many of these items, especially natural features, can be found on base maps made by the municipal planning office. Where possible, relevant updated maps, aerial photographs and satellite images can be used to create the vulnerability map.

Table 4.3:
Sample Checklist for Threatened Objects (Nonexhaustive)

Examples of threatened object	*Responsible department (Fill in the appropriate department for your municipality)*
Bays or lagoons	
Coastlines	
Beaches	
Sand dunes	
Lakes	
Rivers	
Canals	
Forests	
Wetlands	
Special ecosystems (i.e., mangroves)	
Habitat areas for threatened or endangered species	
Environmental sensitive areas	
National parks and nature reserves	
Railways, subways	
Roads	
Highways, small paved roads, dirt roads	
River dykes, drains and levies	
Culverts	
Bridges	
Docks	
Airport terminals	
Drinking water supply	
Wells	
Waste water treatment plants	
Hospitals and medical centers	
Schools	
Day care centers	
Senior citizen centers	
Public places, theatres, sports arenas	

Recreational areas

Agricultural areas

Commercial forestry

Industrial sites

Commercial centers

Hotels

Residential areas

Fire and rescue stations

Emergency response zones

Storage sites for supplies needed for emergencies

Shelters (i.e., for cyclones)

Source: Authors

Some buildings and facilities require special rescue techniques and can be classified as such on the vulnerability map in order to reduce loss of life when a disaster occurs. Some examples are:

- high-risk buildings;
- compound buildings;
- hotels or other buildings with large numbers of visitors;
- senior citizen retirement homes;
- homes with handicapped residents;
- developed areas where the building materials used such as wood or concrete make them especially sensitive to, for example, fires or earthquakes;
- fishing boats, recreational boats, tankers in the harbor;
- underground installations (such as subways, utility cables); and
- mines.

Industries can be broken into different categories such as:

- petroleum refinery;
- chemical plant;
- plastic, rubber, or paint factory;
- steel mill;
- saw mill, paper mill; and
- explosives factory.

Power generating plants can also be mapped as a separate category.

- Electric or hydroelectric power plants
- Nuclear power plants

When the map is complete, there will be sufficient information to begin discussions about action plans for the threatened objects such as:

1. How will the sites be protected?
2. In which order will they be protected?
3. Who will accomplish the mitigation work?
4. Who will check to see if the mitigation work is adequate?
5. How will the vulnerable sites be addressed in the emergency pre-paredness plan?

The vulnerability map can be used to decide where appropriate mitigation measures can be taken such as:

- construction of groynes and sea walls to prevent beach erosion or damage to fishing boats and structures near the coast;
- protecting buildings from high water with sand bags or metal planks;
- stabilizing slopes with rods or by reshaping it by mechanical means;
- coastal embankments to protect roads;
- riverbanks and canals can be widened and strengthened; and
- protecting a harbor by building dykes.

Statistical data and methods are used for calculating the strength of, for example, tide water to determine how strong sea walls shall be. Another mitigation measure is relocating threatened objects. If a lowland area is susceptible to regular flooding, new buildings can be constructed in elevated areas outside the floodplain's risk zone.

4.6 Creating the Vulnerability Map

After the risks have been identified, and one or more scenarios have been made and the risk and vulnerability table has been created, the field work can begin. Vulnerability mapping efforts for an environmental disaster begin with an accurate representation of natural features such as rivers, lakes, landforms, topography, and vegetation type.

Man-made features can then be transposed on the natural landscape. The map will then include such information as land use, road and railway systems, power stations, industrial sites, official buildings, business areas, housing areas, schools, and hospitals. Refer to the list of threatened objects presented earlier. Define what is unique about the areas in order to protect biodiversity and cultural integrity for future generations. In addition, any objects that are essential to the emergency operations should be added to the map.

Environmentally vulnerable

- Soil type and geology
- Hydrology, rivers and lakes
- Forest and bush
- Agriculture
- Pasture and livestock grazing
- Coastal wetlands, coastal ecosystems
- Marine ecosystems

Kosi Flood 2008 (Bihar, India)

Kosi river, the "sorrow of Bihar" breached, (18 August 2008) its embankment at Kusaha, 10 km inside Nepal, bordering Supaul district of Bihar and started flowing along a new course, approximately 15 to 20 km wide and 150 km long, affecting 35 blocks, 407 Panchayats and 980 villages in five districts of Bihar. More than 33 lakh people and their houses/land spread in 110,258 ha area fell in the path of the new course. Hundreds of thousands of people were trapped in the new course with no means to escape. The entire countryside within this 3,000 sq. km swathe was devastated by the rapidly surging flow of the river along its new course. Houses, school buildings, roads, health centers, bridges, telephone towers and railway tracks were flooded or severely damaged/swept away. Several banks closed their branches in the affected areas causing problems in distribution of cash dole. There was complete disruption of road connectivity and, as such, boats became the only source of reaching the people.

Mammoth evacuation: More than a million people were evacuated under conditions of terrible handicaps with the help of 35 columns of army, 855 personnel of NDRF, 4 units of navy, several rescue teams from other states, 3,500 policemen and 5,000 others. More than 1,500 country boats including 561 motor boats were deployed for evacuation. In addition, several SAR teams came from all over the country both government and nongovernment.

(*Continued*)

(Continued)

> *Relief operation:* Marooned people were provided food, water, medicines and shelter by helicopters and boats. Rescued victims were housed in relief camps where they were provided with shelter, food and many other things including clothes and utensils. Toilets were made for them. Hand pumps and water purifiers were installed for clean water supply. During Kosi calamity of 2008, as many as 362 relief camps were set up within a matter of days to house 438,324 inmates. Of this, there were 35 mega relief camps with a population of more than 5,000 each with the help of 12,000 tents procured from army and para military forces. Finally, semi-permanent camps with GI sheets were set up to take care of victims in winter. Arrangements were made for food, water, sanitation, health, education, sports and cultural programmes in these camps.
>
> It has been observed by one of the Fact Finding Mission on the Kosi that the dilapidated state of the Bhimnagar barrage on river Kosi, inside Nepalese territory north of Bihar, India, did not have the capacity to hold its designed discharge of 950,000 cusecs. It was also observed that the silt choked east bank and the west bank canals emanating from the barrage with their combined irrigation capacities reduced by two-third on account of defunct silt ejectors, could only add pressure on the main structure and the already weakened embankments upstream.
>
> Kosi River, used to flow through the district Purnea, with the construction of the embankment the river changed its course westward to Saharasa. The Kosi's embankments have a long and intriguing history. One reference that is worth mentioning relates to the prophetic observations made by a British Engineer Captain F.C. Hirst in 1908, "In recent times, on the left bank of the Kosi, in the Purnea district, private enterprises has copied the work of the makers of the Bir Bund (an embankment), giving temporary relief, which, as will be seen later, is probably a menace to future welfare." Without any regard to such wisdom and accumulated evidences of the negative impact of embankments, the provincial government has turned 'temporary' solution into permanence by building over 3,465 kilometers long embankments to jacket some of its major rivers.

4.7 Seismic Microzonation

Earthquake hazard zonation for urban areas, mostly referred to as seismic microzonation, is the first and most important step toward a seismic risk analysis and mitigation strategy in densely populated regions. In seismic microzonation, one would like to quantify the spatial variation of the subsurface response on a typical earthquake that can be expected in the area. The process involves incorporation of geological, seismological and geotechnical concerns into economically, sociologically and politically justifiable and defensible land-use planning for earthquake effects so that architects and engineers can site and design structures that will be less susceptible to damage during major earthquakes. Microzonation should provide general guidelines for the types of new structures that are most suited an area. It should also provide information on the relative damage potential of the existing structures in the region.

Earthquake microzonation mapping essentially requires (a) Bedrock topography, (b) Subsoil profile, (c) Soil site classification, (d) Peak Ground Acceleration (PGA) and Peak Ground Velocity (PGV) mapping, (e) Liquefaction potential mapping, (f) Geomorphological characterization, and (g) Probabilistic seismic hazard scenario. Figure 4.4 elaborates a typical seismic microzonation framework from seismicity to vulnerability assessment.

Figure 4.4:
Microzonation Framework for Earthquake Hazard Mapping

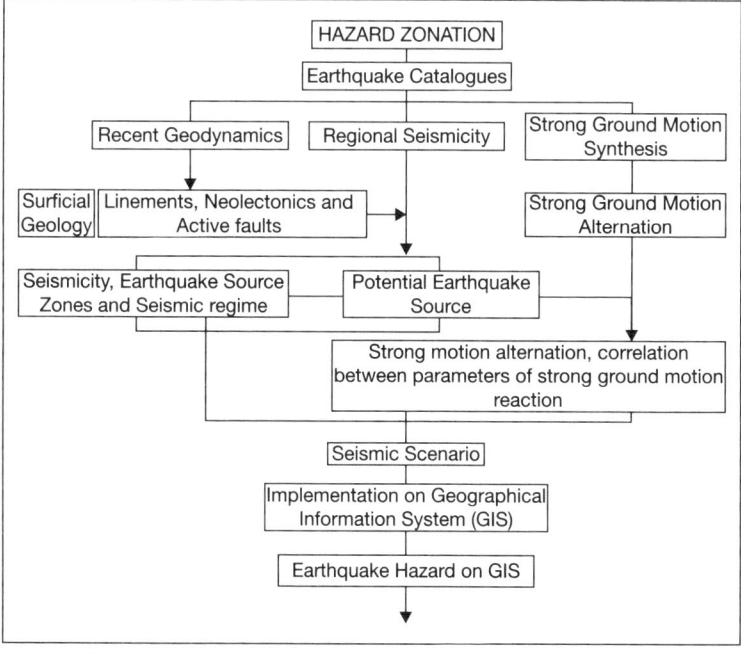

Source: Authors.

4.7.1 GIS-based Thematic Mapping

Geographic Information System (GIS) platform was adopted as a primary working tool in preparing the seismic hazard microzonation study for the Sikkim Himalayan region. Multitasking functionality of GIS makes it ideally suited to seismic microzonation as it enables automation of data manipulation and information of maps. The complex spatial analysis associated with seismic microzonation necessitates GIS technology to data dissemination and its management. GIS stores spatial and aspatial data into two different databases. GIS links the two databases by maintaining one-to-one relationship between records of object location in the topological database and records of the object attribute in relational database by using end-user defined common identification index or code (Hohl 1998; Korte 1997; Marble and Pequet 1983). GIS uses three types of data to represent a map or any geo-referenced data, namely, point type, line type, and area or polygon type. It can work with both the vector and the raster geographic models. The vector model is generally used for describing the discrete features, while the raster model does it for the continuous features (Burrough 1986; Burrough and Mcdonnell 1998; Davis 1996).

The GIS framework allowed us to account for added levels of details and complexity. For the ease of study the data attributes are subdivided into two major groups: (a) geomorphological attributes and (b) seismological attributes. It is very important that the relevant data layers be consistent in their level of detail, in order to successfully combine them and cross analyze them in the pair-wise comparison process. A schematic process flow is shown in the Figure 4.4. Prior to presenting the methodologies used to produce the integrated seismic hazard maps, a review of all the thematic coverage with their analytical detail is described here.

4.7.2 GIS-based Analysis for the Microzonation of Sikkim Region

4.7.2.1 APPLICATION OF GEOGRAPHIC INFORMATION SYSTEM

GIS is used to manage large volumes of data needed for the seismic hazard and risk assessment. As it is well known that the strength of GIS lies in the ability to represent the real world situation closely with layers of maps that can be combined in a manner to identify the impacts of a natural hazard. Geographical data have been traditionally presented in the map form. GIS is a computer-based system that is used to store and manipulate geographic information. GIS can provide better information to support this type of complex decision-making. In general analytical functions and conventional cartographic modeling techniques in GIS are based on Boolean logic, which implicitly assumes that objects in a spatial database and their attributes can be uniquely defined (Sui 1992). The deficiencies of the traditional Boolean logic for the design of spatial databases have been recognized in recent years (Borrough 1986, 1989, quoted in Sui 1992). As an alternative to Boolean logic, Zadeh's fuzzy set theory has been proposed as a new logical foundation for GIS design (Robinson 1988, quoted in Sui 1992).

Fuzzy logic methodologies may provide a scheme for the representation and manipulation of the uncertainty, which is related to the classification of individual locations according to their attribute values. It implements classes or groupings of data with boundaries that are not sharply defined. Fuzzy set theory suggests that the inclusion of an element with in a set is a matter of determination of the degree of belonging. The central idea of fuzzy sets can be aided by the Multi-criteria Evaluation technique and Analytic Hierarchy Process (AHP) to achieve operational simplicity in the evaluation of Hazard model.

McHarg (1969) introduced Multi-Criteria Evaluation technique for a systematic land-use planning. The idea of multicriteria decision making was based on simple matrix system for determining the degree of compatibility. Recent developments in geographical Information Systems have drawn upon concepts of the multicriteria methodology.

4.7.3 Analytic Hierarchy Process (AHP)

Analytic Hierarchy Process (AHP) is a multicriteria decision method that uses hierarchical structures to represent a problem and then develop priorities for alternatives based on the judgment of the user (Saaty 1980). Saaty has shown that weighting activities in multicriteria decision-making can be effectively dealt with via hierarchical structuring and pairwise comparisons. Pair wise comparisons are based on forming judgments between two particular elements rather than attempting to prioritize an entire list of elements (Saaty 1980). Analytic Hierarchy Process (AHP) is used to determine the weights of each individual criteria or thematic layers (Saaty 1990). AHP is a mathematical method to determine priority of the criteria in the decision making process. It is a popular tool used by decision-maker in multi-attribute decisions.

In this method, a matrix of pair-wise comparisons between the factors is built. This matrix is constructed by eliciting values of relative importance on a scale of 1 to 7, 1 indicates that the two factors are equally important, and 8 corresponds that one factor is more important than the other. If one factor is less important than others then it is indicated by reciprocals of the values ranging from 1 to 7 (i.e., 1/1 to 1/7). The process of allocating weights is a subjective and can be done in participatory mode in which a group of decision-maker may be encouraged to reach a consensus opinions about the relative importance of factors.

The matrix developed by pair wise comparisons between the factors can be used to derive the individual normalized weights of each factor. It is performed by calculating the principal eigen vector of the matrix. The weights for each attribute can be calculated by averaging the values in each row of the matrix. Addition of these weights will be '1' and can be used in deriving the weighted sums of rating or scores for each region of cells or polygon of the mapped layers.

Since the values within each thematic map/layer vary significantly, they are classified into various ranges or types, which are known as the features of a layer. These features are then assigned ratings or scores within each layer, normalized to ensure that no layer exerts an influence beyond its determined weight. Therefore, a raw rating for each feature of every layer is allocated initially on a standard scale such as 1 to 10 and then normalized using the relation,

$$x_i = \frac{R_j - R_{min}}{R_{max} - R_{min}} \tag{4.1}$$

where R_j is the raw score, R_{min} and R_{max} are the minimum and maximum scores of a particular layer (Figures 4.5 and 4.6).

4.8 2015 Nepal Earthquake

The Gorkha or Nepal Earthquake of 7.8M occurred on April 25, 2015 at 11:56 NST having epicenter 77 km northwest of Kathmandu and depth of 15 km caused lot of devastation in Nepal. The earthquake of April 25, 2015 had its epicenter in Gorkha district nearby Barpark village followed by another major aftershock of 7.2M on May 12, 2015 with the epicenter at Dolakha district that affected the eastern part of the country. Aftershocks were continuously felt for several months after the disaster. Out of 75, 31

Figure 4.5:
Vector Overlay Operation for Seismic Microzonation

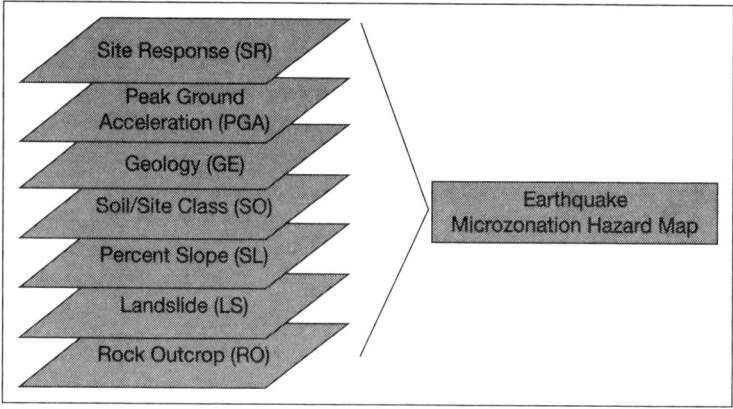

Source: Authors.

Figure 4.6:
Earthquake Hazard Zonation Map of the Sikkim Region

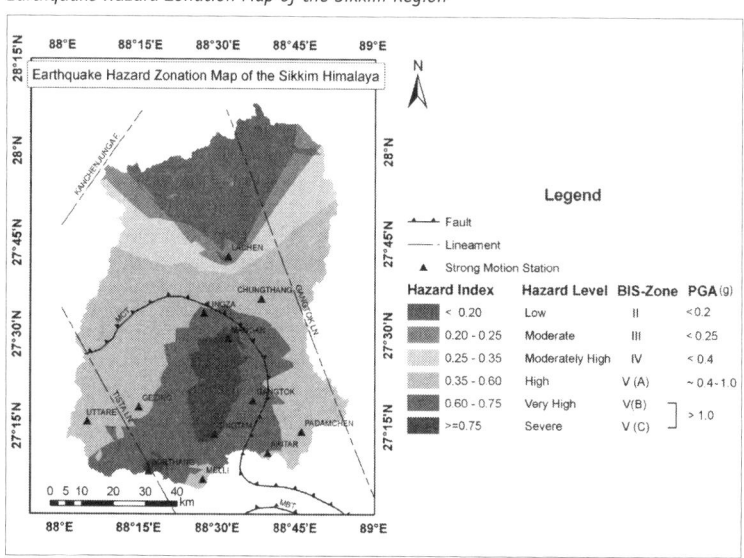

Source: Authors.

districts were reported to have impact from the earthquake in central and western region of Nepal while, 14 out of 75 districts were severely affected by the earthquake in terms of rescue and relief priority (GON, 2015). Immediately after the earthquake, the government declared emergency and requested for international assistance.

According to the government, 500,717 houses were completely destroyed and 269,190 houses partially destroyed till May 25, 2015 and about 8,832 people lost their lives and left 22,309 people injured till June 24, 2015. It was a powerful earthquake after 1934's Nepal–Bihar earthquake and was felt as far as India, Tibet and Bangladesh. The recent earthquake affected about 2.8 million population within the country (OCHA 2015). Over 860,000 people needed immediate assistance to shelter before the monsoon period. On May 4, about US$423 million was appealed for relief operation in Nepal (OCHA, 2015).

4.8.1 Hazard and Vulnerability Profile of Nepal

Nepal is highly susceptible to natural disasters such as floods, landslides, extreme weather patterns, earthquake, avalanche, fire, etc. because of its location, topography, climatic conditions, high rainfall intensity, composition of rocks and soil that has made it more susceptible to disasters. It is second most highly vulnerable country in terms of mortality rate to the risk of multiple hazards (Bronkhorst 2012). Aryal (2007) has mentioned that natural hazards in Nepal has been increasing over time killing more people and increasing the cost. On an average, two lives are lost every day due to natural disasters and the average loss of properties is NRs. 1208 million every year (MOHA 2009). About 23,000 people have been killed by natural disaster from 1983 to 2010, whereas the economic loss incorporated from 1998 to 2008 was over one billion US dollar (DPNET, 2011). Nepal is ranked as 11th most vulnerable country to earthquake in the world according to Disaster Vulnerability and Risk Assessment Study report (UNDP 2004). It lies in active seismic zone where the frequency of earthquake occurrence is enormous. It lies right on the subduction zone of Eurasian and the Indian plate where catastrophic earthquakes tends to occur recurrently. Contemporary cities are known as disaster incubators due to rapid population growth and activities in confined space (Dovers & Handmer 2007; Mitchell 1999; Pelling 2003). Rapid urbanization in cities such as Kathmandu and Pokhara lying on the central part of the country make them more vulnerable to earthquakes.

4.9 Thailand Flood 2011

Thailand is located in the centre of the Southeast Asian Peninsula, covering an area of 513,120 sq. km. It has a coastline of 3,219 km. The northern part of Thailand is mountainous covered by dense forest and the eastern part consists of the Khorat Plateau. The central part of the country is covered predominantly by the flat Chao Phraya river valley, which runs into the Gulf of Thailand. Because of the geoclimatic factors, Thailand is among the most flood-prone nations in the world, according to the global climate risk index 2013. The experience of two catastrophic events within the last decade (the 2004 Indian Ocean Tsunami and the 2011 floods in central Thailand and Bangkok) have revealed existing disaster-related strengths, while at the same time highlighting many remaining challenges. The year 2011 was one of the highest loss years recorded in terms of losses from disasters. The 2011 floods in Thailand caused 892 fatalities and affected over thirteen million people (ASEAN, *Safe Schools Initiative Phase 1 Report*, 2013).

4.9.1 Context of Bangkok

Bangkok is the capital city and nerve center of economic and trading of Thailand. It is located in the Chao Phraya River delta in Thailand's central plains and measures 1,568.7 sq. km. Bangkok is subdivided into fifty districts (35 districts on the eastern side, and 15 districts of the western side of Chao Phraya River), and is adjacent to six provinces, namely Nakhon Pathom, Nonthaburi, Pathum Thani, Chachoengsao, Samut Prakan, and Samutsakhorn. According to the statistical data from Bangkok Metropolitan Administration (BMA), the total population as of 2014 is 5,692,284 persons (2,695,519 male and 2,996,765 female). This makes it the most populated province, and this is likely to continuously increase in the coming years (SED 2015; Wiki 2015; BMA Data Center).

4.9.2 Recent Flood Events

Bangkok is a low-lying area with an average ground elevation of 1.5–2 m MSL; thus, Bangkok is at risk of urban floods each year. The recent 2011 floods in Thailand have been the most devastating in terms of magnitude and economic loss. The damages were severe both in economic as well as social terms. Thailand has faced floods in the past, but the 2011 floods were a jolt to the economic progress of the nation. One of the main reasons for the extreme flooding of 2011 was that the regular monsoon season was accompanied by many other storms. Central Thailand received overwhelming rainfall in its central plains during September, October, and November 2011, which added to the catastrophic impact on every section of the society. Thailand has a total of 78 provinces out of which 28 were severely affected by the 2011 floods (Cohen 2012).

Chapter 5
Mitigation

5.1 Mitigation

5.1.1 Definition

The term 'mitigation' addresses the measures taken to minimize the effect of a possible hazard, and the consequent vulnerable conditions from a future disaster. Hence, the mitigation activities are usually focused on the specific hazard and the elements under threat. A few examples of hazard-specific mitigation measures are sustainable water management in drought-prone areas, relocating people from hazard-prone areas, or retrofitting buildings to reduce possible damage from earthquake. In addition to all these physical measures, mitigation should also aim to reduce the economic and social vulnerabilities toward a resilient society.

5.1.2 Concept of Mitigation

To combat the increasing risks, we need more studies to attempt to understand and help forecast future events. We need to be able to monitor the hazardous systems (e.g., cyclonic storm monitoring and meteorological/weather monitoring) and to be able to quickly communicate the information from the scientists to the general population. All of this helps with the aim to *mitigate* (reduce) the effect of the natural hazards. There are many types of natural events that can impact human processes, such as volcanoes, floods, earthquakes, tornadoes, tsunamis, landslides, avalanches, forest fires, etc. These events can occur either without warning, for example, an earthquake, or they may occur with warning, for example, you can monitor with satellites when and where a cyclone will hit the coastline. Also, some volcanoes change their behavior before an eruption; the sides of the volcano may swell and crack as hot molten rock is pushed up toward the surface. These warnings are called "precursors." Precursors are what scientists look out for when trying to forecast a future event. To help forecast an event and mitigate (reduce) the hazards, the involved scientists need to know three things:

1. The "frequency" of the event, that is, how often the event occurs (on a scale of every month, year, 10 years, 1,000 years, etc.).
2. The "magnitude" of the event, that is, how powerful the event is. This is often related the destructive capacity of the event.
3. There is an important link between frequency (F) and magnitude (M); for example, an event with a high F and low M may not be as devastating as a hazard with low F but high M.

The "scope" of an event describes the area the hazard effects. Will the hazard or damage be contained only locally (e.g., landslides, fires, floods, and earthquakes), or on a larger regional scale (e.g., tsunami's, explosive volcanoes, large-scale earthquakes, and cyclones). The largest, most catastrophic events may even affect the entire globe (e.g., large-scale volcanism, global warming, and meteorite impacts). Knowing these three factors for each natural hazard event will help the population to plan for future events.

5.1.3 The Buzz Words

- All of us are likely to face natural or man-made disasters in our life.
- The impacts of these natural disasters cut across all the sectors of the profession we are in and adversely affect our safety, life, and livelihood. However, these impacts can be scientifically estimated.
- The basic principles of handling such situations lie in not allowing natural hazards to get converted into natural disasters through scientific interventions.
- As disaster comes only when a hazard meets the vulnerable people, this provides us with the clue that reduction of the vulnerabilities of people is the best way to reduce the impact of natural disasters on all of us.
- The modern way of managing a disaster is not after the event has happened, but to manage the root causes of the disaster (much before it appears in front of us).

The hazards risk-management implementation strategy identifies who is responsible for which actions and what funding mechanisms (e.g., grant funds, capital budget, or in-kind donations) and other resources are available or will be pursued and stipulates when the actions are to be completed. This strategy should describe the means by which the community will use its resources to achieve its goals of reducing losses from future hazard events. It should also focus on coordination between the various individuals and agencies involved in the implementation to avoid any duplication or conflicting efforts.

5.1.4 Conventional Mitigation Methods

Structural and Nonstructural Mitigation Detail

The mitigation measures can be engineering or nonengineering, while addressing the possible impact from a hazard. These are often called as hard engineering or soft engineering. The structural solutions are with high investment and recurring maintenance cost, and are often nonsustainable. These may provide initial solutions, but also aggravate the hazard in the near future. Nonstructural measures are designed using the knowledge of traditional practices with the native materials. These are generally of less cost and low maintenance, but may not aggravate the hazard, though they cannot provide long-term protection. Choosing a structural or nonstructural solution depends on the importance of the issue, available fund, and the attitude of the decision-makers. The methodological approach for structural and nonstructural measures has been detailed in Table 5.1.

5.1.5 Mitigation Action

The primary aim of mitigation action is to reduce losses in the event of a future occurrence of a hazard and prioritize reduction of death and injury to the population. The secondary aims are reduction of property and infrastructure damage and economic losses in both public and private sectors, especially when they have adverse effect on the community as a whole. Any mitigation strategy is a bundle of measures from the menu of actions. A wide range of actions may include engineering measures and spatial planning, along with certain economic, management, and societal inputs that are necessary for effective mitigation.

Table 5.1:

Structural and Nonstructural Methods of Hazard/Disaster Mitigation

Structural methods

Retrofitting of existing structures

Reinforcement of new structures
- Design
- Features
- Overdesign
- Structural

Safety features
- Safeguards
- Failsafe design

Engineering phenomenology

Probabilistic prediction of impact strength

Nonstructural methods

(a) short-term

Emergency plans (civil)
- Coordinator (s)
- Police and firemen
- Red Cross and charities
- Volunteer groups
- Medical services

Evacuation plans
- Military forces
- Routes and reception centers for the general public and vulnerable groups: the very young, elderly, sick, or disabled/physically challenged

Protection of impact
- Monitoring equipment
- Forecasting methods and models

Warning process
- General message
- Specialized warning (e.g., ethnic)

(b) Long-term

Building codes and construction norms

Hazard microzonation
- Selected risks
- All risks

Land-use control
- Regulations
- Prohibitions
- Moratoria
- Purchase rule

Probabilistic risk analysis

Insurance

Taxation

Education and training

Source: Authors.

Mitigation actions by planning or development authorities can broadly be categorized into two broad types: "active" and "passive" measures. Active measures are designed for desired actions by offering incentives, often linked with area development and poverty alleviation programs in low-income regions. On the other hand, passive measures are aimed to prevent undesired actions through enforcement of laws, controls, and penalties mostly in areas with higher income. Community-based mitigation actions should ideally be responsive to people's real time needs, mobilizing local resources and materials for long-term development of the community. In generic sense, the common mitigation actions practiced are as follows:

- engineering and construction;
- physical planning;
- societal measures;
- economic measures; and
- management and institutional measures.

In every country, the range of hazards faced are likely to be different than in other countries, and also, the infrastructure, housing, habits, and other elements-at-risk will definitely have their own characteristics. Disaster mitigation actions that have been implemented in Sri Lanka are unlikely to be directly replicated in Thailand. There is no standard solution to mitigate a single disaster risk in different spaces and times.

Mitigation actions can be grouped into six broad categories:

1. *Prevention:* Government administrative or regulatory actions or processes that influence the way land and buildings are developed and built. These actions also include public activities to reduce hazard losses. Examples include:
   a. planning and zoning;
   b. building codes;
   c. capital improvement programs;
   d. open space preservation; and
   e. storm water management regulations.
2. *Property protection:* Actions that involve the modification of existing buildings or structures to protect them from a hazard, or its removal from the hazard area. Examples include:
   a. acquisition;
   b. elevation;
   c. relocation;

    d. structural retrofits;

    e. storm shutters; and

    f. shatter-resistant glass.

3. *Public education and awareness:* Actions to inform and educate citizens, elected officials, and property owners about the hazards and potential ways to mitigate them. Such actions include:

    a. outreach projects;

    b. real estate disclosure;

    c. hazard information centers; and

    d. school-age and adult education programs.

4. *Natural resource protection:* Actions that, in addition to minimizing hazard losses, also preserve or restore the functions of natural systems. These actions include:

    a. sediment and erosion control;

    b. stream corridor restoration;

    c. watershed management;

    d. forest and vegetation management; and

    e. wetland restoration and preservation.

5. *Emergency services:* Actions that protect people and property during and immediately after a disaster or hazard event. Services include:

    a. warning systems;

    b. emergency response services; and

    c. protection of critical facilities.

6. *Structural projects:* Actions that involve the construction of structures to reduce the impact of a hazard. Such structures include:

    a. dams;

    b. levees;

    c. floodwalls;

    d. seawalls;

    e. retaining walls; and

    f. safe rooms.

During the identification of the *appropriate mitigation options*, the planning team must consider the following three questions:

1. Which actions can help to *meet the mitigation objectives*?
2. What *capabilities* are available to implement these actions?
3. What *impacts* may these actions can create on the community?

5.2 Role of Scientists, the Administration, and the Individual for Assessment, Prediction, and Warning

Everyone, including you, has a responsibility to understand the effects of a natural hazard and to respond to assessments, predictions, and warnings. Thus, one of the most important aspects of disaster management and planning is education. Not everyone can be expected to completely understand everything about a potential natural disaster. Therefore, one of the most important links between all those involved is effective communication between various groups of people.

5.2.1 Role of Scientists and Engineers

Hazard assessment: Scientists have the greatest ability to determine where natural hazards exist, and the effects of such hazards when an event occurs.

Prediction: Scientists have access to monitoring of processes that enable prediction. They should be able to communicate probabilities to the appropriate public officials for dissemination to the general public.

Reduction of risk: Scientists and engineers should make information known to public officials about ways to reduce vulnerability and risk, by suggesting zoning regulations and building codes to public officials.

Early warning: Scientists with access to monitoring and hazard information should help develop early warning systems to effectively communicate such warnings to the public officials responsible for communicating the warning to the general public.

Communication: Scientists need to be able to present the information available in a form that is understandable to all concerned.

5.2.2 Role of Administration

Risk assessment: Public officials need to understand hazard assessments, develop risk assessments, and decide where and how resources are to be expended to minimize risk.

Planning and code enforcement: Public officials need to work with scientists and engineers to help reduce vulnerability by making planning decisions (zoning laws) and building codes that help reduce risk and vulnerability.

Early warning: Public officials have the primary responsibility to inform the public about imminent dangers based on the predictions and warnings issued by scientific community.

Response: Public officials have the primary responsibility of maintaining an infrastructure that can deal with the emergencies created by a natural disaster. They need to develop plans for evacuation, emergency response, rescue, and recovery.

Communication: Public officials must be able to communicate effectively with the scientific community and the general public to disseminate information.

5.2.3 Role of an Individual

Understanding of hazards: The general public needs to be aware of the effects of natural hazards on their communities and to have some understanding of what might occur in the event of a disaster.

Understanding of early warning systems: The general public must be informed about what their response should be when a warning is issued.

Communication: In developed countries, people can communicate with public officials, either directly or through the ballot box, to ensure that the public officials make available the necessary information and effectively carry out their responsibilities for hazard and risk reduction.

Thus, a concerted approach to examine the principle on risk governance and its implementation might be useful for disaster mitigation. It provokes the possible adoption of proactive policy formulation (e.g., wise land-use planning), public policies for emergency management as a part of environment management, community development, and capacity-building.

Following components could provide solutions that can produce strategies to combat the incidents:

- central role of administration in mitigation planning;
- integration of mitigation strategies with other socioeconomic and environmental scenario building;

- introducing people perception in mitigation planning for accelerated risk management;
- pragmatic approach to avoid the disruption in environmental functions and their sustainability;
- risk governance and social impact assessment as a tool to understand the framework of regional and biophysical systems;
- change of mindset to accept planning as a process than as a result; and
- post-disaster recovery also provides opportunities for mitigation and community improvement.

5.3 Identify and Analyze State/Local-level Mitigation Cap Abilities

The capability assessment is aimed to review and analyze state and local programs, policies, regulations, funding, and practices currently in place that either facilitate or hinder mitigation in general, including how the construction of buildings and infrastructure in hazard-prone areas is regulated.

This capability assessment will provide information on how and whether the community will be able to implement certain mitigation activities by determining:

1. different types of mitigation actions that may be prohibited by law;
2. probabilistic limitations that may evolve while undertaking actions; and
3. the comprehensive list of administrative, regulatory, technical, and financial resources available to implement the mitigation strategy.

5.3.1 The Process Involved in Capability Assessment

- Check whether the state is able to provide *sufficient resources* to assist implementation of specific, alternative mitigation actions.
- Whether certain mitigation actions *not be available* (for example, the state may prohibit the use of public funds to purchase private property required for the mitigation process).
- Are there state regulations, initiatives, or policies that prevail at the local level that have *negative implications* for implementing the

mitigation effort (for example, a weak building code may be a big hurdle for implementing any mitigation effort). This would be ideally supportive if everyone in the building industry would follow the same code, but it may hinder a coastal community's resilience to withstand the cyclones).

5.3.2 The Procedure to Finalize a Capability Assessment

- The proposed mitigation actions will be evaluated against the backdrop of what is feasible in terms of the government's legal, administrative, and technical capacities. Additionally, there are many types of mitigation activities, some of which will require huge initial investments and recurring maintenance cost, or procedural and policy changes. The jurisdictions should allow these opportunities, depending on the type of activities required.
- There is a necessity to prepare a list of government agencies, departments, and offices having the responsibility for planning, regulating, mapping, building, and/or managing physical assets, as well as for emergency management functions, to understand the people to be pursued for any specific need. The list should incorporate other departments or agencies that do not apparently have a direct impact on mitigation, but rather have an indirect effect on the mitigation program. The list should also include businesses and nongovernmental or nonprofit organizations, as well as operators of critical facilities, colleges, and universities, since they play important roles in pre-/syn-/post-disaster environments.
- Planning team members must interact with government departments or divisions to collect information on all relevant programs, policies, regulations, funding, and practices related to mitigation activities. A comprehensive review of reports, plans, and other community documents is mandatory prior to that consultation to get a basic overview of the talking points during the consultation.
- The final part of this assessment is the analysis of the efficiency of the existing actions and capacities and the gaps that exist in reality, which hinder implementation. This evaluation helps the policy-makers to identify the necessary changes for effective implementation of new actions or to modify the actions in existing ones. Ideally, more extensive analysis is required when the planning team evaluates whether the specific alternative mitigation action meets up to the estimated goal.

5.4 Awareness and Mitigation (National and Local Levels)

The awareness and community building should have some strategic plan, keeping in mind the general psychology and the dynamics in the society. The major components of such understanding are as follows:

- *Knowledge/awareness:* An obvious first step is that people must know there is a problem and that there is a practical and viable solution or alternative. This is important. People are practical; they will always demand clear, simple, and feasible road maps before they start a journey to a strange place. They will also try to identify the personal costs of inaction and the benefits of action in concrete terms.
- *Desire:* People need to be able to visualize a different, desirable future for them. This is different from the rational benefits, as desire is an emotion and is not necessarily related with knowledge. For example, the media advertisements try to stimulate emotions such as lust, fear, envy, and greed to create desire for their product. However, desire can also be created by hoarding a more lucrative future life.
- *Skills:* People need to know what to do and the best possible way for that is by seeing someone else doing it. The best way to do this is to break down the actions into simple steps and use illustrations to make visualization easy. It's amazing how many social marketing campaigns forget this element.
- *Optimism (or confidence on future success):* Strong political or community leadership is definitely a major catalyst of optimism. A considerable number of people are in-active on environmental actions because of their isolation and less access to power. The governance can create leading examples to remove this kind of people's perception.
- *Facilitation:* People are usually with limited resources and less choices. They may need accessible services, infrastructure, and support networks that may not create any kind of frustration, which generates the inaction. The role of campaign may expand the involvement in new services and infrastructure. For example, recycling is not very successful in Southeast (SE) Asia in comparison to the developed nations.
- *Stimulation:* It is observed that with all available knowledge, desire, and understanding, there is still problem with mindset or habit to overcome. Consciousness is the tool human beings use to overcome habit, but we are unconscious most of the time.

In general people act in two situations, either threatening or inspirational, and that is related with stimulation and/or optimism.

An example of essential components of basin-level flood mitigation:

An idealistic procedure of basin management for minimizing the loss from flood can be deciphered through the following steps. The initial objective for that specific purpose can be translated into the several common goals.

Goal 1: Develop basin-level cooperation, coordination, and community participation
Develop an implementable approach for forecast, cooperation, and communication for flood mitigation, which can assure accountability to respond to the local concerns with public awareness and participation.

Goal 2: Build public–private and community partnership
Forge lasting partnerships between government, private organizations, and NGOs/CBOs along with the communities to ensure best practices in flood mitigation.

Goal 3: Provide protection to people and property
Design and implement flood mitigation strategies that minimize the loss of human life, property, and well-being.

Goal 4: Augment environment, economy, and community
Implement innovative flood-mitigation practices that can enhance ecological benefits, economic development, and sociocultural and recreational opportunities.

Goal 5: Integrated oversight and funding
Recommend a method to establish basin-wide, coordinated oversight and secured funding along with other necessary resources to achieve these flood mitigation goals. All these practical steps with developed resources and partnerships can produce a flood-resistant basin and resilient farms and communities.

5.4.1 Creation of Basin-wide Flood Resilience

The framework will contribute to reduce future flood damage and enhance economic growth, conservation, and tourism in the basin by

- creating micro water storage and retention reservoirs with regulated land-use that can reduce downstream runoff by absorbing, retaining, and storing water on the land;

- supporting and linking existing common or shared greenway developments within the basin on both sides of the river and its tributaries to foster sensible floodplain management practices; and
- attracting government and international funding and private-sector resources for the greenway development and management within the basin.

The additional efforts that may have been considered during this exercise are:

- provide financial incentives for land stewardship to combat the problem of frequently flooded, marginal land to increase producer income while reducing taxpayer costs from crop losses and other flood damages;
- encourage cooperation among farmers to implement land management practices for flood mitigation in watershed or subwatershed level;
- improve community infrastructures to reduce flood damages; and
- develop a response and recovery plan as a part of preparedness during times of peak flood conditions to enhance community capacity.

In true sense, all these initiatives are significant for basin-level flood mitigation. This comprehensive effort to mitigate all the possible flood damage also needs creative understanding and innovation to meet local needs can be used as model in other flood basins.

5.5 Community-based Disaster Preparedness and Mitigation

Throughout the world, the discussions about disaster mitigation invariably contain certain call of action, to make the people aware of the risks they face, while implementation is usually left to someone else. In practice, "awareness" often indicate the fact of hazards and the art to live with, or move away from, them or get prepared for responding to risk, rather than making the decisions and changing the behavior that may reduce risk. The clearest evidence for this is that none of the popular tools for risk assessment take into account mitigation activity or seek to measure reduction in risk over time.

When it comes to community disaster response, the engineering emphasis on structures (rather than the people in them) is reproduced in the preoccupation of SAR training with structures rather than people. That would again create a tremendous level of dependency on structures, ignoring the principles of mobilizing spontaneous responders and operating under a flexible "incident command system" that needs to be propagated more widely in order to improve response.

It has been argued that governments and large development agencies tend to adopt a "top-down" approach to disaster mitigation planning, whereby the intended beneficiaries are provided with solutions designed for them by planners rather than selected for themselves. Such "top-down" approaches tend to emphasize physical mitigation measures rather than social changes to build up the resources of the vulnerable groups. They rarely achieve their goals because they act on symptoms, not causes, and fail to respond to the real needs and demands of the people. Ultimately, they undermine the community's own ability to protect itself. An alternative approach is to develop mitigation policies in consultation with local community groups using techniques and actions that they can organize themselves and manage with limited outside technical assistance. Such community-based mitigation programs are considered more likely to result in actions that are a response to people's real needs, and to contribute to the development of the community, its consciousness of the hazards it faces, and its ability to protect itself in the future, even though technically the means may be less effective than larger scale mitigation programs. They will also tend to maximize the use of local resources, including labor, materials, and organizations.

Applying such community-based policies depends on several factors—the existence of active concerned local community groups and agencies able to provide technical assistance and support at an appropriate level, for example, are crucial to success. Nevertheless, opportunities for community-based mitigation actions should always be sought in developing a comprehensive mitigation strategy. They will certainly be cheaper and may be more successful than alternative larger scale programs.

Successful mitigation practices must involve collaboration between the local community and the larger scale development agencies. The local community must be aware of the risk and be concerned to take action to prevent it: in this they may need technical and material assistance and help in building their own capabilities. These forms of assistance may not be available, in which case they need to be provided by external agencies. One of the most effective ways in which such an agency can help promote community protection is by enabling communities to formulate their own project proposals and negotiate with the government and the larger development agencies (or government agencies) for the necessary government actions and the material assistance that they need. This is especially true for technologically based engineering projects, such as large embankments, spillways, and diversion works. For example, the construction of community defenses based solely on hand-labor and local materials alone may result in poor disaster defenses. But local labor supplemented by heavy machinery, and local materials bonded by factory made materials (e.g., cement or wire mesh) provided from external sources can result in lasting defenses that the local community will be able to trust and maintain in the long term. Similarly, a community-based mitigation program may need government action to provide land for safer resettlement of the most vulnerable, which can most effectively be determined by the community itself. The empowerment of the community created by achieving such goals and obtaining assistance from government agencies is likely to be a lasting development benefit.

5.6 Strengthening Community and Government Disaster Risk Management Capacity

Poor and vulnerable communities cannot prevent, prepare for, or manage all of the hazard risks that they face alone. They need the support of their local and national governments. Many countries have disaster management policies and legislation. Some have disaster management authorities and/or coordinating bodies. However, national, central-level government agencies often do not have strong links to the community or even, sometimes, to local governments. The weakest link in their disaster response chain is the community level, and there are numerous examples of disaster early warning and preparedness systems breaking down at this level—with devastating consequences. Whether in collaboration with national platforms or other national disaster management structures, social fund projects can play an important role in developing new models for strengthening community and government relationships in disaster prevention, preparedness, and mitigation. Social fund operations can bring community-level information and perspectives on hazards and vulnerability, as well as lessons learned from their experiences working with communities and local government, into national DRR planning processes. They can introduce demand-driven approaches to DRR consultative and investment processes. Social fund operations will need to develop effective multistakeholder communication strategies in order to achieve these outcomes.

While national governments should provide the overall strategic framework for community-based DRM (CBDRM), the design and implementation of CBDRM measures should occur primarily at the regional, municipal, and community levels. This is due to the fact that most emergencies will be smaller scale, localized events, community members will be the first responders, and more-isolated or underserved communities will need greater self-sufficiency in anticipating and responding to these events (Twigg 2004). Decentralization of disaster preparedness and response responsibilities creates the conditions for more rapid responses that are better informed about local needs and designed for the specific hazard and vulnerability context (Pusch 2004).

5.7 Good Practice in Community-based Disaster Risk Management

Over the past two decades, developmental and humanitarian organizations have learned that the most effective CBDRM programs and projects have many of the following features:

- Local people and their organizations are the main actors in reducing risk and responding to disasters and seek to involve them in defining problems, deciding solutions, implementing activities, and evaluating the results.
- Building linkages between communities and the local and national authorities to promote greater complementarity between their respective roles in disaster risk management.
- Understand the important roles played by women in disaster management and fully include them in decision-making, implementation, and evaluation.
- Analysis of the hazard and risk environment, including the vulnerabilities and capacities of the people affected.
- Attention to the needs and views of particularly vulnerable people who may be marginalized from participation on the basis of their gender, age, disability, ethnicity, socioeconomic status, or other factors.
- Livelihoods are central to poor and vulnerable people's coping strategies, and they incorporate a focus on livelihoods security whenever possible.
- Close linkage between environmental degradation and increased risk from natural hazards. Incorporate appropriate environmental activities to the extent possible.
- Information, education, and communication as a two-way process between communities and other disaster management stakeholders,

combining local knowledge and practice with scientific and technological information to ensure that the disaster early warning, preparedness, and mitigation measures are appropriate to the local context.

- Design locally appropriate and sustainable technological interventions for risk reduction.
- Adequately design and resource baseline data collection, and monitoring and evaluation (M&A) systems.
- Developing good community accountability systems and put them into practice.
- Promote knowledge-sharing, networking, and collaboration between different actors at local, national, and/or international levels to improve good practice.

Chapter 6

Modern Technology in Disaster Management

6.1 Science and Technology Relevance in Disaster Management

Disaster risks and losses can be significantly reduced through the greater use of science and technology, including the natural and social sciences and the applied fields such as environment, health, agriculture, water, and engineering. It is important to take a closer look at the steps that are taken to alleviate the risks and damages associated with them. The UN Secretariat of the ISDR is proactively taking a number of initiatives to analyze how science and technology is organized and how it contributes to DRR. Some of the projects review the importance of science and technology in the HFA, the world's blueprint for action to reduce disaster risks, and it examines the opportunities and barriers to policy-maker adoption of science and technology. Lastly, it makes proposals for how the ISDR can better promote and coordinate science and technology for DRR.

Some technologies that might be better used in some parts of the disaster management cycle, could be less effective, completely limited, or detrimental in others. Some technologies that are only used by some countries could be beneficially used by others. Therefore, it is important to scrutinize in greater detail what different approaches and technologies exist, and when and where they could be applied for the greatest beneficial effect.

The way that people deal with the threat of disasters, including their use of science and technology, is fundamentally dependent on how they view disasters and risks. There are many different conceptual framings of the disaster problem, which has often led to confusion, crossed purposes, and disagreement. For example, the rich and the poor experience disasters differently, and disaster risk will be viewed very differently, for example, by a land developer, a finance ministry official, an emergency manager, a hazard scientist, a community leader, or a subsistence villager.

The conceptual basis for DRR is undergoing significant change, partly as a result of scientific study. The common view of disasters as simply hazards or events is now challenged by the view that disaster risk is the outcome of both natural and human-influenced factors, with risk expressed in terms of potential losses of lives, assets, or incomes.

The HFA 2005–15 puts a strong emphasis on political and social factors in DRR and sets expectations for the better integration of the political/social aspects and the scientific/technical aspects. However, this shift in emphasis has led to a neglect of science and technology in the implementation of the framework and to inadequate stimulation and coordination of science and technology.

The mid-term review of the HFA has revealed unmet demands in countries for science and technology inputs—particularly for risk assessment—practical tools to address specific risks, ways to implement multihazard approaches, methods tailored to adaptation needs, economics-based evidence for advocacy purposes, and greater standardization of methods. There is a desire for initiatives to reduce the barriers to science and technology access and transfer and to support the development of technical capacities in developing countries. These growing demands for practical methods are a natural evolution of advocacy efforts and raised awareness.

6.2 Natural and Social Sciences in Disaster Risk Reduction

The social sciences played a central role in developing new thinking on risk, vulnerability and poverty, the risk process, and human roles in the accumulation of risk. The social sciences also provide core knowledge on risk perception, individual risk–related behavior, and decision-making processes. They are intimately engaged in integrated the modelling of global processes of development, industrial transformation, urbanization, and other aspects of global change. Social science concepts and techniques are essential to risk management practice, especially in community settings.

Many natural and social sciences are very active in work related to disaster risk, particularly the geosciences and the applied fields of agriculture, environment, water management, health, planning, economics, construction, and emergency management. This provides a substantial base of expertise for supporting policy and practice in DRR. By contrast, DRR, as a coherent field of scientific enquiry of its own, is relatively new and small.

Science and technology activities are mostly undertaken by national publicly funded institutions, such as national research institutes, universities, and government agencies. In some cases, a national committee may coordinate activities related to disaster risk and provide advice on priorities and appropriate funding levels. National coordination is particularly valuable owing to the multisectoral, multidisciplinary nature of disaster risk and the need for extensive collaboration on disaster risk-reduction policies and projects.

Science research and applied science have a strong international character, with international publications and peer reviewing, extensive sharing of data and results, foreign training, and much interaction among individual scientists. These processes are critical to shaping and achieving high quality research and to providing solid foundations for public policy and action, including for DRR. They also contribute to the rapid dissemination and application of information.

International collaboration on science and technology, for research, data gathering, and applications, is strongly supported through national academies of science and engineering, international associations of scientists, UN organizations, and various specialist organizations. There are international programs on most hazards and their impacts, as well as a number of multidisciplinary programs on global change and other system issues. A few are devoted solely to DRR, such as the Integrated Research on Disaster Risk (IRDR) program, the Integrated Risk Governance project, and the ISDR Global Assessment Report series.

Over the period 1990–99, the UN International Decade for Natural Disaster Reduction (IDNDR) sought to galvanize international action to reduce the loss of life, property damage, and social and economic disruption caused by natural disasters. The IDNDR was strongly supported by national scientific bodies and international geophysical science organizations, but its achievements were constrained by inadequate attention to the political and social aspects of disaster risk. This gap was a prime target of UN follow-up actions to establish ISDR and to develop the HFA 2005–15.

- In China, flooding in 1931 and 1959 caused millions of fatalities; but now with early warning and evacuation systems, as part of risk reduction policies, such great loss of life no longer occurs.
- In Bangladesh, a national flood warning system provides warnings up to 10 days ahead to millions of villagers, helping them defend against the regular flooding and preserve household assets. Studies indicate savings of about US$40 for every dollar invested.
- In New Zealand, the severe 2011 Christchurch earthquake confirmed the benefits of national building codes and property insurance. But scientific knowledge on soil liquefaction had been neglected in planning laws, resulting in billions of dollars damage to houses and infrastructure.

6.3 Social Sciences and Disaster Risk Reduction

Social sciences provide fundamental inputs to the description and integrated modelling of global processes of development, including industrial transformation, urbanization, environmental loss, and other aspects of global change. Rapid advances are occurring in understanding and tool building, including elaborate computer-based earth systems models.

It is important to recognize that although the social sciences are very influential internationally in DRR, they are not structured, resourced, and coordinated in the way that the natural sciences and related applied fields are. In many cases, they comprise very small capacities at national levels (perhaps with the exception of economics). This can set limitations on the proper contribution of social sciences to particular risk-reduction policies and initiatives.

At the same time, expert social science capacities are increasingly being used within applied and sector-specific science fields such as environment, rural development, agriculture, and health. In this role, they can inform and leverage substantial sector capacities and efforts on DRR.

In this field, the interdisciplinary International Human Dimensions Programme (IHDP) aims to generate innovative social science research to inform and improve societal responses to global environmental changes. One of the IHDP programs, Vulnerability, Resilience and Adaptation, addresses the question: As increasing risks to health, welfare, and safety are distributed very unevenly, how do different human societies cope with and adapt to external stresses and disturbances? One of its core projects, the Integrated Risk Governance project, aims to identify risk formation mechanisms and new paradigms in risk governance, especially for new risks that exceed current coping capacity of the highly inter-connected "socio-ecological system." These programs bring together international alliances of social scientists, natural scientists, economists, and engineers.

Social sciences can and should play a central role in the framing of research issues and the formulation of risk-reduction strategies, as well as to the conduct of particular research projects. One of the most important contributions in this respect has been the influential work on risk, vulnerability, and poverty, and their interconnections. The development of ideas of risk as a process and the delineation of human action in the accumulation of risk were instrumental in the conceptual development from the IDNDR to the HFA. This reframing of disaster risk is driving new themes in research, practice, and risk governance.

Recent work has elaborated finer grained knowledge on local processes of vulnerability and loss, for example, in respect to gender, age, entitlement, etc., and as a function of rural change and urban development. The social sciences have also demonstrated the importance of local communities for effective social and environmental action, and the potential for community-based action on DRR. Social sciences are central contributors to action research and learning projects.

How people perceive and respond to risk information and warnings, in particular concerning low frequency but high impact risks, has long been a topic of behavioral science research. This work continues to inform the development of better means to communicate risk and warning information and to motivate public preparedness and practical risk management action.

The IRDR program is devoted entirely to disaster risk. It stresses an integrated approach across hazards, disciplines, and scales and seeks a close coupling of the natural, socioeconomic, health, and engineering sciences. Social scientists are leading its work on the characterization of vulnerability and risk and the understanding of decision-making in complex and changing risk contexts. A new subprogram has been initiated to undertake the first systematic and critical global assessment of integrated research on disaster risk).

The *World Social Science Report 2010* published by the International Social Sciences Council provides a comprehensive review of the organization and production of social sciences in general. It examines the issues that affect their contributions, such as inequalities in capacity across regions and countries' divides and bridges between disciplines' trends in social science themes, methods, and disciplines' and the relationship between social scientists, policy-makers, and society.

6.4 Policy-making and Science and Technology for DRR

Developing countries can face major problems of access to information and resources along with handicaps to practical application such as policy and institutional shortcomings, scarce resources, lack of necessary data, and limited capacities of vulnerable affected communities.

It has been observed that most people will readily accept scientific and technical information when it is understandable, relevant to the interests of those involved, and affordable. However, there can be many barriers to uptake that need to be overcome, such as lack of political interest, conflicting views on priorities, inadequate institutional mechanisms, and lack of access to knowledge, technical capacity, and funding.

The challenging argument is with the endorsement by policy-makers on the use of science and technology for DRR. The task of making the case to policy-makers for investing in DRR is substantively dependent on the social sciences, not only for their insights into risk processes and impacts but also for their inputs on politics, economics, and administration and their specific contributions to the collection and analysis of disaster-related data. The 2010 World Bank report on the economics of DRR, for example, in addition to considering questions on cost-effectiveness, pointed out the importance of broader policy settings such as information availability, good functioning of markets, good public infrastructure, and effective public institutions.

Investment in DRR requires evidence-based risk management methods. The methods should be well proven and backed by information on the scope of application, costs, implementation process, and expected reductions of risk or losses. Whether as policies or projects, they should be tailored to the specific risk problems faced by countries and communities that arise from specific hazards and socioeconomic circumstances. Generic arguments and examples of good practice have their place, but serious investment needs specific information, analysis, and tools.

Science and technology are accelerants of progress that create new insights and methods, solve old problems, and establish higher standards and better evidence-based policies. People trained in the sciences—both social and natural—are essential to good decision-making and cost-effective implementation. But progress in science and technology does not happen by chance; it needs active leadership, support, and coordination at both national and international levels.

The ISDR system has a UN mandate to promote international efforts on science and technology for DRR. Its institutional mechanisms should be strengthened to address systemic issues in the use or nonuse of science and technology and to support agenda setting, coordination, standardization, validation, and information provision. The key strategy should be to mobilize action by existing scientific and technical organizations, particularly through the Global Platform for DRR. Efforts should be directed to three areas: practical risk reduction, science and technology capability, and advocacy purposes.

Many countries have achieved sustained reductions in risk, for example, through systematic risk assessments, land-use controls, flood management schemes, building codes and their enforcement, hazard monitoring and warning systems, and public education. Drawing on long experience, the 2005 HFA set out a comprehensive guide for the key political and technical areas of action to reduce disaster risk. The efforts on its implementation to date have contributed to rising political awareness and commitment to DRR.

However, the evidence worldwide points to continuing shortcomings in how disaster risk is recognized and managed in practice. Among the underlying causes is an inadequate appreciation of the potential of science and technology to cut risks and losses. Many fields of sciences and technology are important to understand and reduce disaster risk, including the natural and social sciences and the various applied sciences. A major challenge is to coordinate and integrate their potential inputs to produce the comprehensive knowledge and practical tools needed to routinely manage and reduce risks.

6.5 Opportunities for Policy-makers

The starting point for any country is a clear understanding of its risks and past experience of losses, and how they are changing. Which hazards are most prominent? Which parts of the country are affected? Which groups in society suffer most? The answers to these questions require systematic research and assessments, drawing on scientific knowledge concerning the hazards and the exposure and vulnerabilities of populations and assets. This, in turn, requires the availability of scientific and technical institutions and professional cadres knowledgeable in these topics.

Then, with a clearer idea of what is at stake, consideration can be given to potential options for action, weighing up the relative costs and benefits of each option and deciding on the priorities. Specific solutions may need a blend of actions such as infrastructure investment, public information, changes in laws or regulation, new economic or social policies, and sector-specific programs.

Disaster risk and its reduction is a crosscutting issue that spans national and local governments; key sectors such as agriculture, health, and industry; specialist agencies such as emergency management and meteorology; and diverse natural and social sciences. While scientists and engineers often take a lead on disaster risk matters, experience shows that a problem-solving approach that involves a range of experts and lay people is often best, for example, including administrators, planners, community leaders, and representatives of key sectors, government agencies, and the private sector.

The policy-maker can provide critical leadership by keeping the focus on the priority risks and on the target of measurable reductions of risk. He or she should promote political recognition, ensure that the institutional ownership of the problem is clear and accountable, demand cooperation between sectors and between government departments, and support the budgets for the necessary technical capacities. Large investments may be involved, particularly if infrastructure, urban development, and national social support policies are involved.

6.6 Examples of the Power of Science and Technology

One of the great risk-reduction achievements of the last century has been the development of accurate early warning systems for hurricanes, tornadoes, thunderstorms, heavy rainfall, flooding, drought, snowfall, high winds, and other extreme weather hazards. Based on years of public investment in scientific research and data gathering systems, sophisticated computer models can now replicate the physics of the atmosphere and its weather systems and the behavior of water on the land. National meteorological and hydrological services routinely provide timely and detailed forecasts, allowing organizations and individuals to better prepare and protect themselves.

When coupled with effective communication systems, well-informed and well-prepared populations, and good leadership, these warning services can hugely reduce the loss of life and property. Millions of people worldwide owe their lives to these advances. For example, in China, whereas 3,700,000 lives were reported lost in floods in 1931, and 2,000,000 in 1959, now the figure is routinely less than 2,000 per annum. China has implemented comprehensive national and regional "five-year" plans for DRR, based on the scientific understanding of temporal and spatial variation of hazards and risk and including warning and evacuation systems.

In Bangladesh, a unique flood warning system has helped millions of villagers defend against the regular flooding that occurs across the Ganges–Brahmaputra delta. Drawing on international high-tech rainfall predictions made up to 10 days in advance, the national Flood Forecast and Warning Centre prepares and distributes forecasts to district offices and other organizations. Networks of community leaders then disseminate information to villagers on expected flood levels along with advice on evacuation action and other responses, such as storing food and water, protecting household assets, and early harvesting. Recent studies have indicated that approximately US$40 was saved for every dollar invested in the forecasting and warning system, and that the savings were US$400–500 per affected household (locally, about one year's income).

Informed approaches to land-use planning and management, and to construction and building codes, can target the root causes of risk, as is illustrated by the cases below.

The Netherlands, for example, is highly experienced in the use of dykes and other engineering structures to secure its low-lying farmlands and towns against the threat of flooding from rivers and the adjacent seas. However, in recent years, there has been a realization that the thinking must shift away from relying on dykes alone toward a resilience approach that combines traditional engineering actions with other risk-reducing innovations, particularly controlled flooding of specific farmland areas for temporary floodwater storage. The country is now engaged in a decades-long US$3 billion program, "Room for the River," that involves dozens of infrastructure projects to mitigate the likely impacts of climate change and to encourage development that reduces risk and environmental impacts.

In New Zealand, the severe Christchurch earthquake of February 22, 2011, resulted in 185 deaths and an economic loss of about US$18 billion. Once again, there were many lessons on the importance of proactive science-based risk reduction. Most of the deaths occurred through the collapse of three multistory buildings of relatively modern construction, through faults in design, permitting, and construction. Although many old brick buildings collapsed, most buildings were sufficiently intact to allow people to escape largely unharmed, demonstrating the general value of the country's approach to building safety. This was despite the fact that the earthquake greatly exceeded the seismic reference level of the building code. Another positive lesson was the benefit of the mandatory natural hazards component of property insurance, which is now contributing about 80 percent of the recovery costs, a very high proportion by world standards. A costly lesson, however, was the failure to apply available scientific knowledge of liquefaction risk to the city's planning laws. Extensive liquefaction of soils across the city resulted in substantial structural damage to buildings and housing and was a primary source of the large economic losses.

6.7 Remote Sensing, Geographic Information System, and Disaster Risk

In the Indian subcontinent, nearly a billion people are affected by natural and man-made disasters each year. Rapid population growth and unprecedented development, combined with climatic variations and a complex geo-environmental setting, contribute to the increasing effect of disasters. In recent times, South Asia has witnessed large-scale disasters such as the frequent floods in the Indo-Gangetic and Brahmaputra plains, Sri Lanka, the cyclones of the Indian east coast, Gujarat, and Pakistan; the earthquakes of Uttarkashi, Latur, Jabalpur, Chamoli, Gujarat, and Kashmir; and small-scale hazards such as landslides in the Sri Lankan highland and Himalayan range, forest fire, soil erosion, and desertification. These events are stark reminders of the fact that natural disasters can cause a crumbling impact on the economy and take a huge toll of human lives.

6.7.1 Remote Sensing Technology

Present-day Earth observation satellites carry a variety of imaging sensors that are capable of generating an image of the objects on the Earth's surface of various resolutions. With the use of remote sensing (RS) data, it is possible to map, monitor, and measure many environmental parameters related to natural disasters and associated features. For example, weather satellites such as the US National Oceanic and Atmospheric Administration's (NOAA) Geostationary Operational Environmental Satellites (GOES), in combination with land measurement sensors such as NOAA's Advanced Very High Resolution Radiometer (AVHRR), have provided essential information for monitoring disaster events and their aftermath. High-resolution images from the Indian Remote Sensing (IRS) Satellites (Resourcesat-1 and 2, LISS-IV, and Cartosat-1, 2), IKONOS, World View, Geoeye, and Quick Bird provide detailed information that can play a crucial role in disaster management at a large scale. Recent advances of synthetic aperture radar (SAR) interferometry have shown important application potential in studying earthquake, volcano monitoring, glacier movement, landslide, and land-subsidence and for preparing digital elevation models. It has also been demonstrated that airborne scanning laser altimetry, generally referred to as LIDAR, can provide very high-resolution terrain information that can be used for a number of applications related to disaster management. In general, the role of RS or earth observation systems (EOS) in disaster management can be summarized as below:

1. When a disaster occurs, the aerospace/RS media is the quickest mode of information collection and can be utilized for monitoring the event and providing valuable information for relief operations and damage assessment.
2. Some of the disasters have precursors, which can be detected at an early stage by orbiting satellites and can be used for early warning and prediction.
3. Most importantly, RS in combination with GIS can provide database and analytical tools that can be used to produce hazard and vulnerability maps that have immense value in any kind of activities related to disaster management. It can also help to simulate scenarios and assess effects of disastrous events.

High-resolution satellite data products and GIS database can be used for rehabilitation and reconstruction. One of the main advantages of the use of the powerful combination techniques of a GIS is the evaluation of several hazard and risk scenarios that can be used in decision-making about the future development of an area, and the optimum way to protect it from natural disasters.

RS data derived from satellites are excellent tools in the mapping of the spatial distribution of disaster-related data within a relatively short period of time. Many different satellite-based systems exist nowadays, with different characteristics related to their *spatial, temporal, and spectral resolution.*

RS data should generally be linked or calibrated with other types of data, derived from mapping, measurement networks, or sampling points, to derive parameters that are useful in the study of disasters. The linkage is done in two ways, either via visual interpretation of the image or via classification.

The data required for disaster management is coming from different scientific disciplines and should be integrated. Data integration is one of the strongest points of GIS. In general, the following types of data are required:

- data on the disastrous phenomena (e.g., landslides, floods, earthquakes), their location, frequency, magnitude, etc.;
- data on the environment in which the disastrous events might take place: topography, geology, geomorphology, soils, hydrology, land use, vegetation, etc.;

- data on the elements that might be destroyed if the event takes place: infrastructure, settlements, population, socioeconomic data, etc.; and
- data on the emergency relief resources: hospitals, fire brigades, police stations, warehouses, etc.

The amount and type of data that has to be stored in a GIS for disaster management depends very much on the level of application or the scale of the management project. Natural hazards information should be included routinely in development planning and investment project preparation (Table 6.1). Development and investment projects should include a cost/benefit analysis of investing in hazard mitigation measures, and they should be weighed against the losses that are likely to occur if these measures are not taken (OAS/DRDE 1990).

Table 6.1:
Remote Sensing Satellites and Sensors for Disaster Monitoring and Management (Partial List)

Platform	Sensor	Launch	PAN/ Multispectral/ Radar	Pixel size	Swath
Resourcesat-2	LISS-III	2011	M 4	23.5 m	141 km
	LISS-IV		M 3	5.8 m	70 km
	AWiFS		M 4	56 m	740 km
Cartosat-2, 2A, 2B	PAN	2007, 2008, 2009	P	<1.0 m	9.6 km
Cartosat-1	PAN (Stereo)	2005	P	2.5 m	27.5 km
IRS-P4 (Oceansat)	OCM	2009	M 8	360 m × 236 m	1,420 km
RISAT-2	SAR	2009	X bad		
IRS-P4 (Oceansat)	OCM	1999	M 8	360 m	1,420 km
IRS-TES	PAN	2001	P	< 1.0 m	30 km
Resourcesat-1	LISS-III	2003	M 4	23.5 m	141 km
	LISS-IV		M 3	5.8 m	23.9 km
	PAN		P	5.8 m	70.3 km
LANDSAT 7	ETM	1999	M 6	30 m	185 km
			P	15 m	
			T	60 m	
TERRA-ASTER	ASTER	1999	M 14	15–90 m	60 km

(Continued)

Table 6.1:
(Continued)

Platform	Sensor	Launch	PAN/ Multispectral/ Radar	Pixel size	Swath
TERRA-MODIS	MODIS	2000	M 36	250–1,000 m	2,330 km
Ikonos		1999	M 4 P	4 m 1 m	11 km
Orbview-3	Orbview	2003	M 4 P	4 m 1 m	8 km 8 km
Quickbird (Digitalglobe)	QB-P QB-M	2001	P M 4	0.61 m 2.44 m	16.5 km
SPOT-5	PLA MLA	2002	P M	2.5 m 10 m	30 km 60 km
Cosmo Skymed		2007– 10	X band	1–30 m	10–3,040 km
Terra SAR-X	SAR	2007	X band	1–16 m	10–100 km
Envisat	ASAR	2002	C HH VV HV VH	30, 150 m	100, 400 km
RADARSAT-2	SAR		C band 5.3 GHz, HH	3–100 m	20–500 km
RADARSAT-1	SAR	1995	C band 5.3 GHz, HH	8–100 m	50–500 km
SRTM	SAR	2000	C, X	C-30 m	C-225 km, X-50 km
NOAA	AVHRR	1984	M	1.1/4 km	2,600/ 4,000 km

Source: Authors.

Although the selection of the scale of analysis is usually determined by the intended application of the mapping results, the choice of analysis technique remains open. This choice depends on the type of problem, the availability of data, the availability of financial resources, the time available for the investigation, and the professional experience of the experts involved in the survey.

6.7.2 Remote Sensing and GIS Tools

Mitigation of natural disasters can be successful only when detailed knowledge is obtained about the expected frequency, character, and magnitude of hazardous events in an area. Many types of information that are needed in natural disaster management have an important spatial component. Spatial data is data with a geographic component, such as maps, aerial photography, satellite imagery, GPS data, rainfall data, borehole data, etc. Many of these data will have a different projection and coordinate system, and need to be brought to a common map-basis, in order to superimpose them.

We now have access to information gathering and organizing technologies such as RS and GIS, which have proven their usefulness in disaster management. First of all, RS and GIS provide a database from which the evidence left behind by disasters that have occurred before can be interpreted and combined with other information to arrive at hazard maps, indicating which areas are potentially dangerous. The zonation of hazard must be the basis for any disaster management project and should supply planners and decision-makers with adequate and understandable information. RS data, such as satellite images and aerial photographs, allow us to map the variabilities of terrain properties, such as vegetation, water, and geology, both in space and time. Satellite images give a synoptic overview and provide very useful environmental information for a wide range of scales, from entire continents to details of a few meters. Second, many types of disasters, such as floods, drought, cyclones, volcanic eruptions, etc. will have certain precursors. The satellites can detect the early stages of these events as anomalies in a time series. Images are available at regular, short time intervals, and can be used for the prediction of both rapid and slow disasters.

Then, when a disaster occurs, the speed of information collection from air and spaceborne platforms and the possibility of information dissemination with a matching swiftness make it possible to monitor the occurrence of the disaster. Many disasters may affect large areas, and no tool other than RS would provide a matching spatial coverage. RS also allows monitoring the event during the time of occurrence while the forces are in full swing. The vantage position of satellites makes it ideal for us to think of, plan for, and operationally monitor the event. GIS is used as a tool for the planning of evacuation routes, for the designing of centers for emergency operations, and for the integration of satellite data with other relevant data in the design of disaster warning systems.

In the disaster relief phase, GIS is extremely useful in combination with the Global Positioning Systems (GPS) in SAR operations in areas that have been devastated and where it is difficult to orientate. The impact and departure of the disaster event leaves behind an area of immense devastation. RS can assist in damage assessment and aftermath monitoring, providing a quantitative base for relief operations.

In the disaster rehabilitation phase, GIS is used to organize the damage information and the post-disaster census information, and to evaluate the sites for reconstruction. RS is used to map the new situation and update the databases used for the reconstruction of an area, and it can help to prevent such a disaster from occurring again. The volume of data needed for disaster management, particularly in the context of integrated development planning, clearly, is too much to be handled by manual methods in a timely and effective way. For example, the post-disaster damage reports on buildings in an earthquake stricken city may be in thousands. Each one will need to be evaluated separately in order to decide if the building has suffered irreparable damage or not. After that, all reports should be combined to arrive at a reconstruction zoning within a relatively small period of time.

6.7.3 Indian Remote Sensing Programme and Disaster Management

Over last two decades, the Indian Space Program has established a strong space segment for Earth observation. Over the years, it has developed state-of-the-art RS satellites such as IndianIRS-1C/1D, Resourcesat-1 and 2, Cartosat series of satellites, and indigenous launching capability (PSLV series) and a plethora of applications related to environment and natural resource such as groundwater targeting, wasteland mapping, forest cover mapping, landslide mapping, agricultural monitoring, coastal zone mapping, ocean environment studies, urban studies, mineral exploration, geological studies, etc. Currently, large volumes of satellite data are being acquired daily over the Indian subcontinent by a constellation of Indian RS satellites (Cartosat-1, Cartosat-2 series, Resources at 1 and 2, Oceansat-2) and Indian national satellites (INSAT-Series). Apart from this, data from a number of foreign satellites are also acquired at NRSC, Hyderabad.

In addition to technological advancements, other reasons have also been attributed to the need for space inputs in disaster management. India is a vast country and many interior parts have not been adequately surveyed in recent times; therefore, remotely sensed information products are highly desirable for mapping recent land use and terrain features required for developmental planning and disaster management. Therefore, over a period of time, ISRO has developed a very useful disaster management support program under which it delivers data products, inputs for early warnings, and vital communication link for effective disaster management in the country.

6.7.4 International Charter

The International Charter on Space and Major Disasters is a unique initiative that envisages cooperation between space agencies in the use of space facilities in a collaborative manner. It aims at providing a unified system of space data acquisition and delivery to those affected by natural or manmade disasters through authorized users (AUs). Each member agency has committed resources to support the provisions of the charter. In order to support the International Charter, ISRO has set up organizational machinery. Operationally, the charter provides a 24/7 mechanism for the partner agencies to provide space-based data and products to authorized disaster first responders in an effective, timely, and coordinated manner ct no cost to the user. Since its inception and in 2011, the Charter has several disasters related to earthquakes, tsunamis, cyclones, landslides, volcanic eruptions, floods, oil spills, etc. During the recent Sikkim earthquake in India, relevant information was also provided under charter activation.

The International Charter on Space and Major Disasters was initiated by the European space agencies, particularly the French space agency (ESA and CNES), following the UNISPACE III conference held in Vienna, Austria, in July 1999 aiming at providing a unified system of space data acquisition and nonbureaucratic delivery to those affected by natural or man-made disasters. Afterwards, the Canadian Space Agency (CSA) signed the charter on October 20, 2000. Other members who joined the charter later on are the Indian Space Research Organization (ISRO) and NOAA in 2001, the Argentine Space Agency (CONAE) in 2003, the Japanese Aerospace and Exploration Agency (JAXA) and the US Geological Survey (USGS), as well as the Disaster Monitoring Constellation (DMC) via the British National Space Center (BNSC), in 2005, and the China National Space Administration (CNSA) in 2007. As shown in Figure 6.1, the on-duty operators (ODO) and emergency on-call officers (ECO) receive the requests from AUs and process for relevant satellite data on a 24/7 basis. AUs are civil protection and relief organizations or national authorities with a mandate related to disaster management from within the country of a charter member. They request the mobilization of the space and associated ground resources (RADARSAT, Earth Resource Satellite [ERS], ENVISAT, Satellite Pour l'Observation de la Terre [SPOT], IRS, SAC-C, NOAA satellites, LANDSAT, Advanced Land Observing Satellite [ALOS], DMC satellites, and others) of the member agencies to obtain data on a disaster occurrence. Project managers quickly assess the severity of particular disasters and the relevancy of satellite data holdings and capabilities. Satellite image capturing plans and feasibility are determined, and the appropriate space agencies are contacted to conduct satellite-imaging operations.

6.7.4.1 FUNCTIONAL UNITS AND CHARTER OPERATIONS

The charter is made up of six functional units: The AUs, the ODOs, the ECOs, the project manager (PM), the data processing and distribution facilities, and, the value-added resellers (VARs). Figure 6.1 illustrates the operational loop for the charter's functional units, which are described in detail. Although the processes and procedures for using the charter are well-defined, charter activations do not always work smoothly. Sometimes there are delays or oddities in the operational loop because the charter is a relatively new initiative that is not universally well understood by AUs and PMs. Also, the AUs and PMs are sometimes identified and educated in use of the charter during the response phase of a disaster, rather than in advance of the event.

Figure 6.1:
Charter Operation Loop

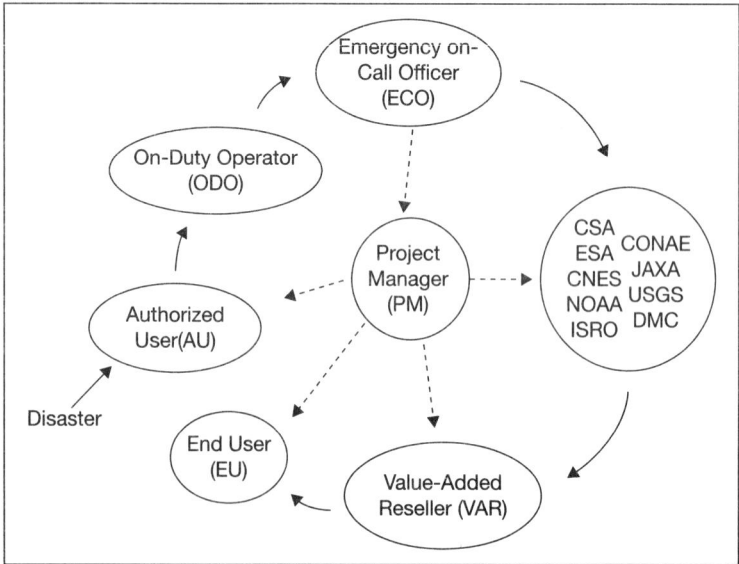

Source: Modified after "Executive Secretariat for the International Charter: Space and Major Disasters."

6.7.5 UNOSAT

The UN Operational Satellite Application (UNOSAT) is a program of UN Institute for Training and Research (UNITAR) and implemented in cooperation with the European Organization for Nuclear Research (CERN). UNOSAT is a people-centered program delivering satellite solutions to relief and development organizations within and outside the UN system to help make a difference in the life of communities exposed to poverty, hazards, and risks, or affected by humanitarian and other crises (UNOSAT 2008). UNOSAT is authorized to request data from the charter members in response to an emergency (Figure 6.2). It has provided value-added processing services for many charter activations over the years. One important achievement is its dedicated support to the planning and coordination of humanitarian relief operations. UNOSAT operates humanitarian, rapid mapping service 24 hours a day, all year-round, ensuring that experts are available whenever needed for rapid acquisition and processing of satellite imagery and data for the creation of maps and GIS layers in support of emergency response and humanitarian relief coordination.

Figure 6.2:
International Charter Member Agencies

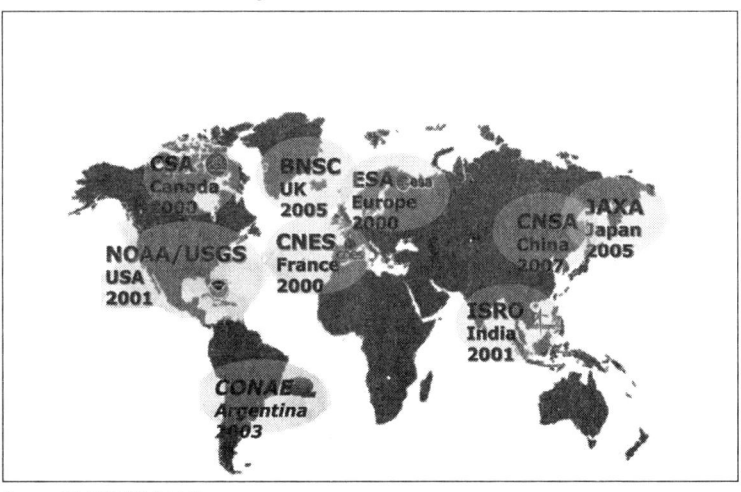

Source: UNOSAT (2008).

6.7.6 Activation of International Charter for Disaster Management in Asian Region (Partial)

Floods in Vietnam (October 17, 2011)
After several days of heavy rain, at least 26 people had been killed due to flooding in parts of Central Vietnam.

Flood in Thailand (October 17, 2011)
Flooding has affected two thirds of Thailand and has killed over 300 people since July. The floods have affected more than 2 million people, with many moving to evacuation shelters.

Flood in Cambodia (October 12, 2011)
Flooding in Cambodia has killed over 200 people since early September. Thousands of people have evacuated as the increasing flood situation has caused massive damage.

Earthquake, Landslide in Sikkim, North East India (September 18, 2011)
The epicentre of India's North Eastern state, Sikkim, was struck by an earthquake of a magnitude of 6.8. Shortly after the quake, two strong aftershocks followed that also affected Nepal and Tibet.

Earthquake in Pakistan (January 19, 2011)
In Dalbandin, several people were injured when the roofs of their houses collapsed, provincial Transport Minister Amanullah Notizai told Reuters, but there were no reports of fatalities in the quake that hit at 1:23 AM (2023 GMT on Tuesday).

6.8 Examples of Remote Sensing and GIS Applications in Disaster Monitoring and Management

6.8.1 Earthquake

The devastating earthquakes of the recent past have emphasised the need to explore technological advancements including space technology for better understanding of the phenomena and its impact in terms of regional and local risk. Although the prediction of seismic events remains as an elusive goal even today after much technological advancement, the main value of scientific investigation lies in the potential for identification of hazard-prone areas where proper preventive measures can be taken well before the fatal shock strikes. Second, comprehensive seismic hazard assessment over a range of spatial and temporal scales will provide a more systematic approach in prioritizing areas for retrofitting of vulnerable structures, relocating populations at risk, protecting lifelines, preparing for disasters, and educating the public.

The most spectacular observation from space has been made possible by deployment of dense geodetic network of GPSs in seismic active regions and satellite Interferometric Synthetic Aperture Radar (InSAR) as well as subpixel correlation (of optical images) based deformation mapping to understand the kinematics of fault rupture and transient stress built up. A review of the literature reveals significant progress in the application of Earth observation in earthquake studies in the following areas.

6.8.2 Neotectonics and Morphotectonics to Identify Active Faults

In the initial phase of earthquake hazard studies, the most significant contribution is made by monitoring the subtle neotectonic changes often expressed in the form of diagnostic landform and relief changes. Monitoring and mapping of such subtle changes on relief and landform is very important, and in one case study, the evidence of crustal deformation and its manifestation on surface geomorphology in the recent past (1972–2003) have been observed in the seismically active Himalayan Frontal Thrust (HFT) zone at Dehradun, India, by comparative analysis of the Landsat images of 1972 and IRS Satellite.

6.8.3 Seismicity-induced Landslide

Earthquakes in hilly regions often trigger landslides that can be the main cause of damage and destruction as observed during the Kashmir, China, and recent Sikkim earthquake that caused several landslides that killed people and damaged houses and infrastructure. Using remotely sensed images, landslides, landslide debris blocked dams can be mapped and their damage potential can be assessed. Using temporal information, it is even possible to make volumetric estimation of the landslide as well as the reservoir.

6.8.4 Liquefaction Study Using Optical Remote Sensing

Liquefaction is a soil behaviour phenomenon in which a saturated soil looses a substantial amount of strength due to high pore-water pressure generated by and accumulated during strong earthquake ground shaking. The liquefaction phenomenon can be very damaging or hazardous when accompanied by strong ground displacement or ground failure as was observed during the Bhuj earthquake of 2001 (Champati Ray et al. 2001). Using image processing on pre- and post-LISS III datasets, it was possible to delineate earthquake-induced liquefaction areas indicated by liquid emanation. In the event of an earthquake, the liquefaction probability can be estimated based on the geological characteristics of soil, earthquake strong motion in terms of PGA, and ground water table. For the Bhuj earthquake, the liquefaction probability has been modelled, which corresponds well with the actual observation as revealed on satellite image and in the field.

6.8.5 Landslides

RS data can provide authentic and timely information on the extent of landslides in a cost-effective and timely manner. It also provides information on causative factors such as rock type, geomorphology, geological structure (fold, fault, joints, etc.) land use, vegetation cover, etc. Aerial photographs and high-resolution satellite images have been used extensively by several agencies to map, monitor, and predict landslide hazard-prone areas in different mountainous regions of the world. In the recent past, the geosciences division at the Indian Institute of Remote Sensing (IIRS) in Dehradun, India, has experimented with various quantitative methods for spatial prediction of landslides in the Himalayan region using remotely sensed data products and other ancillary information in GIS environment (Champati Ray 1996; Das et al. 2010). The results of such studies indicate that RS data products can be utilized for deriving various information related to landslide studies, and the information can be utilized for landslide damage assessment and prediction in GIS.

In the recent past, attempts have been made to map, monitor, and make landslide hazard zonation maps using data integration techniques in GIS showing areas prone to landslides. These maps are of immense value in planning hilly area development and minimizing loss due to landslides in the region. Subsequent to the Malpa and Okhimath tragedy in India in 2008, at the behest of the cabinet secretary (Government of India), the Department of Space launched a major project, titled National Landslide Project, on landslide hazard zonation along major tourist and strategic routes covering 1,800 km of road length in UP and HP Himalayas. The project was carried out with the collaboration of 11 government organizations and it generated landslide hazard zonation maps along with management plans on a 1:25,000 scale. A similar project was carried out by IIRS, NESAC, and CBRI for landslide hazard assessment in the north Eastern (India) region along the highway from Shillong to Aizwal via Silchar.

6.8.6 Cyclones and Storm Surges

The South Asian coastal areas encircling the Indian Ocean have emerged as one of the most prominent regions experiencing cyclonic hazards in the recent past. For example, the Orissa super cyclone (1999) and the tsunami (2004) have wreaked havoc by taking thousands of lives, and it will take ages for the community to recover fully from such catastrophes. With approximately 8,000 km of coastline exposed to tropical cyclones, India is country highly vulnerable to cyclonic disasters. Over a period of time, it has developed expertise in cyclone prediction and warning with the help of ground-based radar and space-based observation. A network of cyclone-detection radars covering the entire east west coasts is being used for cyclone warning within a range of 400 km, beyond which NOAA and INSAT datasets are used. For precise location of the cyclone, all half-hour INSAT images are used on operational basis.

The EOS plays an important role in assessing the cyclonic disaster and subsequent emergency relief operation. During the super cyclone of Orissa, information on the spatial extent of inundated areas was derived using IRS LISS-III and Radarsat Synthetic Aperture data (SAR) data. The map showing inundated areas draped over a topographical map was delivered to the stakeholders. The information thus generated was of immense help to various agencies involved in the relief operation. Temporal information on water-inundated areas was also generated for the subsequent period, which was found to be very useful for the relief operation and damage assessment. Additionally, crop damage assessment along with block-wise statistics could also be generated using IRS Wide Field Sensor (WiFS) data (Rao 2000).

The tropical cyclones constitute one of the most destructive natural disasters that affect India, especially its east coast. Its impact is greatest over coastal regions, which bear the brunt of strong winds, heavy rainfall, and flooding. RS data has been utilized for tracking, monitoring, and forecasting of cyclones, assessment of damage, and preventive measures. INSAT data have been regularly utilized to monitor the track of cyclones and forecasting its crossing point on land. IRS 1C/1D data have been utilized to assess the damage caused by cyclones. IRS WiFS data was utilized to delineate area under inundation. NDVI images generated using WiFS data can also provide damage to vegetation. LISS III data was found to be quite useful in assessing damage caused to agriculture and horticulture areas. Panchromatic (PAN) data provided input on structural damage caused to large buildings. An important aspect of damage assessment is to provide input within 2–3 days of the event so that necessary relief measures can be taken up. Many times, optical data are not useful due to cloud cover during the event. It is necessary use radar data in such cases.

However, about one week will be required to acquire such data. It was also observed that mangroves get affected marginally. Generally, defoliation occurs and mangroves recover in 3–4 months. This also suggests that mangrove afforestation should be taken up in cyclone-prone areas to protect life and property in coastal regions. RS data can be used in identifying sites for such afforestation.

6.8.7 Flooding

Different types of flooding (e.g., river floods, flash floods, dam-break floods or coastal floods) have different characteristics with respect to the time of occurrence, magnitude, frequency, duration, flow velocity, and areal extension. Many factors play a role in the occurrence of flooding, such as the intensity and duration of rainfall, snowmelt, deforestation, land-use practices, sedimentation in riverbeds, and natural or man-made obstructions. In the evaluation of flood hazard, the following parameters should be taken into account: depth of water during flood, duration of flood, flow velocity, rate of rise and decline, and frequency of occurrence. Satellite data has been successfully and operationally used in most phases of flood disaster management (CEOS 1999). Multichannel and multisensor data sources from geostationary operational environmental satellite (GOES)/ polar operational environmental satellite (POES) are used for meteorological evaluation, interpretation, validation, and assimilation into numerical weather prediction models to assess hydrological and hydrogeological risks (Barrett 1996). Quantitative precipitation estimates (QPE) and forecasts (QPF) use satellite data as one source of information to facilitate flood forecasts in order to provide early warnings of flood hazard to communities. Earth observation satellites can be used in the phase of disaster prevention by mapping geomorphologic elements, historical events, and sequential inundation phases, including duration, depth of inundation, and direction of current.

Earth observation satellites are also used extensively in the phases of preparedness/warning and response/monitoring. The use of optical sensors for flood mapping is seriously limited by the extensive cloud cover that is mostly present during a flood event. Synthetic aperture radar (SAR) from ERS and RADARSAT has been proven very useful for mapping flood inundation areas, due to their bad weather capability. In India, ERS-SAR has been used successfully in flood monitoring since 1993, and Radarsat since 1998 (Chakraborti 1999). A standard procedure is used in which speckle is removed with medium filtering techniques and a piece-wise linear stretching. Color composites are generated using SAR data during floods and pre-flood SAR images. For the disaster relief operations, the application of current satellite systems is still limited, due to their poor spatial resolution and the problems with cloud covers. Hopefully, the series of high-resolution satellites will improve this. RS data for flood management should always be integrated with other data in a GIS. A large number of hydrological and hydraulic factors need to be integrated, especially on the local scale. One of the most important aspects in which GIS can contribute is in the generation of detailed topographic information using high-precision, digital elevation models, derived from geodetic surveys, aerial photography, SPOT, LiDAR (Light detection and Ranging), or SAR (Corr 1983). These data are used in two and three dimensional (2D and 3D) finite element models for the prediction of floods in river channels and floodplains (Gee et al. 1990).

The optical and microwave data from IRS, LANDSAT, ERS, and Radarsat series of satellites have being used to map and monitor flood events in near real time in operational mode. The information on inundation and damage due to floods is furnished to user agencies to effectively carry out relief operation, damage assessment, and rehabilitation. The capability of EOS in flood management is well demonstrated through various case studies carried out by the NRSC (Chakraborti 2000;), particularly through a pilot project in Assam carried out by a task force setup by the Department of Space, GOI, involving NRSA, RRSSC, SAC, EOS/ISRO HQ, and DMS/ISRO in close coordination with the state government, and central departments (Department of Agriculture and Cooperation [DAC], Central Water Commission [CWC], IMD, and the Assam Remote Sensing Application Centre).

In order to improve the flood forecasting and warning by incorporating RS inputs, a pilot study, titled Spatial Flood Warning System, was undertaken for the Godavari basin (Rao 2000). Under this project, a digital terrain model generated from differential global positioning system (DGPS), hydrologic modelling, and a decision support system were used for flood forecast; the result is highly encouraging and holds good potential to use in operational mode.

6.8.8 Drought

The space-based technology can be used in various ways for drought mitigation. For identification of drought-prone areas, various types of information on land use, land cover, waste lands, forest cover, and soil can be obtained from remotely sensed data. At present, under various programs of the Department of Space and state RS agencies, such information is being generated at various levels. Medium-range weather prediction plays an important role in drought prediction. The National Centre for Medium Range Weather Forecasting (NCMRF) uses satellite-based sea surface temperature, normalized difference vegetation index, snow covered area, snow depth, surface temperature, soil moisture, etc. for weather forecasting. Also INSAT-based visible and thermal data are being used for these purposes (Rao 2000). Satellite RS offers effective means of drought surveillance. One way to minimize the impact of drought is to continually monitor the conditions in drought-prone regions, particularly in the wake of poor rainfall, and provide early warning to prevent or minimize the effects of impeding hydrological or agricultural drought. For effective drought monitoring and assessment, a project titled National Agricultural Drought Assessment and Monitoring System (NADAMS) was sponsored by the Department of Agriculture and Cooperation and the Department of Space (DOS) and was taken up by the National Remote Sensing Centre (NRSC) in collaboration with the Indian Meteorological Department (IMD), the CWC, and concerned state government agencies. The main emphasis was on the assessment of agricultural drought conditions in terms of prevalence, relative severity level, and persistence through the season. Satellite-derived vegetation index (VI), which is sensitive to vegetation stress, was used as a surrogate indicator to monitor drought conditions on a real-time basis. Initially, NDVI derived from NOAA-AVHRR was used for drought monitoring; subsequently WiFS data from the IRS was used for deriving information on sowing, surface water spread, taluk-/*mandal*-level crop condition. The NADAMS provides detailed assessment of drought conditions for providing short-term relief. As a part of long-term strategy, a nationwide project, titled Integrated Mission for Sustainable Development (IMSD), was taken up in collaboration with the Department of Space centers and state RS application centers. The project essentially aims at generating locale-specific action plans for the development of land and water resources on microwatershed basis in drought-prone areas using IRS data.

6.8.9 Forest Fire

RS and GIS offer effective tools to study forest fires. At the national and regional scales, NOAAA/AVHRR data are mostly used for burnt area mapping. The thermal data of NOAA-AVHRR were used to map hot pixels on a daily basis. In India, around 50 percent of the total forest cover of 63 million hectares is prone to forest fire and annually 6 percent of the forest area gets affected by forest fire, a majority of which is started by humans. Satellite RS has tremendous potential in determining fire risk, fire detection, fire monitoring, and fire damage assessment. Various case studies from Garhwal Himalayas and Gir protected area have demonstrated the capabilities of IRS satellite products in forest fire management (Roy 2000). In India, the temporal data from IRS series have been effectively used to map and monitor the incidence of fire in some of the wildlife sanctuaries and national parks of the country. Information on the extent of the burnt area and the types of vegetation affected has been derived forest range-wise. Past satellite data of the Mudumalai wildlife sanctuary and national park have been analysed for time series forest fire events and been integrated to generate a fire frequency map. In another attempt, the fire risk zones have been identified by integrating vegetation type/density, road and settlement network, slope, drainage, and past history of forest fire occurrence using logistic models (Jain et al. 1996). These types of maps are required to study the impact of forest fires on the forest vegetation, regeneration status, and delineated forest risk zones. Under a new initiative, known as Indian Forest Fire Response and Assessment System (INFFRAS), near real time, daily, daytime, and nighttime active forest fire alerts are provided during the fire season (February–June). Fire alerts are prepared using satellite data from the moderate-resolution imaging spectroradiometer (MODIS) sensors aboard the Terra and Aqua platforms (for daytime observations), as well as the operational linescan system (OLS) sensor on the Defense Meteorological

Satellite Program (DMSP) satellites (for nighttime observations). The fire alerts report only active forest fires observed by these satellites. This is made available to users on a no-cost basis.

6.9 Impact of Developmental Activities

Rapid industrialization has lead to the development of new ports or the expansion of existing facilities. The groins, seawalls, breakwaters, and other protective structures have secondary effects resulting in downstream erosion. Erosion has been observed north of Visakhapatnam, Paradip, Ennore, and north of Madras and near Nagapattiam and Kanyakumari ports on the East Coast of India, while deposition has been observed south of these ports. These changes are attributed to the construction of artificial barriers such as breakwaters, jetties, etc. Erosion noticed at the Kavaratti and Agatti Islands is mainly attributed to dredging and other human activities. The knowledge about suspended sediment movement helps in understanding water flow near the shore. In one such study, a sediment plume of circular shape emerging from the Kochi harbor was identified. This plume indicated a sharp contact with the sediments along the coast indicating two different water masses. It is also observed that sediment concentration is more on the northern side of the plume compared to the southern side. Such behaviour was observed in all seasons. This clearly indicated that the plume is acting as obstruction to the sediment movement. This has resulted in erosion on the southern coast and deposition on the northern coast. This behaviour is also seen in the shoreline-change maps. The recently available IRS P4 ocean color monitor (OCM) data are extremely useful to study sediment dispersal and sediment transport studies due to their two-day repetitivity.

6.10 Unmanned Arial Vehicle/Drone in Disaster Management

6.10.1 Definition of Unmanned Arial Vehicle

"UAVs are to be understood as uninhabited and reusable motorized aerial vehicles." These vehicles are remotely controlled, semi-autonomous,

autonomous, or have a combination of these capabilities. Comparing an unmanned aerial vehicle (UAV) to a manned aircraft, it is obvious that the main difference between the two systems is that on the UAV no pilot is physically present in the aircraft. This does not necessarily imply that a UAV flies by itself autonomously. The term UAV is commonly used in the computer science, robotics and artificial intelligence, as well as the photogrammetry and RS communities. Additionally, synonyms such as remotely piloted vehicle (RPV), remotely operated aircraft (ROA) or remotely piloted aircraft (RPA) and unmanned vehicle systems (UVS) can also infrequently be found in the literature. RPV is a term to describe a robotic aircraft flown by a pilot using a ground control station. The first use of this term may be addressed to the US Department of Defence during the 1970s and the 1980s. The terms ROA and RPA have been used by National Aeronautics and Space Administration (NASA) and Federal Aviation Administration (FAA) in the USA in place of UAV. Furthermore, the term Unmanned Aircraft System (UAS) is also being used (Colomina et al. 2008). The FAA has adopted the generic class UAS, which was originally introduced by the US Navy. Common understanding is that the terminology UAS stands for the whole system, including the Unmanned Aircraft (UA) and the Ground Control Station (GCS).

6.10.2 UAV Photogrammetry

The new terminology UAV photogrammetry (Eisenbeiss 2008c) describes a photogrammetric measurement platform that operates remotely controlled, semiautonomously, or autonomously, without a pilot sitting in the vehicle. The platform is equipped with a photogrammetric measurement system, including, but not limited to, a small- or medium-size still-video or video camera, thermal or infrared camera systems, airborne LiDAR system, or a combination thereof. Current standard UAVs allow the registration and tracking of the position and orientation of the implemented sensors in a local or global coordinate system. Hence, UAV photogrammetry can be understood as a new photogrammetric measurement tool. UAV photogrammetry opens various new applications in the close-range domain, combining aerial and terrestrial photogrammetry, but also introduces new (near-) real-time applications and low-cost alternatives to the classical manned aerial photogrammetry.

Existing UAVs can be used in large-scale and small-scale applications. The system price may vary within some orders of magnitude, depending on the complexity of the system. With costs between US$1,000 and several millions of dollars, the system can be similar or even higher priced compared to a standard manned aircraft system.

The extraction of the terrain, ortho images, and textured 3D models from UAV images or other sensor data can be applied to all kinds of hazards and catastrophic or environmental disasters, in any country, to instruct and to coordinate urgent response measures, such as building collapse, aircraft accidents, SAR operations, fire combat, crop damages, landslides, and volcano outbursts. Further applications include 3D documentation of the environment and cultural heritage sites, surveying of power lines, pipeline inspection, dam monitoring, and recording of cadastral data. The main advantage is to provide detailed area information and area profiles for further detailed mission planning, and to facilitate quick response times. Typical customers of such applications include not only civil security services but also mapping agencies, real estate companies, and environmental organizations.

The combination of photogrammetric aerial and terrestrial recording methods using a mini UAV opens a broad range of applications, such as surveillance and monitoring of the environment and infrastructural assets. In particular, these methods and techniques are of paramount interest for the documentation of cultural heritage sites and areas of natural importance, which are facing threats from natural deterioration and hazards. Today, UAVs can be used as precise, automated, and computer-controlled data acquisition and measurement platforms, thanks to the recent developments of low-cost sensors such as off-the shelf digital cameras, GPS/INS (Inertial Navigation System)-based stabilization, navigation units, and laser scanners.

As far as emergency mapping UAV applications are concerned, several organizations have already begun using the consolidated capability of UAVs to produce maps and provide high-resolution imagery as part of disaster response or DRR programing. The main interest is in damage assessment information and in population count estimation.

Drones, or UAVs, are already being used in humanitarian response around the world. An unprecedented number of small and lightweight UAVs were launched in the Philippines after Typhoon Haiyan in 2013. They were used in Haiti following Hurricane Sandy in 2012. And more recently, they were flown in response to the massive flooding in the Balkans and after the earthquake in China. This increasing use of UAVs for humanitarian purposes explains why the UN recently published an official policy brief on the topic. And, several UN groups like the Office for the Coordination of Humanitarian Affairs (OCHA) are actively exploring the use of UAVs for disaster response.

6.10.3 How to Make Maps with Drones

Inexpensive drones are capable of making sophisticated maps. Small, portable drones are quickly deployable. They carry lightweight digital cameras that can capture good quality images. These cameras can be set to take pictures at regular intervals, and digital memory is cheap and plentiful. After landing, the pictures can be knit into georectified orthomosaics—that is to say, they can be geometrically corrected to a uniform scale, adjusted so that they adhere to a common geographical coordinate system, and knit together. Lightweight GPS units enable drones to make spatially accurate maps. Because there is no need for the information in real time, drones do not have to carry data links that add weight and complexity. Such drones can be used at a local level to create maps rather than having to rely on centralized mapping authorities. They complement other mapping methods and fill in imaging gaps left by satellite mapping and traditional surveying.

6.10.4 Types of Maps Generated by UAVs/Drones

Drones can produce several different types of maps: geographically accurate, ortho-rectified 2D maps; elevation models; thermal maps; and 3D maps or models. If properly produced, these data products can be used for the practice of photogrammetry, which is more simply defined as the science of making measurements from photographs.

6.10.5 2D Maps

2D maps are still the most commonly created products from imagery collected by a UAV. The simplest way to create a mosaic from aerial imagery is by using photo stitching software, which combines a series of overlapping aerial photographs into a single image. However, without geometric correction, a process that removes the perspective distortion from the aerial photos, it is hard to accurately gauge distance. Images that have been simply stitched are continuous across boundaries, but do not have perspective distortion corrected. Geometric correction is only one step in making a usable map.

6.10.6 Orthophoto

An orthomosaic is a series of overlapping aerial photographs that have been geometrically corrected (orthorectified) to give them a uniform scale. This process removes perspective distortion from the aerial photos, making the resulting "mosaic" of 2D photographs free of distortion. Ortho-rectified photos can be used to produce GIS-compatible maps for archaeological applications, construction, cadastral surveying, and other applications.

6.10.7 3D Maps

3D models, which permit researchers to make volume calculations from a set of aerial images, are increasingly common outputs from UAV technology, as new hardware and software have made it easier than ever to produce them. Instead of flat, 2D outputs created by standard photo-stitching techniques, 3D models resemble video games that let you navigate virtual worlds from within. "Other data products that can be made from UAV-collected imagery include Digital Elevation Models (DEM), Normalized Difference Vegetation Index (NDVI) maps, and thermal maps, which require specialized payloads and processing software."

6.10.8 Digital Elevation Models

Digital elevation models are distinct from 3D models; they are more akin to topographical maps. They represent only the underlying terrain; surface features such as buildings, vegetation, and other man-made aspects are removed, revealing the underlying surface. In a digital elevation model, a given point in the plane has a unique height, so features with cavities, like buildings, cannot be adequately represented.

6.10.9 NDVI Maps

NDVI maps, most commonly used for agricultural applications, are made from specialized normalized difference vegetation index (NDVI) images, which are taken with cameras that can see in both the visual and the near-infrared spectrum. NDVI imagery is used to assess whether a certain area has green vegetation or not, based on the amount of infrared light reflected by living plants. Standard point-and-shoot cameras, such as the Canon A490, can be modified to capture the wavelengths required for the imagery used to create NDVI images, considerably bringing down the cost of gathering this data.

6.10.10 Thermal Maps

Thermal maps image the temperatures of a given mapping area and are useful for applications such as detecting structural damage to roads, identifying the source of groundwater discharge, spotting hidden archaeological ruins, and detecting roe deer fawns that may be harmed by mowing operations. Specialized thermal imaging cameras, such as those made by FLIR, are light enough to be mounted on a UAV and are increasingly being adopted by civilian pilots interested in gathering thermal imagery. Many of these systems remain quite expensive, and some are subject to export restrictions.

6.11 Limitations of UAV/Drone Applications

UAVs, especially low-cost UAVs/drones, limit the sensor payload in weight and dimension, so that often low weight sensors like small or medium format amateur cameras are selected. Therefore, in comparison to large format cameras, UAVs must acquire a higher number of images in order to obtain the same image coverage and comparable image resolution. Moreover, low-cost sensors are normally less stable than high-end sensors, which results in a reduced image quality. In addition, these payload limitations require the use of low weight navigation units, which implies less accurate results for the orientation of the sensors. Furthermore, low-cost UAVs are normally equipped with less powerful engines, limiting the reachable altitude. Existing commercial software packages applied for photogrammetric data processing are rarely set up to support UAV images, as through no standardized workflows and sensor models are being implemented.

In addition to these drawbacks, UAVs do not benefit from the sensing and intelligent features of human beings. Thus, UAVs cannot react like human beings in unexpected situations, for example, unexpected appearance of an obstacle. In general, there are no sufficient regulations for UAVs given by the civil and security authorities (Colomina et al. 2008). Low-cost UAVs are not equipped with air traffic communication equipment's and collision avoidance systems, like manned aircrafts. Therefore, due to the lack of communication with the air traffic authorities, UAVs are restricted to the flight in line-of-sight and to operate with a back-up pilot. The flight range of the UAV is also, in addition to the line-of-sight regulation, dependent on the skill of the pilot to detect and follow the orientation of the UAV-system. To take full advantage of the impressive flying capabilities of UAVs, like the fully automated operating rotary wing UAVs, there needs to be a well-trained pilot, due to security issues. The pilot should be able to interact with the system at any time and maneuvers.

6.12 Application of Information and Communications Technology

Warning dissemination could take advantage of new information and communications technologies (ICT), which includes the Internet and mobile services. Use of ICT for warning dissemination is, however, context specific, with consideration of available communication infrastructure, social culture, literacy, etc.

6.12.1 Websites and Dashboard

The emergency management websites and related dashboards allow the sharing of observed and forecast data, hazard and risk information, and warnings in visual form through infographics, data tables, geospatial layers, maps, etc. The interactive websites also provide web-GIS services along with the statistical charts and graphs.

6.12.2 Emails

Emails may be customized according to the information and format required by the user groups. Programing tools may be used to automate email alerts for flood warning. The option for subscribing to email alerts could be added as feature of the website of the designated warning centers or disaster management departments.

6.12.3 SMS Alert and Cell Broadcasting

Mobile communication could provide push and pull services for warning dissemination to and fetching by users. Messages, however, may be limited by the number of characters that can be used for SMS or cell broadcasting. Also, these communication channels would require close collaboration with mobile phone operators to ensure that warnings are given priority for sending.

6.12.4 Interactive Voice Response

Interactive Voice Response (IVR) is a useful dissemination tool, particularly for users with reading disability, as well as for dissemination in the local language. Messages could be recorded in various local languages, each assigned with a specific number for users to choose from and access. Voice messages can also be pushed to registered mobile phone numbers as incoming calls. The system may be complemented by a call center for receiving and responding to users that require more information or seek clarification.

6.12.5 Social Media

Social media, such as Facebook and Twitter, have become powerful communication tools, which may also be used for warning dissemination. Many national meteorological and hydrological services and warning agencies have taken advantage of social media as complementary dissemination channels.

6.12.6 Common Alerting Protocol

Common alerting protocol (CAP) is an open source standardized digital message format for simultaneously disseminating alerts and warnings for various hazards and emergencies over different communications systems, such as sirens, phone/fax, Internet-based systems, and radio/television. Its capabilities include (OASIS 2010):

- templates for framing messages;
- support for digital images and audio;
- messaging in different languages for different receivers;
- phased/delayed timing of message effectivity and expiration;
- message update and cancellation;
- digital signature compatibility; and
- targeted geographic dissemination.

CAP provides a harmonizing platform for warning sources and dissemination systems for all hazards, which are otherwise independent for each hazard (Figure 6.3).

Figure 6.3:
The Common Alerting Protocol

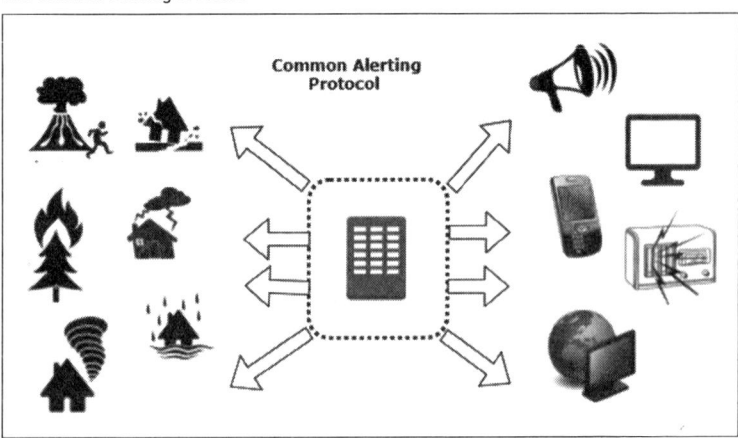

Source: Modified after World Meteorological Organization (WMO).

6.12.7 Warning Communication

Warning communication, in contrast with warning dissemination, refers to users' understanding of the received message, prompting users to take appropriate actions. It, thus, depends on the presentation and dissemination of warning information, and the users' awareness and understanding of risks.

Communication is important because:

- forecasts and warnings have value only when users understand and use them;
- users are able to provide feedback when they understand and know how to use forecasts and warnings;
- users appreciate easy-to-understand forecasts and warnings, thus aiding the credibility of the NHS communication skills; and
- education of and outreach to communities at risk are, hence, essential.

6.12.8 HAM Radio

Despite all technological advancement, the amateur radio (HAM) continues to play an important role during any disaster. As witnessed during some of the most recent disasters, for example, the Orissa Super Cyclone, Gujarat earthquake in 2001 and recent tsunami when all other communication networks failed, the HAM radio provided the only means of communication to carry out rescue and relief.

What is HAM radio? HAM means Hertz Armstrong and Marconi (HAM). HAM radio is a kind of wireless set. Amateur radio operators are often called ham radio operators or simply "HAMs." Amateur radio transmitters and receivers are used to communicate with other amateur radio operators.

Amature means nonprofessional, that is, the radio activity is carried out by nonprofessionals. People take it as a hobby and one HAM is able to communicate with other HAM anywhere in the world. Every HAM is identified by a call sign, for example, VU2RDM, VU2NYD, VU2PEZ, etc.

World over, HAM has been popularized among various sections of people as a hobby, and a country such as Japan that is prone to various natural hazards have a large number of HAM volunteers. The activity is also becoming popular in India where there is a growing awareness of its utility.

ICT for Flood early Warning: Connecting Remote Communities at Risk in Bangladesh

Flood is a regular occurrence in Bangladesh due to its location in the floodplains of the Ganges-Brahmaputra-Meghna basin. The Flood Forecasting and Warning Centre (FFWC) of the Bangladesh Water Development Board (BWDB) generates 5-day deterministic and 10-day probabilistic forecasts as operational flood forecast products, and 8-day satellite-altimetry-based forecast and 3-day flash flood forecast products on experimental basis for pilot locations.

FFWC issues operational forecast products by fax and email to disaster management committees that translate these products into risk information and disseminate to communities at risk through display boards, community meetings, and word of mouth. With more than 100 million of its over 156 million population owning mobile phones, and about a million Internet subscribers, FFWC has adopted ICT technologies for forecast and warning dissemination. Location-specific water level forecasts and flood warnings are sent by text and voice messages, as well as posted in the FFWC Facebook page. FFWC has recently upgraded its website, making dynamic bulletins and infographic and map products available to users. The development of an online interactive web portal (dashboard) at BWDB District Flood Information Centres and Union Parishad Digital Centres is ongoing, with support from Cordaid and regional integrated multihazard early warning system (RIMES), for data collection and analysis and dissemination of flood risks. Concurrently, a mobile phone application is also being developed to increase user access to flood forecast and risk information, facilitate user feedback, and allow user participation in water level monitoring. Another ongoing initiative is the integration of the voice message warning dissemination system into the national IVR, which currently provides flood situation updates in major rivers. These efforts are in line with the country's vision of a Digital Bangladesh by 2021.

Chapter 7

Overview of Disaster Management Mechanism in Asian Countries

7.1 History of Disaster Management Mechanism in Asian Countries

While the Asian countries have been successful in achieving economic growth and poverty reduction (Figure 7.1), the region cannot avoid exposure to a variety of disasters. Indeed, Asia, particularly the area of the ASEAN member states (AMSs), is the most prone region to disasters in the world (Sawada and Oum 2012). According to Table 7.1, during the past decade Asia experienced more than 150 natural disasters (40 percent of world total) annually that affected more than 200 million people annually (about 90 percent); and caused more than US$41.6 billion in annual

Figure 7.1:
Effect of Natural Disasters on People and Total Economic Loss in ASEAN Countries, 1970–2012

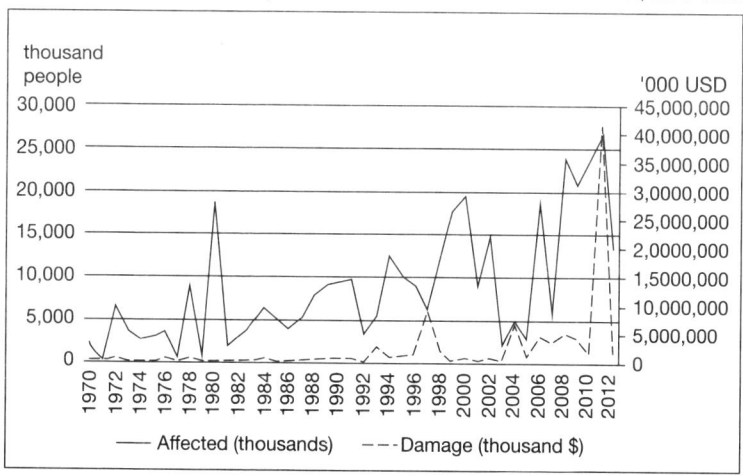

Source: Emergency Events Database, CRED.
Note: Disasters data cover drought, seismic earthquake, flood, mass movement, storm, volcano, and wildfire.

Table 7.1:
Natural Disaster Occurrence and Impacts: Regional Figures (Average from 2001 to 2010)

	Africa	Americas	Asia	Europe	Oceania	Global
(1) Number of Natural Disasters per year						
Climatological	9	12	11	17	1	50
Geophysical	3	7	21	2	2	35
Hydrological	44	39	82	24	6	195
Meteorological	9	34	40	14	7	104
Total	65	92	153	58	16	384
(2) Number of Victims per year (in millions)						
Climatological	12.29	1.22	63.45	0.27	0.00	77.23
Geophysical	0.08	1.02	7.77	0.01	0.04	8.92
Hydrological	2.18	3.31	100.82	0.35	0.04	106.70
Meteorological	0.35	2.72	35.88	0.11	0.04	39.10
Total	14.91	8.27	207.92	0.74	0.12	231.95
(3) Damages (in USD billions)						
Climatological	0.04	1.90	3.45	3.23	0.48	9.10
Geophysical	0.69	4.75	17.38	0.57	0.69	24.08
Hydrological	0.28	3.15	11.15	5.57	1.24	21.39
Meteorological	0.08	40.47	9.62	4.03	0.56	54.77
Total	1.10	50.27	41.61	13.40	2.97	109.35

Source: Annual Disaster Statistical Review (2011) and CRED, IRSS & UCL (2012).

damage (39 percent). MunicRe's 2010 NatCatSERVICE data reports that only 9 percent of the total property losses due to natural disasters in Asia were covered by private insurance.

Indeed, at the recent high-level forums in East Asia such as Fourth East Asia Summit (EAS) in Cha-am HuaHin, Thailand, held on October 25, 2009, the Fifth EAS on October 30, 2010, in Hanoi, Vietnam, the Sixth EAS in Bali, Indonesia, in November 2011 and the Seventh EAS in Phnom Penh, Cambodia, in November 2012, the leaders noted and reiterated the need to enhance disaster management cooperation for the region. The Special ASEAN-Japan Ministerial Meeting in April 2011 also emphasized the need to strengthen such cooperation through sharing of exercises and lessons learned as well as conducting training and capacity-building programs for disaster preparedness, emergency response, relief, and reconstruction efforts. The chair's statement at the 18th ASEAN Summit held in Jakarta, Indonesia, on May 7–8, 2011 noted the potential transboundary impact of accidents at nuclear plants in the aftermath of the Fukushima incident. They agreed that ASEAN should engage in appropriate information-sharing and promote transparency on relevant nuclear-related issues in the region and to achieve goal of building disaster-resilient societies and towards a safer community by the year 2015. The Fourth ERIA Governing Board Statement on June 3, 2011, also recognized that knowledge-sharing and exchange of technologies on DRM on a regional basis is essential.

In a study by Sawada and Zen (2014), 12 countries were covered for natural disasters (Cambodia, China, Indonesia, Japan, Korea, Laos, Malaysia, Mongolia, the Philippines, Taiwan, Thailand, and Vietnam) and 8 for economic disasters (China, Indonesia, Japan, Korea, Malaysia, the Philippines, Taiwan, and Thailand) and the result has been shown in Table 7.2.

Keeping in mind the disaster-prone conditions, a majority of ASEAN member states are raising their collective efforts to cope with the challenges through the Current Regional Effort in Disaster Management. Since its inception back in 1976, ASEAN has been recognizing and adopting disaster management as one of its eight principles and objectives. The declaration stated that "natural disasters and other major calamities can retard the pace of development of member states, therefore they shall extend, within their capabilities, assistance for relief of member states in distress." One of early cooperation is done through ASEAN Experts Group on Disaster Management (AEGDM) (AEGDM). The 11th AEGDM Meeting in Chiang Rai in August 2000 considered the elevation of AEGDM to an ASEAN committee or a senior officials meeting on disaster management that would report to the ASEAN Standing Committee (ASC) or to the ASEAN Ministerial Meeting on Disaster Management. Within this form, ASEAN member countries could regularly meet to monitor the programs and projects they adopt. The idea then was put as recommendation in the 12th AEGDM meeting to establish the ASEAN Committee on Disaster Management (ACDM). The ACDM was established in 2003, reporting to the ASC. It consists of heads of national agencies responsible for disaster management of ASEAN member countries. The ACDM assumes overall responsibility for coordinating and implementing the regional activities. ACDM has vision of a region of disaster-resilient nations, mutually assisting and complementing one another, sharing a common bond in minimizing adverse effects of disasters in pursuit of safer communities and sustainable development. And, its mission is to enhance cooperation in all aspects of disaster management,

Table 7.2:
Incidence of Disasters in East and Southeast Asia

	Currency	Inflation	Banking	Geo-physical	Meteoro-logical	Hydro-logical	Climato-logical	Bio-logical
No. of Countries	8	8	8	12	12	12	12	12
Year								
1980	1	1	1	7	0	0	6	4
1981	0	1	2	6	8	7	4	3
1982	0	0	2	8	9	9	6	5
1983	2	0	4	8	9	0	6	5
1984	3	1	3	6	6	9	3	3
1985	0	1	4	8	0	9	5	5
1986	1	0	4	5	8	9	5	3
1987	0	0	4	4	7	6	7	3
1988	0	0	2	8	6	8	5	4
1989	1	0	0	7	0	7	3	3
1990	1	0	0	6	1	9	6	6
1991	0	0	0	4	9	8	5	3
1992	0	0	3	6	0	8	5	5
1993	0	0	2	6	9	9	4	1
1994	1	1	3	4	7	8	3	
1995	0	0	3	6	6	7	3	2
1996	0	0	3	6	7	0	2	5
1997	5	0	8	5	7	6	5	5
1998	2	1	7	3	6	6	4	7
1999	0	1	7	5	6	0	6	5
2000	3	0	6	5	9	0	2	6
2001	0	0	6	4	8	1	0	2
2002	0	0	2	6	9	8	4	9
2003	0	0	0	6	0	0	3	1
2004	0	0	0	8	1	1	5	8
2005	0	0	0	5	9	2	7	8
2006	0	0	0	6	7	2	4	6
2007	0	0	0	8	7	0	5	6

Source: Disaster Statistical Review (2011) and CRED, IRSS & UCL (2012).

including prevention, mitigation, preparedness, response, and recovery through mutual collaborative activities (ASEAN DRR Portal 2013). In 2004, the ASEAN Ministerial Meeting on Disaster Management (AMMDC) was set up aimed at reviewing and enhancing regional cooperation on disaster management. In the same year, the ASEAN Regional Program on Disaster Management (ARPDM) was also established. It aims to create cooperation among member countries with regard to capacity-building and sharing of information and resources. It also creates engagement external partnerships and public education, awareness, and advocacy. The coordinating unit responsible for the tasks is ASEAN Agreement on Disaster Management and Emergency Response (AADMER) established in 2005 and put on effect since 2009. It is the first HFA-related binding instrument in the world. The operational coordination body and engine of AADMER is the ASEAN Coordinating Centre for Humanitarian Assistance on Disaster Management (AHA Centre) headquartered in Jakarta. Given its young age, the AHA Centre currently is still developing its programs; thus, the scope is limited to logistics and rapid assessment in preparedness and response, technical support for early warning, risk assessment and monitoring, and capacity-building. The AHA Centre shall work on the basis that the party will act first to manage and respond to disasters. In the event that the party requires assistance to cope with such situation, in addition to direct request to any assisting entity, it may seek assistance from the AHA Centre to facilitate such request (AADMER Article 20.2). Apart from ASEAN context, cooperation in DRR is also performed by other international entities, including the UN, International Red Cross, and the Asia-Pacific Economic Community (APEC). The Asian Disaster Preparedness Center (ADPC) was set up in Bangkok in 1986 as a follow up of a feasibility study conducted jointly by two agencies of the UN: the Office of the UN Disaster Relief Coordinator (currently the UN Office for the Coordination of Humanitarian Affairs) and the World Meteorological Organization. In 2005, APEC established the APEC Task Force of Emergency Preparedness (TFEP) to coordinate and promote responses to emergencies and disasters, which in 2010 became the Emergency Preparedness Working Group (EPWG).

At the national level, each country has a national body responsible for managing DRR that also acts as a national focal point for regional cooperation. Most works are devoted to technical and logistics aspects, from the stage of preparedness, response, to reconstruction. The financing aspect is a bit lagged behind; it is focused on financing the logistics to respond to the emergency situation, and to some extent on reconstruction.

Disaster resiliency is an important core component of sustainability for ASEAN because ASEAN and East Asia have experienced various crises and disasters during the past two decades, and those shocks were utilized to improve resiliency in the region. Not to mention, strengthening regional cooperation in the fields of financial, trade, energy security, food security, and disaster management will pave the way for smooth development in the region. The region has experienced diverse forms of disasters, including floods, typhoons, earthquakes, epidemics, and the financial crises of the late 1990 s, which necessitates better regional organization for quick action. This is the very reason why there is a need for more effective insurance mechanisms against various kinds of disasters. When we consider the actual form of such insurance mechanisms, there are numerous issues involved, such as whether it would be an institutionalized system such as a disaster fund or something more flexible such as a coordination forum. It is worth pursuing reforms that undertake comprehensive preparations against the risks of a variety of disasters in Asia. In regional cooperation, the existing schemes shall be improved to cover better system of financing and transfer. In developing countries, the cost of preparation—response—post-disaster is typically following a bell-shaped graph and also reflects cycle-related fiscal needs. The government and individuals spend a small portion on preparation efforts; therefore, when the disaster occurs, they are burdened by large financial consequences, some obligations usually filled by donors out of humanitarian considerations, and at later stage, reconstruction may face delayed and under-budgeted program. With appropriate design, the bell-shaped financing burden can be changed into an upside down curve—even though not completely. The situation can be changed if there is sufficient fiscal allocation for preparation in pre-disaster. Thus, in the occurrence of disaster, claims will close a large part of fiscal needs and can be used for the later stage as well. Given the financial constraints of many developing economies in the regions, development partners can contribute to assist the programs. The participation rate for disaster-linked insurance can be increased by contribution from the government for paying the premium. Once the coverage is sufficient in terms of fair calculation of premium, the insurance company can sell CAT bonds. Apart from government contribution, international development partners can take the role to provide soft loans for the government or grants. The contribution can have a positive impact on the institution itself; it can be a way to enforce some constructive liabilities, for example, safe standard for building in the covered areas, obligation to build sufficient mitigation system, capacity-building, etc. The donors can also expect declining costs of contribution in the events of disaster and decreasing fatalities and damages. Another policy to consider is to support the acquiring and publicly providing of the hazard map and data. Rashcky and Chantarat (2013) suggested that regional cooperation develops a regional center for disaster risk data, modelling, and insurance. Reliable spatiotemporal-rich data on exposures and disaster losses are largely unavailable in ASEAN countries. These necessary risk data and modeling are critical in enhancing risk-based pricing and supervision, stimulating development of new insurance products, and helping the governments to identify appropriate risk financing strategies for effective and timely disaster responses. The center shall have the objectives to enhance the development of regional risk market infrastructure and to promote cross-border knowledge exchange and capacity-building on natural disaster risk financing and transfer practices.

The ISDR Asia Partnership (IAP) is an informal multistakeholder forum of Asian Governments and stakeholders to facilitate DRR through the implementation of the HFA 2005–15 and the Sendai Framework for Disaster Risk Reduction 2015–30. The IAP has been the main consultation forum for the Asia Ministerial Conferences (Regional Platform) in Asia. The forum includes regional intergovernmental organizations, governments, civil society organizations, the UN and international organizations, and bilateral and multilateral donors. The forum meets twice a year.

The Asian Ministerial Conference on Disaster Risk Reduction (AMCDRR) 2016 also *recognized* the opportunity to build on past achievements by reaffirming the commitments to DRR and resilience and *recognized* the urgent need to accelerate the implementation of the agreed global frameworks. It is our primary responsibility to put in place national and local-level strategies to ensure the achievement of the seven global targets of the Sendai Framework.

The AMCDRR 2016 declarations commit to the principle of a people-centered and whole-of- society approach and the need to strengthen national and local multistakeholder platforms, enhance participation and partnership of the major groups and stakeholder groups, and further support the coordination role of the UN International Strategy for Disaster Reduction (UNISDR).

7.1.1 Salient Points of the AMCDRR Declarations

1. Pursue with a sense of urgency the paradigm shift from disaster management to DRR.
2. Ensure that policies and practices reflect an understanding of disaster risk. More specifically, collect and share risk information for pre-disaster

risk assessment, risk prevention and reduction through development, and appropriate preparedness for effective response to disasters.

3. Strengthen national and local governance of DRR to ensure coherence among policies, institutional arrangements across sectors, with representation of stakeholders in line with national circumstances and policies. Increase public and private investment in capacity-building, science and technology, innovation, critical infrastructure, and services, to contribute to achievement of community resilience.

4. Increase investment in DRR for resilience including in multihazard early warning systems and dissemination channels; contingency planning that engages all people to further strengthen disaster preparedness. In this regard, reaffirm that an effective and meaningful global partnership and further strengthening of international cooperation, including fulfillment of respective commitments and official development assistance by developed countries, are essential for effective disaster risk management.

5. Encourage meaningful participation and support representation of women, children, youth, and persons with disabilities in leadership role for DRR.

6. Improve preparedness for disaster recovery by strengthening institutional frameworks, establishing standards, and enhancing capacities to ensure that disaster recovery integrates risk reduction measures to build back better.

7. Use the International Day for Disaster Reduction and the World Tsunami Awareness Day to raise awareness, promote better understanding of risks, and develop tools to address them.

8. Adopt indicators of the Sendai Framework being developed by the UN General Assembly Open-Ended Intergovernmental Working Group (OEIWG) and ensure that they are anchored in national priorities and aligned with the indicators of the 2030 Agenda for Sustainable Development.

9. Collaborate for promoting disaster-resilient infrastructure involving governments, multilateral development agencies, financial institutions, the private sector, and major infrastructure investors in Asia.

10. Strengthen inclusive collaboration at the local level to build on community initiative, knowledge, and resources, and leverage national policies and programs to achieve resilience.

11. Promote application of science and technology, and research for evidence-based DRR policies, practices, and solutions, including through international cooperation.

12. Enhance regional cooperation including strengthening the role of intergovernmental organizations for coherent implementation of the Sendai Framework and the 2030 Sustainable Development Agenda, fostering innovative partnerships, and North–South, South–South and Triangular Cooperation in all areas related to DRR (AMCDRR, New Delhi, India, 2016).

7.2 Institutional Framework of Disaster Management in Asian Countries (Nepal)

It is commonly accepted that most of the world's worst disasters occur between the Tropic of Cancer and Tropic of Capricorn. Asia is the most affected continent with respect to natural disasters. About 39 percent of the total reported disasters from 1992 to 2001 struck Asia, accounting for 74.5 percent of the total number of casualties (IFRC 2002).

Institutional structure confers the benefits of unity of command and definite roles for actors in field situations that prevents confusion. Territorial span of control prevents overlap and the three-tier structure separates the distinct functions: operational, tactical and strategic. The effectiveness of the policy lies in the empowerment of the concerned offices and the fixing of accountability at all levels. Emergency powers have been delegated to the nodal officers (NOs) at operational and tactical levels over the line departments and also over any institution or individual that may be of assistance.

In the present context, we will discuss the institutional framework of Nepal and India as two representative country to understand the Asian perspectives.

7.2.1 Nepal: Different Disaster Management Acts, Frameworks, and Guidelines

Many guidelines, policies, and laws have been formulated regarding disaster management in Nepal. Recently, different strategies, disaster management acts, and other flagships has been proposed to systematize the disaster management sector in Nepal. The existing laws, acts, and policies in Nepal are given further. Only a few of them will be discussed according to the need in emergency response and management related to the Gorkha earthquake.

1. Natural Calamity Relief Act (NCRA) 1982
2. National Strategy for Disaster Risk Management (NSDRM) 2009
3. Disaster Management Act 2014 (in process)
4. Guidance Note: Disaster Preparedness and Response Planning 2011
5. National Disaster Response Framework (NDRF) 2013
6. Contingency Plan for the Coordination of Shelter Preparedness and Response in Nepal 2014
7. Nepal Disaster Management Reference Handbook (NDMR) 2015
8. Rescue and Relief Standards 2007
9. Prime Minister Disaster Response Fund Guideline 2006, amendment 2007
10. National Strategic Action Plan on Search and Rescue 2014
11. National Platform on Disaster Risk Reduction 2008/2012

7.2.2 Natural Calamity Relief Act 1982

The Natural Calamity Relief Act (NCRA) mainly focuses on the emergency response and relief which had been last updated in 1992. The committee formed consists of 22 representatives of government services, and representatives of the Nepal Red Cross Society and Nepal Scout. The administrative functions are governed by the Ministry of Home Affairs (MOHA) and chief district officers (CDO) in district-level office. There are four tiers of relief committees within the Act. The Central Natural Calamity Relief Committee (CNCRC) is the central body. The Regional Natural Calamity Relief Committees (RNCRC) and District Natural Calamity Relief Committees (DNCRC) are next two tiers of the disaster management structure. The DNCRC play a key role of coordination between different levels of administrative roles. At the local operational end of disaster management structure, the Local Natural Calamity Relief Committee (LNCRC) looks after assessments, evacuation, organizing volunteers, and distributing relief products.

The roles of subcommittees are limited to being response-related rather than proactive which is inefficient (*Practical Action* 2010a, 2010b). The Act title in itself clarifies that this Act is solely based on the relief activities, with only relief-centric approach used by the government. The importance of integrated disaster management plans and preparedness has not been recognized in the NCRA Act (*Practical Action* 2010a, 2010b).

The overall responsibility of the government has been thoroughly described in the Act regarding banning any activities that hinder the response process, mobilizing governmental as well as nongovernmental agencies and resources, protecting and evacuating people and property for certain time period, controlling the relief items and distributing them, deploying international aid agencies, and use of resources (COE-DMHA 2012).

7.2.3 National Strategy for Disaster Risk Management 2009

The National Strategy for Disaster Risk Management (NSDRM) is the national framework of Nepal based on HFA 2005. It was prepared to incorporate all the sections of the disaster management cycle and is also dedicated for the guarding and promotion of heritages and infrastructures. NSDRM focuses on all aspects of DRM cycle, namely, risk management phases such as prevention, mitigation, and preparedness along with crisis management inclusive of response and recovery (Koirala 2014).

The main objective of this national framework is to act as a guideline in the developmental process for national development and reduction of disasters. To accomplish disaster resiliency, five major priority actions with 29 strategic activities have been formulated to meet the goal. The NSDRM consists of different DRM planning directives such as mainstreaming DRM concept into development, rights to life safety and social security, gender and social inclusion, decentralized the working process, personal safety and security, one-window policy, and cluster approach and holistic approach. In emergency response, the one-window policy will be applied by government authorities for relief distribution and the cluster approach will be mandatory for international organization.

7.2.4 One-window Policy and Cluster Approach

The coordination process from central to local level comprising of responsibility with the Disaster Management Committee (DMC) or Disaster Management Authority (DMA) as a single entity is known as one-window policy. Under this policy, any institutions such as the government, NGOs, or International non-governmental organizations (INGOs) work to prevent any duplication in response activities or any other disaster management activities to increase the overall performance.

The cluster approach UN Inter-Agency Standing Committee (IASC) was started in 2005 in Nepal. Several sector-experts with specific expertise form a network to share and coordinate together for better response and recovery through joint effort. The cluster approach in the national level emphasizes on humanitarian cooperation, accountability, and partnership among variable activities in DRM. It also helps in clarity of work division for each sector involved to meet the humanitarian needs.

Along with NSDRM, the Nepal RedCross Society (NRCS) contingency plan for a major earthquake in Kathmandu Valley has been designed to predict the worst case scenario of destruction with a magnitude 8.3 on the Richter scale. The plan covers Kathmandu, Lalitpur, and Bhaktapur districts in Kathmandu Valley. The estimated death was 22,000, with 60 percent of the houses destroyed and 600,000–900,000 homeless (COE-DMHA 2012).

7.2.5 Organizational Structure and Approaches for DRM

The proposed organization structure of DRM in Nepal includes the National Council for Disaster Management (NCDM), National Disaster Management Authority (NDMA) and regional, district and local disaster management committees (Figure 7.2). Different hierarchies will comprise of separate specification of tasks with NCDM dealing with endorsing DRM policies and national and sectoral plans; taking policy decisions; and supervising the DRM cycle and its activities in all areas of DRM. NDMA is established as a national focal point for implementation, coordination,

Figure 7.2:
Organizational Structure of NCDM

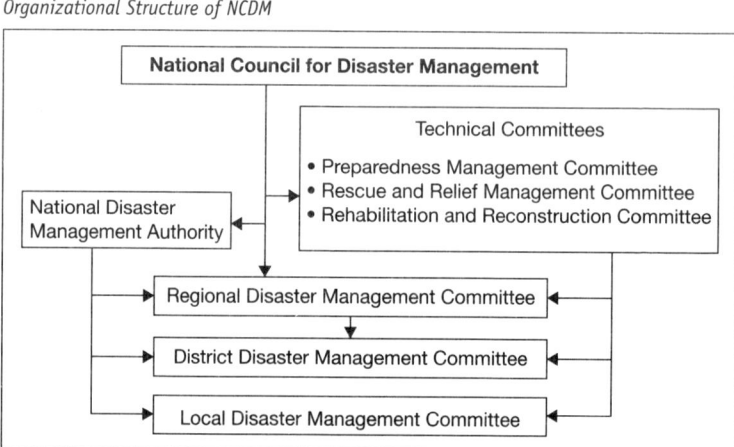

Source: Modified after NSDRM (2009).

facilitation, and monitoring of disaster management strategies. Allocated disaster management committees will help in decentralization of power and task division for ease. NDMA works for direct cooperation and coordination among ministries, departments, and different stakeholders at various levels for proper management of a task in different stages of the DRM cycle.

NDMA has the responsibility to ensure sectoral action plan and program for NSDRM, ensure adequate budget allocation for implementation, and allow resource mobilization from government funds for M&E of the implantation of the plan. The working strategy of NDMA consists of building support of plan and sector specific action plan for DRR; supporting all kinds of initiatives and M&E DRR and its indicators.

7.2.6 National Emergency Operation Center

The National Emergency Operation Center (NEOC) is the emergency operation center under MOHA positioned in Kathmandu with other EOCs in 5 developmental regions, 42 districts, and 5 municipalities. The standard operating procedure (SOP) of EOC is found in the national and district levels.

7.2.6.1 ROLE OF NEOC

NEOC plays a very important role in emergency response to command and coordinate with different agencies to tackle crisis situation. According to the disaster at hand, the establishment of district emergency operation centers (DEOCs) in high-risk areas and SOP preparation takes place. Disaster Information Management System (DIMS), notification system, and training is another key role. Moreover, it also develops human resource and expertise in emergency response.

The chief of NEOC is the disaster response focal point signifying ministries and departments of GON. Being the main coordinating body, it assess the logistics within the country for management of relief operation.

The CNDRC, RDRC, and DDRC are the different tiers of committees to maintain humanitarian operation during a disaster. Each has different responsibilities to coordinate and cooperate smoothly. The CNDRC and DDRC monitor and evaluate the rescue and humanitarian assistance operations. The RDRC monitors the performance of DDRCs and evaluates information. The CNDRC recommends amendments on the NDRF, which the government reviews and implements in the future. Similarly, the NEOC maintains levels of operation: NEOC, REOC, and DEOC. The main functions are the establishment of a command post within the NEOC, media management, deployment of disaster response focal points, coordinate relief, disseminate information, and regular information to the public (COE-DHMA 2012).

7.2.7 National Disaster Response Framework 2013

The National Disaster Response Framework (NDRF) is the overall disaster framework which integrates all government organizations and committees within the country. NDRF explains about the national system of disaster response, international assistance and coordination structure, special operation arrangements and national framework. It also sketches disaster response operations to be handled by responsible lead agencies with operational activities and their timeline. Table 7.3 gives an overview of different responsibilities during emergency response of different government bodies.

Table 7.3:
Overview of Operational Activities in Emergency Response

Timeline	Operational Activities	Responsible Lead Agency
0–1 hour	Provision on immediate information of earthquake	National Seismological Centre (NSC)
	Provide information of incidents related to the disaster	DDC, Municipality and VDC Offices
	Immediate SAR activities	DDRCs and Security Force
	Public reporting on the extent of disaster and rescue efforts	MOHA/NEOC
	Setup of emergency information system and information dissemination	Ministry of Information and Communication technology (MOIC), MOHA
	Situation report preparation	Chief District Office (CDO)/DDRC
	DDRC meeting	District Administration Office (DAO)/DDRC
0–7 hours	Emergency meeting at MOHA	MOHA/NEOC
	Activation of NEOC following SOP	MOHA/NEOC

Timeline	Operational Activities	Responsible Lead Agency
	Develop emergency communication system and provide information to the public regularly	NEOC/REOC/DEOC
	CNDRC meeting	MOHA/NEOC
	Coordination with national and international stakeholders for potential support	MOHA/CNDRC
	Act as disaster response focal point	NEOC
	SAR team deployment to affected areas	DEOC and security forces
0–24 hours	First aid provision to injured and their treatment	Red Cross/hospitals
	CNDRC meeting and emergency declaration	CNDRC/Cabinet
	Initial rapid assessment (IRA)	DDRC/DEOC
	Activation of UN Cluster via UN Humanitarian Coordinator (HC)	Activation of UN Cluster
	Appeal for national and international assistance	Cabinet
	Establishment of media management center within NEOC/REOC/DEOC	NEOC/REOC/DEOC
	Airport security and air traffic management	Civil Aviation Authority Nepal (CAAN)
	Restore road networks	Dept. of Road
	Management of transportation facilities for search and rescue materials and equipment	Department of Transport Management (DOTM)
	Availability of basic food items for survivals and IDPs	Ministry of Commerce and Supply (MOCS)
	Registration and facilitation of international humanitarian communities (IHC) and relief consignments	MOFA/DOI
24–48 hours	Coordination and deployment of relief consignments	NEOC/DEOC
	Maintenance of law and order around warehouse, IDP camps, personnel, humanitarian convoy, protect property and security in affected areas	MOHA
	Registration and tracking of affected families; issuance of Victims ID card and maintain database at districts and central level	DDRC
48–72 hours	Rapid assessment of existing hospitals, schools, health facilities along with bridges, roads and others for structural damages	MoHA
	Debris management (collection and disposal) including dead animals	Municipality/VDC
	Collection and management of animal carcass	Municipality/VDC

(*Continued*)

Table 7.3:
(Continued)

Timeline	Operational Activities	Responsible Lead Agency
48–72 hours	Distribution of immediate lifesaving relief materials to the survivors as per the agreed standards	CDO/DDRC
72 hours–7 days	Proper management of dead body including cremation and issuance of death certificates	DDRC
	Special protection arrangement at the camps and distribution sites for the protection of women, children, elderly, disabled	Ministry of Women, Children and social welfare (MOWCSW)
	Multicluster initial rapid assessment (MIRA) activation	UN HC
	Distribution of nonfood items such as cooking utensils, fuel, clothes, blanket, family kits, baby kits, hygiene kits	NRCS
7 days–2 weeks	Setup of temporary shelter in the pre-determined safe and open/evacuation sites for displaced families	Ministry of Urban Development (MOUD)
	Monitoring of malnutrition of children in the camps	MOHP
	Monitoring and reporting of humanitarian response and relief	DDRC
	Ensure safe environment for survivors particularly the vulnerable groups, and control gender based violence	DWDO
	Psychological rehabilitation to person with mental trauma	MOHP
	Social support to unaccompanied children, disabled and elderly who have lost their supporting family members; and reunion of their lost members or relatives	MOWCSW/DWDO
2 weeks–1 month	Cash or other assistance to affected populations	DDRC
	Re-start schools to help children feel secure and help them to go back to normal life	Dept. of Education
	Provide minimum essential services and security at the hospitals, schools, etc. in the camp sites	DUDBC
	Conduct early recovery assessment and plan development covering key sector including shelter, livelihood, etc.	DUDBC
	Document the lesson learnt on search and rescue efforts, immediate relief, camp management and rehabilitation efforts and prepare an analytical report	MOHA

Source: NDRF (2013).

7.2.8 Coordination Mechanism between Operation Centers

During a large-scale disaster, contact among NEOC, the Onsite Operation Coordination Centre (OSOCC) and Multi-National Military Coordination Centre (MNMCC) should be maintained. NEOC functions as a secretariat of CNDRC, and information flows through NEOC between OSOCC and MNMCC.

During a disaster response, designated ministries and their departments execute their assigned functions immediately. Other national and international agencies also carry out their functions immediately. NEOC is the main coordinating body during emergency response. Meetings are conducted immediately after the disaster by the secretary of MOHA along with members from NEOC, the joint secretary, and the disaster management division to analyze the situation through incident reports. Humanitarian response includes coordination among agencies, restoration of infrastructures, provision of basic needs, protection, and management of road, debris, dead body, etc. that should be carried out by responsible agencies. A military command post is established in NEOC for proper rescue and relief management. The command post consists of the chief of disaster management division, brigadier general from the Directorate of Military Operation, deputy inspector general of police (DIG), and armed police force (NDRF 2013).

7.3 Institutional Framework of Disaster Management in Asian Countries (India)

7.3.1 Brief Background of Disaster Management in India

India has been traditionally vulnerable to natural disasters on account of its unique geo-climatic conditions. Flood, droughts, cyclones, earthquakes, and landslides are regular phenomena in India. The multihazard scenario depicted in the *Vulnerability Atlas of India* (Building Material and Technology Promotion Council, New Delhi, India) shows that out of total geographical area of 3,287,263 sq. km, about 60 percent of the landmass is prone to earthquakes of various intensities, over 40 million hectare area is prone to floods, about 8 percent of the total area is prone to cyclones, and 68 percent of the area is susceptible to droughts. During 1990–2000, an

Table 7.4:
Overview of Disaster Management Strategic Action Plan, Policies, and Framework

S.N.	Acts/ Frameworks/ Policies	Salient Features
1	Natural Calamity Relief Act 1982	• Baseline legislation for other laws, policies, and regulation • Focus on emergency response and relief in disaster
2	Local Self Governance Act 1999	• Concept of local authority for decentralized framework • Emphasizes the interrelationship among environment, development, and disaster
3	National Strategy for Disaster Risk Management (NSDRM) 2009	• National framework to promote, grow, and protect national heritages and physical infrastructure by GON • Vision to make Nepal a disaster-resilient country with five priority actions and 29 strategic activities • One-window working policy and cluster approach for disaster management activities
4	Nepal Disaster Risk Reduction Consortium (NDRRC) 2009	• National strategy for risk management priorities • Five flagships focusing on risk management
5	Guidance Note: Disaster Preparedness and Response Planning 2011	• gives detail process of response of different organization and scheduled meetings • Facilitate process of preparing disaster preparedness and response plans at district and VDC/municipality level
6	National Disaster Response Framework (NDRF) 2013	• Sketches disaster response operations and timeline to be handled by responsible lead agencies with operational activities • 49 different actions
7	Disaster Management Act 2014 draft (not publicized)	• Covers full DRM cycle • Decentralize disaster management at local level • Comprises natural and made-made disasters
8	Contingency Plan for the Coordination of Shelter Preparedness and Response in Nepal 2014	• Support GON meet immediate shelter needs of affected population after disaster • Appropriate, adequate, timely, and effective management of shelter needs of population
9	National Strategic Action Plan on Search and Rescue 2014	• Enhance the rescue capacity in emergency response • Common language to be used • Component of SAR, disaster management, medical, and logistics to be added

Source: Authors.

average of about 4,344 people lost their lives, about 30 million people were affected by various disasters every year, and the average annual damage has been estimated to be approximately ₹2.70 billion as per the World Bank estimates; during 1996–2001 and the Bhuj earthquake alone in Gujarat in January 2001, the loss amounts to US$13.8 billion (Table 7.4).

India, due to its geoclimatic and socioeconomic condition, is prone to various disasters. During the last 30 years, the country has been hit by 431 major disasters resulting in enormous loss to life and property. According to the Prevention Web statistics, 143,039 people were killed and about 1.5 billion were affected by various disasters in the country during these three decades. The disasters caused huge loss to property and other infrastructures, costing more than US$48 billion.

In India, the cyclone occurred on November 25, 1839 had a death toll of three lakh people. The Bhuj earthquake of 2001 in Gujarat and the super cyclone of Orissa on October 29, 1999, are still fresh in the memory of most Indians. The most recent natural disaster of a cloud burst resulting in flash floods and mudflow in Leh and the surrounding areas in the early hours of August 6, 2010, caused severe damage in terms of human lives as well as property. There was a reported death toll of 196 persons, 65 missing persons, 3,661 damaged houses, and 27,350 hectares of affected crop area. The disaster-wise total numbers of deaths between 1999 and 2012 are shown in Figure 7.3.

Figure 7.3:
Disaster-wise Death Toll in India from 1999 to 2012

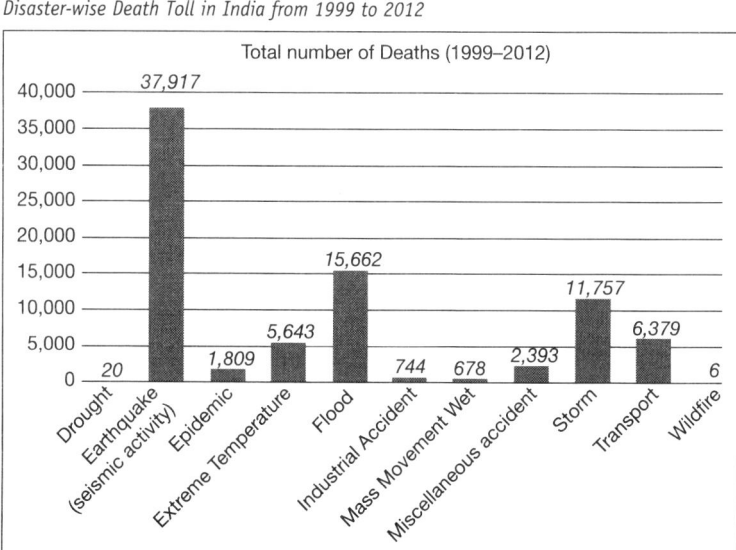

Source: EM-DAT (2016).

7.3.2 Vulnerability Profile of India

India has been vulnerable, in varying degrees, to a large number of natural as well as human-made disasters on account of its unique geoclimatic and socioeconomic conditions. It is highly vulnerable to floods, droughts, cyclones, earthquakes, landslides, avalanches, and forest fires. Out of the 35 states and union territories in the country, 27 of them are disaster-prone. Almost 58.6 percent of the landmass is prone to earthquakes of moderate to very high intensity; over 40 million hectares (12 percent of land) are prone to floods and river erosion; of the 7,516 km long coastline, close to 5,700 km is prone to cyclones and tsunamis; 68 percent of the cultivable area is vulnerable to drought; and the hilly areas are at risk from landslides and avalanches.

Vulnerability to disasters or emergencies of chemical, biological, radiological and nuclear (CBRN) origin has increased on account of socioeconomic development. Heightened vulnerabilities to disaster risks can be related to expanding population, urbanization and industrialization, development within high-risk zones, environmental degradation, and climate change. During the last two decades of the 19th century (1982–2001), natural disasters in India had claimed a total death toll of around 107,813 people (on an average more than 5,390 every year). As mentioned previously, India with its extended coast line is exposed to five to six tropical cyclones on average, both from the Arabian Sea and Bay of Bengal, annually.

7.3.3 Hazard Profile of India

1. India is one of the 10 worst disaster-prone countries of the world. The country is prone to disasters due to number of factors, both natural and human induced, including adverse geoclimatic conditions, topographic features, environmental degradation, population growth, urbanization, industrialization, nonscientific development practices, etc. The factors either in original or by accelerating the intensity and frequency of disasters are responsible for heavy toll of human lives and disrupting the life-supporting system in the country.

   The basic reason for the high vulnerability of the country to natural disasters is its unique geographical and geological situation. As far as the vulnerability to disaster is concerned, the four distinctive regions of the country, that is, the Himalayan region, alluvial plains, hilly part of the peninsula, and coastal zone have their own specific problems. While on one hand the Himalayan region is prone to disasters such as earthquakes and landslides, the plain is affected by floods almost every year. The desert part of the country is affected by droughts and famine, while the coastal zone susceptible to cyclones and storms.

2. The natural geological setting of the country is the primary, basic reason for its increased vulnerability. The geotectonic features of the Himalayan region and adjacent alluvial plains make the region susceptible to earthquakes, landslides, water erosion, etc. Though peninsular India is considered to be the most stable portion, but occasional earthquakes in the region show that geotectonic movements are still going on within its depth.

3. The tectonic features, characteristics of the Himalayas are prevalent in the alluvial plains of Indus, Ganga and Brahmaputra too, as the rocks lying below the alluvial pains are just extension of the Himalayan ranges only. Thus this region is also quite prone to seismic activities. As a result of various major river systems flowing from Himalaya and huge quantity of sediment brought by them, the area is also suffering from river channel siltation, resulting into frequent floods, especially in the plains of UP and Bihar.

4. The western part of the country, including Rajasthan, Gujarat, and some parts of Maharashtra, are hit very frequently by drought situations. If the monsoon worsens, the situation spreads in other parts of the country too. The disturbance in the pressure conditions over oceans results into cyclones in the coastal regions. The geotectonic movements going on in the ocean floor make the coastal region prone to tsunami disaster too.

5. The extreme weather conditions, huge quantity of ice and snow stored in the glaciers, etc. are other natural factors that make the country prone to various forms of disasters.

6. Along with the natural factors discussed previously, various human-induced activities such as increasing demographic pressure, deteriorating environmental conditions, deforestation, unscientific development, faulty agricultural practices and grazing, unplanned urbanization, construction of large dams on river channels, etc. are also responsible for accelerated impact and the increase in frequency of disasters in the country.

Figure 7.4:
India: Losses Due to Disasters

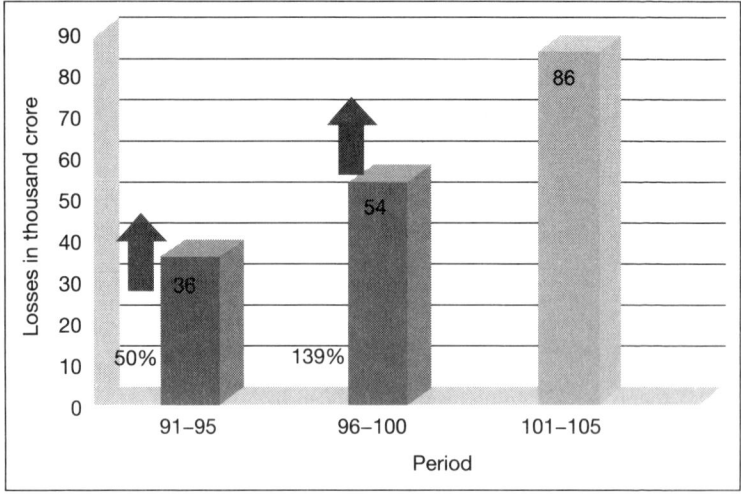

Source: Modified after World Bank.

Losses due to disasters have been shown in Figure 7.4. It shows that economic loss is accounted for 2 percent of the GDP due to disasters as per the study of the World Bank (Figure 7.4).

7.3.4 Institutional Framework of Disaster Management in India

The institutional structure for disaster management in India is in a state of transition. The new setup, following the implementation of the National Disaster Management Act 2005, is evolving, while the previous structure also continues. Thus, the two structures co-exist at present. The NDMA has been established at the national level, and the State Disaster Management Authority (SDMA) at state and district authorities at the district level are gradually being formalized. In addition to this, the National Crisis Management Committee, part of the earlier setup, also functions at the national level (Figure 7.5).

Within this transitional and evolving setup, two distinct features of the institutional structure for disaster management may be noticed. First the structure is hierarchical and functions at four levels—national, state,

Figure 7.5:
Institutional Framework for Disaster Management in India

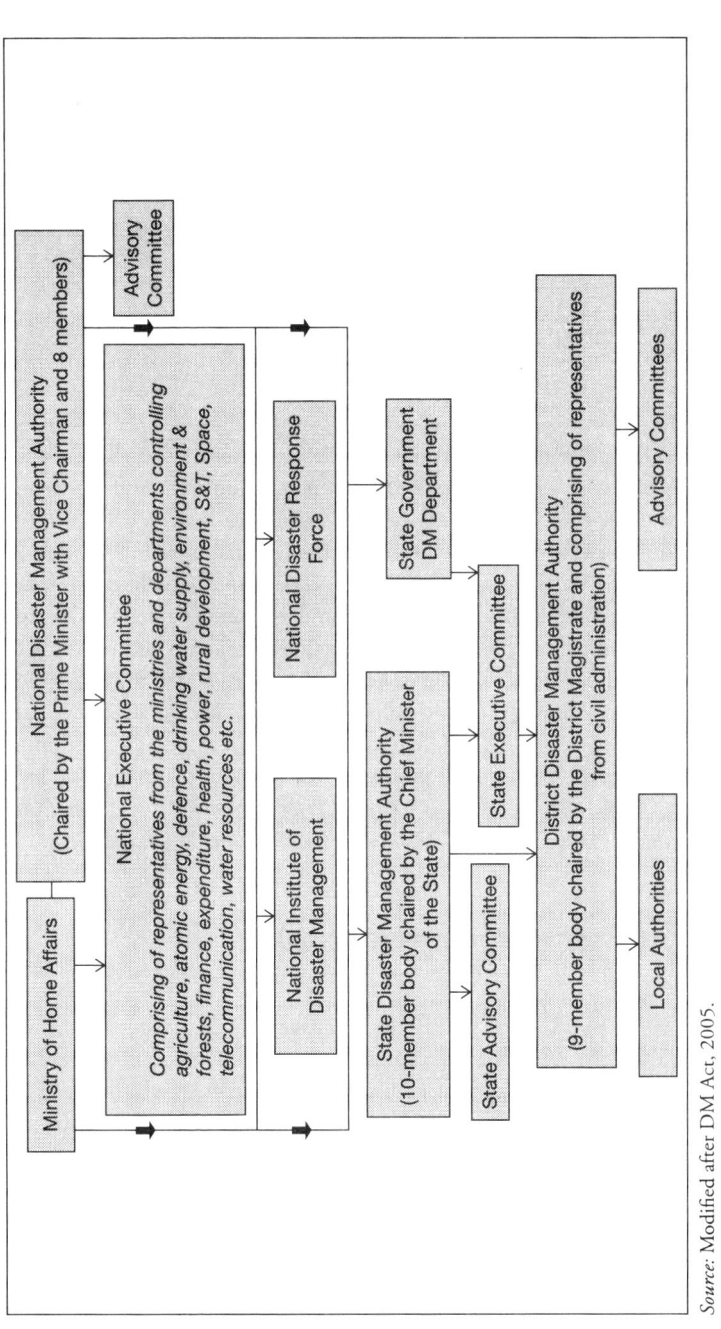

Source: Modified after DM Act, 2005.

district, and local. In both the setups—one that existed prior to the implementation of the Act and other that is being formalized post-implementation of the Act—there have existed institutionalized structures at the national, state, district, and local levels. Each preceding level guides the activities and decision-making at the next level in hierarchy. Second, it is a multistakeholder setup, that is the structure draws involvement of various relevant ministries, government departments, and administrative bodies.

7.3.5 Paradigm Shift in Disaster Management Approach

In the last decade, there has been a paradigm shift in disaster management in India, a distinct move from the earlier approach of post-disaster relief and rehabilitation to pre-disaster preparedness, mitigation, and risk reduction. The 10th Five Year Plan document emphasize that "while hazards, both natural or otherwise, are inevitable, the disasters that follow need not be so and the society can be prepared to cope with them effectively whenever they occur" and called for a "multi-pronged strategy for total risk management, comprising prevention, preparedness, response and recovery, on the one hand, and for initiating development efforts aimed towards risk reduction and mitigation, on the other." It is stated that only then we can look forward to "sustainable development." The country is also committed to mainstream DRR into the process of development planning at all levels for sustainable development, as stated in *Hyogo Framework of Action 2005–15: Building the Resilience of Nations and Communities to Disasters*.

7.3.6 Disaster Management Act 2005

This Act provides for the effective management of disaster and for matters connected therewith or incidental thereto. It provides institutional mechanisms for drawing up and monitoring the implementation of the disaster management. The Act also ensures measures by the various wings of the government for prevention and mitigation of disasters and prompt response to any disaster situation.

The Act provides for setting up of an NDMA under the chairmanship of the prime minister, SDMAs under the chairmanship of the chief ministers, and District Disaster Management Authorities (DDMAs) under the chairmanship of collectors/district magistrates/deputy commissioners. The

Act further provides for the constitution of different executive committees at national and state levels. Under its aegis, the National Institute of Disaster Management (NIDM) for capacity-building and National Disaster Response Force (NDRF) for response purpose have been set up. It also mandates the concerned ministries and departments to draw up their own plans in accordance with the national plan. The Act further contains the provisions for financial mechanisms such as creation of funds for response, the National Disaster Mitigation Fund and similar funds at the state and district levels for the purpose of disaster management. The Act also provides specific roles to local bodies in disaster management.

Further with the enactment of 73 rd and 74th amendments to the constitution and emergence of local self-government, both rural and urban, as important tiers of governance, the role of local authorities becomes very important. The Disaster Management Act 2005 also envisages specific roles to be played by the local bodies in disaster management.

7.3.7 Institutional Framework in India

7.3.7.1 NATIONAL DISASTER MANAGEMENT AUTHORITY (NDMA)

The NDMA was initially constituted on May 30, 2005, under the chairmanship of the prime minister vide an executive order. Following the enactment of the Disaster Management Act, 2005, the NDMA was formally constituted in accordance with Section-3(1) of the Act on September 27, 2006, with the prime minister as its chairperson and nine other members, and one such member to be designated as the vice-chairperson.

7.3.7.2 MANDATE OF NDMA

The NDMA has been mandated with laying down policies on disaster management and guidelines that would be followed by different ministries and departments of the Government of India and state governments in taking measures for DRR. It has also to laid down guidelines to be followed by the state authorities in drawing up the state plans and to take such measures for the management of disasters. Details of these responsibilities are given as under:

1. Lay down policies on disaster management.
2. Approve the national plan.

3. Approve plans prepared by the ministries or departments of the Government of India in accordance with the national plan.
4. Lay down guidelines to be followed by the state authorities in drawing up the state plan.
5. Lay down guidelines to be followed by the different ministries or departments of the Government of India for the purpose of integrating the measures for prevention of disasters or the mitigation of its effects in their development plans and projects.
6. Coordinate the enforcement and implementation of the policy and plan for disaster management.
7. Recommend provision of funds for the purpose of mitigation.
8. Provide such support to other countries affected by major disasters as may be determined by the Central Government.
9. Take such other measures for the prevention of disaster, mitigation, preparedness, and capacity-building for dealing with the threatening disaster situation or disaster as it may consider necessary.
10. Lay down broad policies and guidelines for the functioning of the NIDM.

7.3.7.3 COMPOSITION OF NDMA

Besides the nine members nominated by the prime minister, chairperson of the authority, the organizational structure consists of a secretary and five joint secretaries, including one financial advisor. There are 10 posts of joint advisors and directors and 14 assistant advisors, undersecretaries, and assistant financial advisor and duty officer along with supporting staff.

7.3.7.4 NATIONAL EXECUTIVE COMMITTEE

A National Executive Committee (NEC) is constituted under Section 8 of the Disaster Management Act, 2005, to assist the national authority in the performance of its functions. The NEC consists of the home secretary as its chairperson, *ex-officio*, with other secretaries to the Government of India in the ministries or departments having administrative control of agriculture, atomic energy, defense, drinking water supply, environment and forest, finance (expenditure), health, power, rural development science and technology, space, telecommunication, urban development, and water resources. The chief of integrated defense staff of the chiefs of staff committee, *ex-officio,* is also its member.

NEC may, as and when it considers necessary, constitute one or more subcommittees for the efficient discharge of its functions. For the conduct of NEC, Disaster Management National Executive Committee (Procedure and Allowances) Rules, 2006, has been issued which may be visited at the website of the ministry of home affairs.[1] NEC has been given the responsibility to act as the coordinating and monitoring body for disaster management, to prepare a national plan, monitor the implementation of a national policy, etc. vide Section 10 of the Disaster Management Act.

7.3.7.5 STATE DISASTER MANAGEMENT AUTHORITY

The Disaster Management Act, 2005, provides for constitution of the SDMAs and District Disaster Management Authority (DDMA) in all the states and UTs. As per the information received from the states and UTs, except Gujarat, and Daman and Diu, all the rest have constituted SDMAs under the Disaster Management Act, 2005. Gujarat has constituted its SDMA under its Gujarat State Disaster Management Act, 2003. Daman and Diu has also established an SDMA prior to the enactment of the Disaster Management Act 2005.

7.3.7.6 STATE EXECUTIVE COMMITTEE

The Act envisages the establishment of the State Executive Committees (SEC) under Section 20 of the Act, to be headed by chief secretary of the state government with four other secretaries of such departments as the state government may think fit. It has the responsibility for coordinating and monitoring the implementation of the national policy, the national plan and the state plan as provided under Section 22 of the Act.

7.3.7.7 DISTRICT DISASTER MANAGEMENT AUTHORITY

Section 25 of the Disaster Management Act, 2005, provides for the constitution of a DDMA for every district of a state. The district magistrate/ district collector/deputy commissioner heads the authority as its chairperson besides an elected representative of the local authority as co-chairperson, except in the tribal areas where the chief executive member of the District Council of Autonomous District is designated as co-chairperson. Further in districts where Zila Parishads exist, its chairperson shall be the co-chairperson of DDMA. Members of this authority include the CEO of the district authority, superintendent of police, chief medical officer of the district and two district level officers designated by the state government.

---

[1] See www.mha.nic.in.

The district authority is responsible for planning, coordination, and implementation of disaster management and taking such measures for disaster management as provided in the guidelines. The district authority also has the power to examine the construction in any area in the district to enforce the safety standards and also to arrange for relief measures and respond to the disaster at the district level.

7.3.7.8 INSTITUTIONAL FRAMEWORK FOR METROPOLITAN CITIES

In the larger cities (say, with population exceeding 2.5 million), the recommendation of the second Administrative Reforms Commission of India has suggested that the mayor, assisted by the commissioner of the municipal corporation and the police commissioner, be directly responsible for crisis management.

7.3.7.9 HIERARCHICAL STRUCTURE OF AUTHORITY AND COMMITTEE

In this structure, NDMA is the authority for formulation of policy and guidelines for all disaster management work in the country. The state authorities further lay down the guidelines for departments of the state and the districts falling in their respective jurisdictions. Similarly, district authorities direct the civil administration, departments, and local authorities such as the municipalities, police departments, and civil administration. The SECs are responsible for the execution of the tasks envisaged by the authorities. The structure thus discussed is summarized in Figure 7.4.

7.3.8 National Institute of Disaster Management

In the backdrop of the International Decade for Natural Disaster Reduction (IDNDR), a national center for disaster management was established at the Indian Institute for Public Administration (IIPA) in 1995. The center was upgraded and designated as NIDM on October 16, 2003. It has now achieved the status of a statutory organization under the Disaster Management Act, 2005. Section 42 of Chapter VII of the Disaster Management Act, 2005, entrusts the institute with numerous responsibilities, namely, to develop training modules; undertake research and documentation in disaster management; organize training programs; undertake and organize study courses, conferences, lectures and seminars to promote and institutionalize disaster management; and undertake and provide for publication of journals, research papers, and books.

The Institute has four academic divisions, namely, geohazard division, hydromet hazard division, policy planning and cross-cutting issues division, and response division.

7.3.9 Disaster Response Force (SDRF)

The states/UTs have also been advised to set up their own specialist response force for responding to disasters on the lines of the NDRF vide a MOHA letter dated July 26, 2007, and March 8, 2011. The Central Government is providing assistance for training of trainers. The state governments have been also advised to utilize 10 percent of their State Disaster Response Fund (SDRF) and Capacity Building Grant for the procurement of SAR equipment and for training purposes of the response force.

7.3.10 Civil Defense

Aims and objectives of Civil Defense Act: The civil defense (CD) policy of the Government of India, until 1962, was confined to making the states and UTs conscious of the need of civil protection measures and to keep in readiness civil protection plans for major cities and towns under the Emergency Relief Organization (ERO) scheme. The legislation on CD, known as the Civil Defense Act, was enacted in 1968, and is in force throughout the country. The Act defines CD and provides for the powers of the Central Government to make rules for CD, spelling out various actions to be taken for CD measures. It further stipulates for the constitution of CD corps, appointment of members and officers, functions of members, etc. The Act has since been amended in 2010 to cater to the needs of disaster management so as to utilize the services of CD volunteers effectively for enhancement of public participation in disaster management-related activities in the country. The CD organization is raised only in such areas and zones which are considered vulnerable to enemy attacks. The revision and renewal of categorized CD towns is done at regular intervals, with the level of perceived threat, external aggression, or hostile attacks by anti-national elements or terrorists to vital installations.

7.3.11 Fire Services

Fire services are mandate of the municipal bodies as estimated in Item 7 of Schedule 12 under Article 243 W of the Constitution. The structure across is not uniform. Presently fire prevention and firefighting services are organized by the concerned states and UTs. The MOHA, Government of India, renders technical advice to the states and UTs and central ministries on fire protection, fire prevention and fire legislation.

The Government of India, in 1956, formed a "Standing Fire Advisory Committee" under the MOHA. The mandate of the committee was to examine the technical problems relating to fire services and to advise the Government of India for speedy development and upgradation of fire services all over the country. This committee had representation from each state fire services, as well as the representation from Ministries of Home, Defense, Transport, and Communication and the Bureau of Indian Standards. This committee was renamed as the "Standing Fire Advisory Council" (SFAC) during the year 1980.

Fire Services in Gujarat, Chhattisgarh, Punjab, Maharashtra, Himachal Pradesh, Haryana, and Madhya Pradesh excluding Indore are under the respective concerned municipal corporations. In other remaining states, it is under the Home Department. While some states have enacted their own Fire Act, some others have not. As of today, there is no standardization with regard to the scaling of equipment, the type of equipment, or the training of their manpower. In each state it has grown according to the initiatives taken by the states and the funds provided for the fire services.

Presently the only basic lifeline of fire and emergency services that is fully committed to the common public is the municipal in some states and state fire services. The airport authority, big industrial establishments, CISF, and armed forces, however, also have their own fire services and many a times, in case of need, rush in aid to the local fire services. Apart from the lack of being a proper government department with a complete developmental plan, state fire services have their own organizational structure, administrative setup, funding mechanism, training facilities, and equipment.

7.3.12 National Disaster Response Force

Constitution and role of NDRF: NDRF has been constituted under Section 44 of the Disaster Management Act, 2005 by upgradation/conversion of eight standard battalions of the Central paramilitary forces, that is, two battalions each from the Border Security Force (BSF), Indo-Tibetan Border Police (ITBP), Central Industrial Security Force (CISF), and Central Reserve Police Force (CRPF) to build them up as a specialist force to respond to disaster or disaster-like situations.

At present, NDRF consists of 12 battalions, three each from the BSF and CRPF and two each from CISF, ITBP, and SSB. Each battalion has 18 self-contained specialist SAR teams of 45 personnel each including engineers, technicians, electricians, dog squads, and medical/paramedics. The total strength of each battalion is 1,149. All the 12 battalions have been equipped and trained to respond to natural as well as man-made disasters. Battalions are also trained and equipped for response during CBRN emergencies.

Based on vulnerability profile of different regions of the country, these specialist battalions have been presently stationed at the 10 places shown in Figure 7.6.

Figure 7.6:
NDRF Battalions Deployment in Strategic Locations

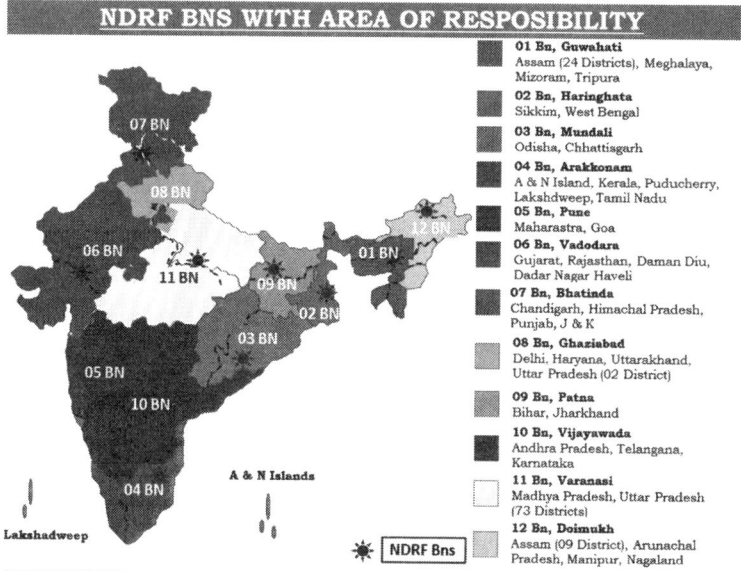

Source: Modified After http://ndrf.gov.in/ndrf
Disclaimer: This figure has been redrawn by the author and is not to scale. It does not represent any authentic national or international boundaries and is used for illustrative purposes only.

7.3.13 Incident Response System

India is vulnerable to a variety of natural and man-made disasters that hinder the country's growth. The management of response in disasters requires the existing administrative setup, civil society and its various institutions to carry out a large number of tasks. The activities involved in response management would depend on the nature and type of disaster. It has been observed that in times of disaster, apart from lack of resources, lack of coordination among various agencies and an absence of role clarity amongst various stakeholders pose serious challenges. If the response is planned and the stakeholders are trained, there will be no scope for ad hoc measures and the response will be smooth and effective. The objective of these guidelines is to pre-designate officers to perform various duties as well as train them in their respective roles. Realization of certain shortcomings in response and a desire to address the critical gaps led the Government of India to look at the world's best practices and adopt the Incident Command System (ICS).

The ICS incorporates all the duties that may be performed in case of any disaster or event. It envisages a complete team with various sections to attend to all possible requirements. If the ICS is put in place and stakeholders are trained in their respective duties and roles, it will help reduce chaos and confusion during actual incident management, and everyone involved will know what all needs to be done, who will do it, where are the resources, and who is in command, etc.

The ICS is a flexible system and all its sections need not be activated in every situation at the same time. Only required sections may be made operational as and when required. This system envisages that the roles and duties shall be laid down in advance, the personnel earmarked and trained in their respective roles and duties.

This system consists of a number of useful features such as (a) management by objectives (MBO), (b) unity and chain of command, (c) transfer of command, (d) organizational flexibility, (e) manageable span of control, (f) area command, (g) unified command, (h) common terminology, (i) personnel accountability, (j) integrated communications, (k) planning and comprehensive resource mobilization, deployment, and demobilization, (l) incident action plan, (m) information management, (n) proper documentation of the entire response activities through forms and formats, (o) responder's safety, (p) media management, and (q) agency coordination.

Keeping in mind the Disaster Management Act, 2005, and the existing administrative structure of the country, the ICS required some modifications and adaptation to the Indian context. In India, the main stakeholders in any incident response are the administrators of the various government departments at the national, state, district, UT, and metropolitan city level. The roles of NGOs, CBOs, PRIs, volunteers of CD, NDRF, SDRF personnel, and communities, etc. also need to be carefully integrated in the response structure. NDMA, therefore, decided to adapt the ICS duly indigenized so that it is in consonance with the administrative structure of the country in order to strengthen and standardize the response system in India.

7.3.13.1 NEED FOR IRS

The Disaster Management Act, 2005, has heralded a paradigm shift in disaster management from a post-event response to one of pre-event prevention, mitigation, and preparedness. Though India has a long history of battling disasters and providing adequate response, it was clearly realized that there were a number of shortcomings such as

1. lack of accountability because of ad hoc and emergent nature of arrangements and no prior training for effective performance;
2. lack of an orderly and systematic planning process;
3. unclear chain of command and supervision of response activity;
4. lack of proper communication, inefficient use of available resources, use of conflicting codes and terminology, and no prior communication plan;
5. lack of predetermined method/system to effectively integrate interagency requirements into the disaster management structures and planning process;
6. lack of coordination between the first responders and individuals, professionals, and NGOs with specialized skills during the response phase; and
7. lack of use of common terminology for different resources resulting in improper requisitioning and inappropriate resource mobilization, etc.

7.3.13.2 DEFINITION AND CONTEXT

The Incident Response System (IRS) is an effective mechanism for reducing the scope for ad hoc measures in response. It incorporates all the tasks that may be performed during disaster management irrespective of their level of complexity. It envisages a composite team with various Sections to attend to all the possible response requirements. The IRS identifies and designates officers to perform various duties and get them trained in their respective roles. If IRS is put in place and stakeholders trained and made aware of their roles, it will greatly help in reducing chaos and confusion during the response phase. Everyone will know what needs to be done, who will do it and who is in command, etc. IRS is a flexible system and all the Sections, Branches and Units need not be activated at the same time. Various Sections, Branches and Units need to be activated only as and when they are required.

7.3.13.3 IRS Organization

The IRS organization functions through incident response teams (IRTs) in the field. In line with our administrative structure and Disaster Management Act 2005, responsible officers (ROs) have been designated at the state and district level as overall in charge of the incident response management. The RO may, however, delegate responsibilities to the incident commander (IC), who in turn will manage the incident through IRTs. The IRTs will be pre-designated at all levels; state, district, subdivision, and tehsil/block. On receipt of early warning, the RO will activate them. In case a disaster occurs without any warning, the local IRT will respond and contact the RO for further support, if required. An NO has to be designated for proper coordination between the district, state, and national level in activating air support for response. Apart from the RO and NO, the IRS has two main components: (a) command staff and (b) general staff (Figure 7.7).

7.3.13.4 Command Staff

The command staff consists of an incident commander (IC), information and media officer (IMO), safety officer (SO), and liaison officer (LO). They

Figure 7.7:
IRS Organization

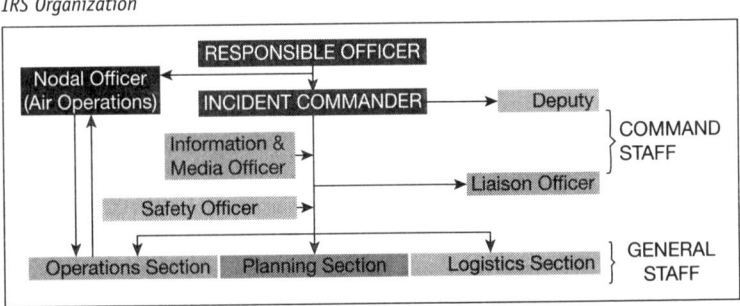

Source: National Disaster Management Guidelines—Incident Response System (2010).

report directly to the IC and may have assistants. The command staff may or may not have supporting organizations under them. The main function of the command staff is to assist the IC in the discharge of his or her functions.

7.3.13.5 GENERAL STAFF

The general staff has three components which are as follows:

Operations section: The operations section (OS) is responsible for directing the required tactical actions to meet incident objectives. Management of disaster may not immediately require activation of branch, division, and group. Expansion of the OS depends on the enormity of the situation and number of different types and kinds of functional groups required in the response management.

Planning section: The planning section (PS) is responsible for collection, evaluation, and display of incident information, maintaining and tracking resources, preparing the incident action plan (IAP) and other necessary incident-related documentation. They will assess the requirement of additional resources, propose from where it can be mobilized, and keep the IC informed. This section also prepares the demobilization plan.

Logistics section: The logistics section (LS) is responsible for providing facilities, services, materials, equipment, and other resources in support of the incident response. The section chief participates in the development and implementation of the IAP and activates and supervises branches and units of his or her section. In order to ensure prompt and smooth procurement and supply of resources as per financial rules, the finance branch has been included in the LS.

7.3.13.6 FEATURES OF IRS

Management by Objectives
MBO covers four essential steps in IRS. These steps should be taken for the management of every incident, regardless of its size or complexity:

1. Understand government policy and directions including relief code, evacuation procedures, etc.
2. Establish incident objectives.
3. Select appropriate strategies.
4. Performa tactical moves (assigning the right resources, and monitoring the performance, etc.).

Unity of Command and Chain of Command

In IRS, unity of command means that every individual has a designated supervisor. Chain of command means that there is an orderly line of authority within the ranks of the organization with a clear reporting pattern right from the lowest level to the highest. In the IRS, the chain of command is established through a prescribed organizational structure, which consists of various layers such as sections, branches, divisions, etc. This feature eliminates the possibility of receiving conflicting orders from various supervisors. Thus it increases accountability, prevents freelancing, improves the flow of information, and helps in smooth coordination in operational efforts.

Transfer of Command

The command of an incident initially is vested in the highest ranking authority in the area where the disaster occurs. The transfer of command in any incident may take place for the following reasons:

1. When an incident becomes overwhelming for the IC and IRT.
2. More qualified and experienced senior officers arrive at the scene.
3. The incident situation changes over time, where a jurisdictional or agency change in command is operationally required.
4. Normal turnover of personnel in the case of long or extended incidents.

The various processes in IRS of briefing, debriefing, and documenting through forms and formats, proves very useful during transfer of command. The IAP, assignment list, details of actions already taken, resources deployed, available, and ordered, etc. give an immediate and comprehensive view of the incident status to the newcomer.

Organizational Flexibility

The IRS organization is a need-based, flexible organization. All the components need not be activated simultaneously. It would depend upon the nature and requirements of the incident. Each activated section, branch, or unit must have a person in charge to perform its role. In some cases, because of lack of personnel, a single supervisor may be made in charge of more than one group, unit, or section. It should be clearly understood that in such cases the groups, units, and sections do not get merged/amalgamated. Their functioning would continue to be independent. Only the supervisory work/responsibility will be performed and discharged by the same individual. The organizational elements that are no longer required should be deactivated to reduce the size of the organization and to ensure appropriate use of resources.

Span of Control

Span of control refers to the number of elements (section, branch or unit) that one supervisor can directly manage effectively. Ideally, a supervisor should have five organizational elements under his control. However, if the elements increase to more than five or are reduced to less than three, necessary changes in the IRS organizational structure should be carried out.

Area Command

Area command is an expansion of the incident response function, primarily designed to manage a very large number of incidents that has multiple IRTs assigned or an area being isolated because of geographical reasons. It is established for overseeing the response and to ensure that conflicts, jurisdictional or otherwise, do not arise amongst the deployed responding teams.

Unified Command

Unified command (UC) is a team effort that allows all agencies with jurisdictional responsibility for the incident, either geographical or functional, to manage an incident by establishing a common set of incident objectives and strategies under one commander. This is accomplished through the UC framework headed by the governor/Lt governor (LG)/administrator/chief minister (CM) and assisted by the chief secretary (CS) without losing or abdicating agency authority, responsibility, or accountability.

Common Terminology

In IRS, common terminology is applied to organizational elements, position titles, resources, and facilities, which are as follows:

1. *Organizational elements:* There is a consistent pattern for designating each level of the organization (e.g., sections, branches, divisions, and units, etc.).
2. *Position titles:* Those charged with management or leadership responsibility in IRS are referred to by specific position titles such as commander, officer, chief, director, supervisor, leader, in-charge, etc. It provides a standardized nomenclature for requisitioning personnel to fill various levels of positions.

3. *Branch:* The organizational level having functional or geographic responsibility for major segments of incident operations. The branch is found in operations and logistics sections. It is based on various functional requirement of the section.

4. *Division:* Divisions are used to divide an incident into geographical area of operations. It is positioned in the IRS organization between the branch and the groups. Divisions are established when the number of resources deployed exceeds the span of control of the OSs chief. It is also activated for closer supervision when an area is very distant or isolated.

5. *Group:* Group refers to only the functional responsibilities for major segments of incident operations. A group consists of different functional teams (single resource, strike team, and task force).

6. *Resources:* Resources are grouped into two categories: (a) primary and (b) support. The primary resources are meant for the responders and the support resources are meant for the affected people. All resources are, however, designated according to the "kind" and "type." "Kind" would mean the overall description of the resource such as bus, truck, bulldozer, medical team. "Type" would mean the performance capability of the resource, which may be large, medium, or small. This helps in ordering the exact and correct resource by the ordering unit. It also helps the deploying agencies to send the correct requirement.

7. *Facilities:* Different kinds of facilities have to be established to meet the specific needs of the incident. IRS tries to standardize them by using common terminology such as incident command post, staging area, incident base, camp, relief camp, helibase, helipad, etc.

7.3.13.7 Symbols for Different IRS Facilities

In the IRS, different symbols are used for identification of different facilities established for response management. They are given in Figure 7.8.

7.3.13.8 Accountability

In IRS, through a clear-cut chain of command, it is ensured that one individual or group is not assigned to more than one supervisor. Through other procedures and use of various forms, accountability of personnel and resources are ensured. It makes the response effort absolutely focused and leaves no room for unsupervised activity. It helps maintain a complete record of all activities performed and resources deployed.

Figure 7.8:
Symbols Used for Identification of Different Facilities

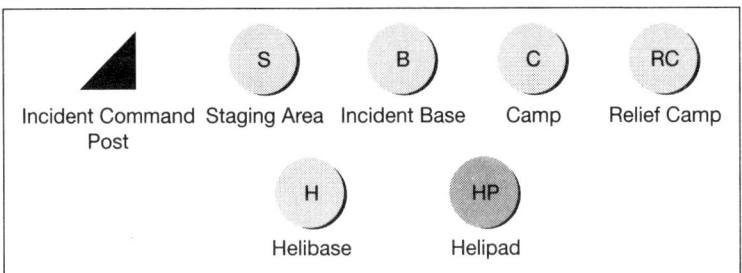

Source: National Disaster Management Guidelines—Incident Response System (2010).

In line with the federal structure of the country, it should be clearly understood that response to any disaster will be carried out by the concerned states and districts. The Government of India will play a supporting role by way of assistance in the form of resources, manpower (NDRF, armed, and paramilitary forces), equipment, and funds. At the national level, the NCMC or NEC will coordinate and provide the required resources. NDMA will also help in monitoring the coordination of response (Figure 7.9).

7.4 Asia Regional Plan for Implementation of the Sendai Framework for Disaster Risk Reduction (2015–30)

7.4.1 Overview of the Regional Plans

Asia has enjoyed fast economic growth over recent decades. This has contributed to the progress in achieving the Millennium Development Goals (MDGs). In the next 15 years, many countries in the region aim to continue this progress and generate higher national income on the way to achieving sustainable and equitable development. Asia is exposed and vulnerable to a wide range of natural and man-made hazards. In many respects, it is the global epicenter for disasters. In 2015, the Nepal earthquake killed more people than any other disaster (8,831). The drought in the Democratic People's Republic of Korea affected the food security of more than 18 million. Four of the top five most disaster-hit countries were in Asia: China (26

Figure 7.9:
Organization Structure of Incident Response System

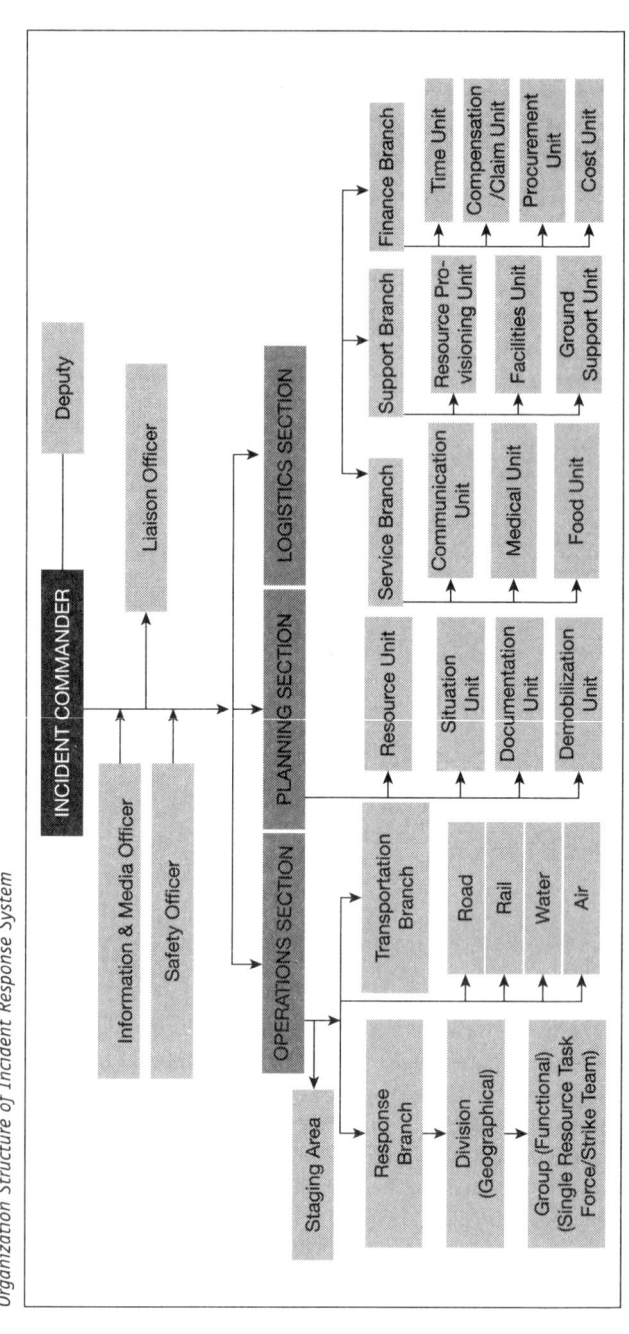

Source: National Disaster Management Guidelines—Incident Response System (2010).

disasters); India (19), Philippines (15), and Indonesia (11). In terms of economic losses, China, India, and Nepal were among the five worst-hit countries in the world. These figures are consistent with longer term trends over recent decades. During the implementation of the HFA 2005–15, it was evident from national and regional progress reports that countries in Asia made initial progress in reducing disaster risk at all levels. In particular, the region moved forward on dedicated legislation, policies, and establishment of institutions to reduce disaster risk; establishment of tsunami, cyclone, and other hydrometeorological early warning systems; improvements in information generation and dissemination, awareness-raising, and school education on DRR; and strengthened disaster preparedness and response capacity at all levels. This has led to a decrease in mortality risk, particularly to hydrometeorological hazards. The region has, however, struggled to reduce the underlying risk factors. Rapid and poorly managed urbanization, natural resource exploitation, and increasing social inequalities, amongst others, increased risk levels. Underlying risk drivers are either not well identified and understood or not adequately addressed because of capacity constraints, lack of priority, or scarcity of resources. This has resulted in development practices that are not fully risk-informed. The overall impact of disasters on economies and communities is still rising. They are further challenged by the impact of climate change. Exposure of populations and assets in the region has increased faster than vulnerability has decreased. Much of this is because of rapid economic growth and private and public investments in hazardous areas such as tsunami- and cyclone-prone coasts, flood-exposed river basins and earthquake-prone cities. This has generated new risk and led to a steady rise in disaster losses with a significant economic, social, health, cultural, and environmental impact across Asia. For instance, the socioeconomic impact of the 2011 earthquake/tsunami/nuclear disaster in Japan, the 2011 floods in Thailand, and the 2013 typhoon Haiyan in the Philippines are still being felt in the daily lives of millions of people. Disasters bring challenges to achieving sustainable development. The Sendai Framework aims to achieve "[t]he substantial reduction of disaster risk and losses in lives, livelihoods and health and in the economic, physical, social, cultural and environmental assets of persons, businesses, communities and countries" by 2030. This expected outcome will be monitored via indicators against seven targets.

To reduce

1. mortality,
2. the number of affected people,

3. economic losses, and
4. damage to critical infrastructure;
  and increase
5. the number of national and local DRR strategies,
6. level of international cooperation, and
7. availability of and access to multihazard early warning systems and disaster risk information.

The Sendai Framework set the goal to prevent the creation of new risk, reduce existing risk, and strengthen resilience. It highlights the importance of: understanding, assessing, and monitoring disaster risk; strengthening disaster risk governance and coordination across relevant institutions and sectors and the full and meaningful participation of relevant stakeholders at appropriate levels; investing in disaster risk prevention and reduction through structural and nonstructural measures that are essential to enhance the economic, social, health, and cultural resilience of persons, communities, countries, and their assets, as well as the environment, also through technology and research; and enhancing multihazard early warning systems, preparedness, response, recovery, rehabilitation, and reconstruction. Countries in Asia have made initial progress in implementation of the Sendai Framework and have contributed to the development of indicators to measure its seven global targets. This underpins the Sendai Framework's stronger focus on risk reduction as a major contributor to sustainable development. Regional-level plans and activities need to support national actions.

The Sendai Framework for Disaster Risk Reduction 2015–30 (Sendai Framework) provides the way forward to prevent and reduce disaster risk in order to achieve resilient and sustainable development. Under the leadership of national governments, and embracing all actors at all levels, the Sendai Framework offers a solution to saving lives, livelihoods, and assets as well as for reducing the fiscal burden on governments to bail out the aftermath of failed "development." The Sendai Framework's primary focus on risk-reduction and resilience is a common element highlighted in all the 2030 development agendas adopted by all member states of the UN, such as the Addis Ababa Action Agenda on Financing for Development, the SDGs, the Paris Agreement on Climate Change, the Agenda for Humanity, and the New Urban Agenda. The Sendai Framework highlights the need for agreed regional and subregional strategies and mechanisms for cooperation and for the progress to be reviewed by regional and global platforms. At the Sixth Asian

Ministerial Conference for Disaster Risk Reduction (AMCDRR) in June 2014 and the ISDR Asia Partnership (IAP)2 deliberations after the Third UN World Conference on Disaster Risk Reduction (3WCDRR), Asian countries and stakeholders agreed to develop an "Asia Regional Plan for implementation of the Sendai Framework" (Regional Plan) to facilitate cooperation and collaboration for building risk resilience in Asia. This regional plan aims to provide:

1. *Broad policy direction* to guide the implementation of the Sendai Framework in the context of the 2030 sustainable development agendas in the region;
2. *Long-term road map* spanning the 15-year horizon of the Sendai Framework outlining a chronological pathway for implementation of priorities to achieve seven global targets;
3. *Two-year action plan* with specific activities that are prioritized based on the long-term road map and in line with the policy direction.

The Asia Regional Plan seeks to guide and support the national implementation of the Sendai Framework, but it is not a replacement of national plans. It does so by identifying priorities at regional activities to support national and local actions; enhancing exchange of good practice, knowledge, and information among governments and stakeholders, and strengthening regional cooperation to support the implementation of the Sendai Framework.

This regional plan is one of the main outcome documents of the AMCDRR in November 2016 in New Delhi, India. The regional plan has been developed through the IAP and an advisory working group (AWG) mechanism set up by the UNISDR secretariat in Asia-Pacific. Three informal AWGs were established to ensure that the actions for all priority areas of the Sendai Framework are developed with the engagement of key partners—the primary implementers of the Sendai Framework from the public, private, and local levels. This aligns with the outcome of the Sixth Asian Ministerial Conference and further deliberations in the IAP. The AWG comprised individuals with expertise in public policy, private-sector engagement, and local-level implementation respectively. They consulted governments, stakeholder groups, and partners and provided the necessary implementation focus of the regional plan through inputs to the policy direction, the road map, and in particular the 2-year action plan.

7.4.2 The Policy Direction for Implementation of the Sendai Framework in Asia

1. *Coherence and integration:* The Addis Ababa Action Agenda, the 2030 Agenda for Sustainable Development, the Sendai Framework, and the Paris Agreement on Climate Change (COP21), all strive towards making development resilient and sustainable. Effective DRR is an indispensable element towards this end. The integration of risk reduction in development will build resilience and protect development gains. Risk reduction and resilience is a common element across the various frameworks and agreements. This will help establish a more collaborative environment between the DRR community and the development sectors. Further, the outcomes of the World Humanitarian Summit and the Habitat III underline the significance of DRR in their respective sectors, particularly through local actions. The incorporation of DRR into the 2030 development agendas will provide an opportunity to break down silo approaches within and between respective sectors. The Sendai Framework states:

   > The development, strengthening and implementation of relevant policies, plans, practices and mechanisms need to aim at coherence, as appropriate, across sustainable development and growth, food security, health and safety, climate change and variability, environmental management and DRR agendas. DRR is essential to achieve sustainable development. (Para. 19.h)

   The respective international frameworks/agendas mutually reinforce and depend on each other. For instance, the effective implementation of the Addis Ababa Action Agenda and the Paris Agreement will enable the Sendai Framework to contribute to the overall 2030 Agenda for Sustainable Development. Consequently, countries in the region need to ensure that all planning on DRR is an integral part of the implementation and monitoring of the SDGs, while recognizing the need to achieve more specific targets and indicators of the Sendai Framework.

2. *Guiding principles:* The Sendai Framework provides principles for resilient development that emphasize the primary responsibility of the state to prevent and reduce disaster risks; shared responsibilities across all levels, sectors, and stakeholders with an "all-of-society" engagement approach; the integration of gender, age, disability,

and cultural perspective in all policies and practices; a multihazard approach; risk-informed public and private investments; Building Back Better (BBB) in post-disaster recovery, rehabilitation, and reconstruction; effective and meaningful global partnerships; and adequate, sustainable, and timely provision of support from developed to developing countries. Countries should adopt and/or adapt these principles in their national policies and strategic action plans for implementation of the Sendai Framework. The first four of the Sendai Framework targets focus on substantial reduction of disaster losses and the remaining three focus on increasing capabilities. To achieve these targets, countries will need to prioritize actions outlined under the Sendai Framework's four "priorities for action" Target (e) requires the development and/or revision of risk-informed national and local DRR strategies by 2020. As such, countries should prioritize, as appropriate, the improvement of multihazard risk assessment. The strengthening of national disaster loss databases is also crucial so that the baselines for targets (a, b, c, and d) can be set. Effective databases will provide evidence to inform policies and investments to reduce disaster risk and losses, particularly in the most affected sectors.

3. *Enabling environment:* For governments, stakeholders, and organizations to implement the Sendai Framework, actions will be required that create an enabling environment for risk–resilient development.

At the regional level, a common understanding and approach is needed to tackle regional and transboundary issues. Assessment and monitoring of regional and transboundary hazards and emerging risks have improved. However, much remains to be done to enhance the understanding of regional and transboundary risks. The main intergovernmental regional organizations (IGO) in Asia—ASEAN, SAARC, the Economic Cooperation Organization (ECO), the Tri-lateral Cooperation Secretariat, Shanghai Cooperation Organization (SCO), APEC, the Bay of Bengal Initiative for Multi-Sectoral Technical and Economic Cooperation (BIMSTEC), the UN Economic and Social Commission for Asia and the Pacific (UNESCAP), and the Asian and Pacific Centre for the Development of Disaster Information Management (APDIM)—need to strengthen their capacity to foster the implementation of the Sendai Framework from a regional perspective. This could include building human and institutional capacity to carry out science-based, transboundary risk assessments in the region; developing common policies, tools, and political commitments towards implementing and monitoring the Sendai Framework; fostering data, information, and knowledge exchange in the region; and so on. At the same time, other nonintergovernmental and informal networks need to promote regional cooperation through science, research, and knowledge exchange. To enhance the regional mechanisms to ensure effective follow-up and implementation of the regional plan, there is a need to strengthen the biennial AMCDRR and its consultation mechanism the IAP as the principal means of regional DRR governance.

At the national level, countries will need to align their DRR strategies and plans with the Sendai Framework. Over the coming years, in order to expedite the implementation of the Sendai Framework in the region, additional emphasis will be given to (a) promoting policy coherence among DRR and development in-country, (b) making DRR a development practice to achieve resilient public investment and the SDGs, (c) encouraging private-sector engagement towards risk-sensitive investments, and (d) building capacity and leadership to implement the Sendai Framework at the national and local levels. These focus areas also need to be supported by provisioning of adequate capacity and resources at the local-level, knowledge and information from the scientific and academic community, and practical guidance and tools. Adopting an inclusive approach—via multi-sector/stakeholder DRR platforms, both at national and local levels—is particularly important. It should embrace the leadership of persons with disability, women, children and youth, and the significant contribution of the business sector.

At the local level, it is important to raise awareness of the Sendai Framework, through local campaigns. Target (e) provides an opportunity to promote local practices through the development and adoption of local DRR strategies/plans by 2020. These strategies and plans should be based on improved local risk assessment and capacity to monitor hazards, exposure, and vulnerabilities. Strengthening leadership and capacity of local authorities, communities, civil society, volunteers, organized voluntary work organizations, and community-based organizations (CBOs) will be critical for them to work together through an inclusive approach. Peer learning and cooperation between local governments play an important role in this regard. At the same time, it is important to allocate resources to local governments as appropriate to implement DRR measures.

7.4.3 The Road Map for Implementation of the Sendai Framework in Asia

The 15-year timeframe of the Sendai Framework together with other international frameworks including the SDG's provides an opportunity for greater coherence and integration in terms of planning, implementation, and monitoring. By providing guidance towards the achievement of the Sendai Framework's seven global targets and presenting a set of intended results in chronological order, the road map contributes to this process of mutual reinforcement in Asia. The road map fits within the overarching policy direction and will steer the 2-year action plan. The milestones are based on a shared vision across Asia towards risk prevention and resilience building that will achieve the seven global targets of the Sendai Framework. The milestones are sequenced in a way to enable this. The earlier milestones—such as the development and establishment of countries' DRM status, disaster damage, and loss databases, and the national and local strategies and plans (target e)—are relevant for other targets, namely, a reduction in mortality, people affected, economic losses, and loss of critical infrastructure and services. The regional plan will support progress against all Sendai targets, specially to target (f), that is, enhanced international cooperation. It also facilitates achievement of target (g), that is, the strengthening of transboundary early warning systems and the sharing of disaster risk information. The key milestones of the road map are as below.

By 2016

1. Translation and dissemination of key messages and essence of the Sendai Framework in national languages to increase awareness.
2. All countries have identified their Sendai Framework focal point.
3. 20 percent of countries take stock of their current status of DRR.
4. The AMCDRR agreed on implementation of the Asia Regional Plan for implementation of the Sendai Framework which was one of the outcomes of the Conference.

By 2018

1. Technical guidance by UNISDR to national indicators is finalized with a link to SDG targets and indicators.
2. Fifty percent of countries have prepared a design to establish a national mechanism to collect, analyze and disseminate information on disaster losses and risk aiming to achieve appropriate level of disaggregation for gender, age and disability.
3. Forty percent of countries have revised/ developed their national strategies and/ or plans for DRR in line with the Sendai target (e).
4. Fifty percent of countries have reviewed their initial progress in implementation of the Sendai Framework through the Sendai Monitor.
5. Forty percent of countries have established multisectoral and multistakeholder national and local platforms to foster the dialogue and cooperation between governments, science and technology community and other stakeholders for risk-sensitive development and innovative risk management.
6. 10 percent of countries have developed regulatory or policy frameworks to reinforce risk considerations and risk reduction measures into development initiatives, particularly in the infrastructure sector.

By 2020

1. All countries have established methodologies to collect disaster loss data and risk profiles, with gender, age, disability disaggregated data.
2. All countries have revised/developed their national strategies and/or plans for DRR with increased focus on local actions, achieving Sendai Framework Target (e).
3. Thirty percent of countries have developed regulatory or policy frameworks to reinforce risk considerations and risk-reduction measures into development initiatives, in particular in the infrastructure sector
4. Sixty percent of countries have established multisectoral and multistakeholder national and local platforms.
5. Review regional targets under the Asian Regional Plan from 2022 to 2028, every biennium.

By 2022

1. Fifty percent of countries have developed regulatory or policy frameworks to reinforce risk considerations and risk-reduction measures into development initiatives, in particular in the infrastructure sector.
2. Sixty percent of countries have improved their early warning systems including improved monitoring and forecast systems evacuation procedures, analyses of risk, availability and access to early warning information.
3. All countries have established multisectoral and multistakeholder national and local platforms.

By 2030

1. All countries have demonstrated reduction in disaster-related mortality, affected population, economic losses and damages to critical infrastructure and basic services.
2. All countries have reviewed their progress of implementation through the Sendai Monitor 3. A regional review report of the Sendai Framework is available.
3. All countries have risk-sensitive development planning and practice.
4. All countries have improved their governance and accountability for risk-resilient investment in both the public and private-sector.
5. Sub-regional and regional cooperation mechanisms established for trans-boundary DRR efforts.

7.4.4 Two-year Action plan (2017–18)

The two-year action plan is in line with the policy direction and contributes to the achievement of the milestones in the long-term road map. The action plan is derived from priorities shared by governments and stakeholders during the development of the Asia Regional Plan and in consultations through the IAP. The two-year action plan will be reviewed and updated in line with the biennial Asian Ministerial Conferences and through the IAP forum.

7.4.4.1 REGIONAL-LEVEL ACTIONS

These actions cover transboundary issues, regional cooperation mechanisms and so on. They will be implemented through the cooperation of countries and by regional and sub-regional actors such as inter-governmental organizations (IGOs); UN regional entities; other regional and international organizations; networks and consortiums; and stakeholders and partners. UNISDR, as the secretariat will have the role of facilitating and advocating for implementation of these actions in line with the Sendai Framework. The actions are as follows:

1. Strengthen the Regional Platform for DRR—that is, the Asian Ministerial Conference, the ISDR Asia Partnership and its secretariat—by means of enhanced high-level engagement of governments, intergovernmental organizations and partners to support the implementation and monitoring of the Sendai Framework.
2. Establish a collaborative mechanism of UNISDR, intergovernmental organizations (IGOs) and UNESCAP Regional Coordination Mechanism to support the implementation and monitoring of Sendai Framework in line with the SDGs.
3. Strengthen existing regional mechanisms to reduce the risk of and enhance early warning and preparedness for transboundary disasters.
4. Strengthen regional cooperation including through public-private partnership for the application of science, technology and innovation in policy making for DRR.
5. Promote and support gender-sensitive DRR actions at national and local level including universal access to sexual and reproductive health-care services, prevention and response to gender-based violence and women's leadership.
6. Strengthen the role of inter-governmental organizations (IGOs) and cooperation among them for information and knowledge exchange and concerted support to national actions.
7. Establish regional cooperation for disaster resilient infrastructure development in the region with an aim of "preventing future risks."
8. Strengthen regional cooperation between private-sector organizations and chambers of commerce to promote public private partnership for DRR.
9. Enhance stakeholder groups' engagement in the Sendai Framework by means of implementation and monitoring of their voluntary commitments.

10. Implement the "United Nations Plan of Action on Disaster Risk Reduction for Resilience" and integrate DRR into UN country level operations through regional UN Development Group processes.
11. Strengthen UNISDR to facilitate, coordinate and advocate for the implementation of the Asia Regional Plan by countries, partners, inter-governmental organizations, UN, international organizations and stakeholders.
12. Promote the provision of human, financial and technical resources by multilateral and bilateral donors to support actions from the Asia Regional Plan. Raise public awareness of Tsunami taking the opportunity of the World Tsunami Awareness Day on November 5.

7.4.4.2 NATIONAL AND LOCAL-LEVEL ACTIONS

The following national and local actions address the priorities of Asia in the next 2 years and are guided by the Sendai Framework. Bearing in mind that countries will develop or revise their national DRR strategies and plans in the coming years, this set of actions will complement such plans and provide guidance on the setting of priorities. National and local activities can be implemented by governments, stakeholders, and partners. The actions are as follows.

Priority 1

Understanding disaster risk

1. Establish/strengthen consistent and appropriate level of disaggregation for gender, age, and disability data disaster loss baseline data at the national level with local data.
2. Assess disaster risk, vulnerability, capacity, exposure, hazard characteristics, and their possible impacts (risk profiling).
3. Establish risk information systems and promote the collection, analysis, management, use, and dissemination of the relevant data and information, using space and in situ information, including (GIS, and information and communication technology and innovations.
4. Incentivize businesses to strengthen business continuity, and conduct location-specific risk assessments of their operating environment and supply chain with a clear understanding of their hazard exposure, vulnerability, and risks to hazards.
5. Integrate disaster risk information into business investment planning and management across private-sector entities, in particular micro, small and

---

(Continued)

(Continued)

medium enterprises (MSMEs), through business associations, chambers of commerce, and national and local platforms on DRR.

6. Mobilize public–private partnership initiatives for awareness raising, advocacy, and education to strengthen private-sector attitudes towards risk-sensitive development.

7. Consolidate/adapt guidelines for local risk assessments and their use in local development planning, ensuring the complementarities of proven traditional, indigenous, and local knowledge and practices, and science and technology.

8. Institutionalize efforts to build the knowledge and capacity of local and national government officials, civil society, communities, and volunteers to monitor hazards, exposure, and social vulnerabilities.

9. Raise public awareness of disasters by taking the opportunities provided by the International Day for Disaster Reduction and the World Tsunami Awareness Day.

Priority 2
Strengthening disaster risk governance to manage disaster risk

1. Review/revise existing national and local DRR strategies/plans in line with the Sendai Framework, Paris Agreement, SDGs and New Urban Agenda to ensure disaster risk-sensitive development.

2. Establish/strengthen multistakeholder and multisectoral national and local platforms that are gender responsive and inclusive, with the participation of local community representatives and other stakeholders.

3. Improve the legal, policy, and regulatory environment to incentivize businesses to reinforce DRR.

4. Build corporate governance and risk-sensitive investment beyond corporate social responsibility, through business associations, chambers of commerce, and national and local platforms on DRR.

5. Develop guidelines for coherent implementation of the 2030 development agendas at the local and national level involving the relevant national authority.

6. Promote implementation of health aspects of the Sendai Framework for Disaster Risk Reduction 2015–30, including the Bangkok Principles, with a view to ensuring more systematic cooperation, coherence, and integration between disaster and health risk management.

7. Undertake an inventory of available local DRR strategies/plans, and work towards achieving the Sendai Framework Target (e) on "risk-informed local DRR strategies/plans" by 2020.

8. Foster local leadership and forums such as the "Asian Local Disaster Risk and Resilience Forum."
9. Ensure women's full and effective participation and equal opportunities for leadership at all levels of decision-making in DRR.

Priority 3

Investing in DRR for resilience

1. Establish/strengthen regulatory frameworks to reinforce risk consideration in structural and nonstructural investments.
2. Promote national mechanisms for disaster risk transfer and insurance as appropriate.
3. Increase resilience and integrate DRR into business models, corporate practices, and supply chains through intensified partnerships between the private sector, local government, and other stakeholders.
4. Promote appropriate financial mechanisms to integrate DRR considerations and measures to support the building of disaster-resilient communities by engaging development funds, banks, private foundations, and stakeholders.
5. Allocate resources to local governments, as appropriate, to implement DRR measures.
6. Institutionalize community-based DRM (CBDRM) to strengthen resilience of households and communities.
7. Strengthen education on disaster and climate risk reduction, and accelerate the implementation of comprehensive school safety.
8. Invest in the development of resilient health systems, and design and implementation of inclusive policies to ensure access to social safety nets and primary healthcare services, including maternal, newborn and child health, and sexual and reproductive health.

Priority 4

Enhancing disaster preparedness for effective response and to "Build Back Better" in recovery, rehabilitation, and reconstruction

1. Prepare/update disaster preparedness and contingency plans at local, national, and regional levels with a multistakeholder and multisectoral approach, ensuring comprehensive and accessible service and referral mechanisms to promote specific needs of women and children, the elderly, people with disabilities, and other at-risk populations.
2. Strengthen multihazard early warning systems to ensure last-mile accessibility.

(Continued)

(*Continued*)

> 3. Integrate disaster risk concerns and measures in post-disaster recovery and reconstruction planning and practice, such as developing/revising building codes and standards in recovery and reconstruction practices at the national and local levels.
> 4. Expand private-sector engagement in preparedness for response, recovery, and to "Build Back Better" through more systematic public–private cooperation.
> 5. Develop and implement disaster management plans that would include early warning, response coordination, evacuation plans, and stockpiling of necessary materials to implement rescue and relief activities.
> 6. Build the capacity of local authorities, including strengthening leadership of women, persons with disability, and youth to develop local disaster recovery plans that include retrofitting, reconstruction, building code enforcement, post-trauma programs, resilient livelihoods, shelter provision for displaced population, etc.
> 7. Mainstream ecosystem-based approaches through transboundary cooperation to build resilience.

7.4.5 Implementation and Monitoring of the Asia Regional Plan

The Asian regional plan is intended to guide the implementation of the Sendai Framework at national and local levels. The application of the policy direction, pursuing the roadmap, and monitoring the 2-year action plan requires some level of dedicated support in terms of advocacy, monitoring, and the provision of technical support. In this regard, cooperation at the regional level through North-South, South-South, and triangular cooperation will be important, including through the exchange of experiences and home-grown approaches of countries in the region, such as the sufficiency economy philosophy of Thailand. The overall roles and responsibilities of international, regional, and subregional organizations and entities will be in line with the Sendai Framework (General Assembly Resolution A/RES/69/283). The IAP forum—under the stewardship of UNISDR's Regional Office for Asia Pacific and with the engagement of governments (via Sendai focal points), intergovernmental organizations, the UN, international organizations, civil society organizations and various other stakeholders—constitutes a regional technical support mechanism. Among its responsibilities, the regional technical support mechanism will periodically assess the progress of the regional plan; identify key gaps in DRR in the region from the national perspective, provide or mobilize technical expertise and guidance at the regional level, and advocate for accelerated implementation of the Sendai Framework. UNISDR is mandated to facilitate implementation, follow-up, and monitoring of the Sendai Framework. As part of its support to the regional technical support mechanism, UNISDR's Regional Office for Asia-Pacific requires strengthening with adequate resources and capacities. This may include resourcing through various funding mechanisms, including increased, timely, stable, and predictable contributions to the UN Trust Fund for Disaster Reduction as well as enhancing the role of the trust fund in relation to the implementation of the Sendai Framework. The following factors will support effective implementation of the Asia Regional Plan:

- Regional intergovernmental organizations playing a stronger role in the regional platform for DRR and enhanced regional cooperation to enable policy and planning for the implementation of transboundary risk assessments; mutual learning and exchange of good practices and information.
- The entities of the UN system—including the regional commission, funds and programs, and the specialized agencies—promoting and supporting the implementation of the regional plan through technical assistance upon request.
- UNISDR's Regional Office for Asia-Pacific providing a range of support, including (a) advocating for and facilitating the implementation, follow-up, and review of the Asia Regional Plan through partners and the IAP forum, including the AWG for AMCDRR; (b) supporting the implementation of the Sendai Framework monitoring system at the national level; (c) providing technical support to countries to establish disaster-loss baselines; (d) convening the Asian Ministerial Conference with governments, regional organizations, and partners; (e) mobilizing science and technical work for DRR through an enhanced role of its Asian Science and Technology Advisory Group in close cooperation with the Science and Technology Advisory Group of the UNISDR; and (f) facilitating engagement of the various stakeholder groups, including the private-sector, civil society organizations, and local actors in the implementation of the Asia Regional Plan.
- The various stakeholder groups providing support to States, in accordance with national policies, laws, and regulations, in the implementation of the Sendai Framework at all levels. This will build on their past DRR actions. Stakeholders have consistently provided and reviewed "Voluntary Statements of Action" at previous AMCDRRs. The Asia Regional Plan and all "Voluntary Statements of Action" from stakeholder groups will constitute two principal regional instruments for tracking progress.

- The periodic review of the Asia Regional Plan by governments and partners through the annual meetings of the IAP forum. There will be no separate monitoring system or mechanism specifically for the Asia Regional Plan. The progress will be reviewed through the overall monitoring of the Sendai Framework. The monitoring of the Sendai Framework will follow the internationally agreed set of indicators and monitoring system, expected to be agreed at the OEIWG in November 2016, and will be endorsed by the member States at the UN General Assembly. The Asian Ministerial Conferences from 2018 onwards will provide an opportunity for the regional plan to be reviewed and revised, based on the emerging priorities and needs. Experience sharing and peer learning among countries within the region and beyond will be promoted toward standardized data and best practices.

7.5 Illustrative List of Activities Identified as an Immediate Nature

7.5.1 Drinking Water Supply

1. Repair of damaged platforms of hand pumps/ring wells/spring-tapped chambers/public stand posts and cisterns
2. Restoration of damaged stand posts including replacement of damaged pipe lengths with new pipe lengths and cleaning of clear water reservoir (to make it leak proof)
3. Repair of damaged pumping machines, leaking overhead reservoirs, and water pumps including damaged intake—structures and approach gantries/jetties

7.5.2 Roads

1. Filling up of breaches and potholes, use of pipe for creating waterways, and repair and stone pitching of embankments
2. Repair of breached culverts
3. Providing diversions to the damaged/washed out portions of bridges to restore immediate connectivity

4. Temporary repair of approaches to bridges/embankments of bridges, repair of damaged railing bridges, repair of causeways to restore immediate connectivity, and repair of damaged stretch of roads to restore traffic

7.5.3 Irrigation

1. Immediate repair of damaged canal structures and earthen/masonry works of tanks and small reservoirs with the use of cement, sand bags, and stones
2. Repair of weak areas such as piping or rat holes in dam walls/embankments
3. Removal of vegetative material/building material/debris from canal and drainage system

7.5.4 Health

Repair of damaged approach roads, buildings and electrical lines of public health centers (PHCs)/community health centers.

7.5.5 Community Assets of Panchayat

1. Repair of village internal roads
2. Removal of debris from drainage/sewerage lines
3. Repair of internal water supply lines
4. Repair of street lights
5. Temporary repair of primary schools, Panchayatghars, community halls, anganwadi, etc.

7.6 Mainstreaming Disaster Risk Reduction: Tools and Techniques

Realizing the regional imperatives, national needs, and prospects of mainstreaming, the following mechanisms and instruments need to be developed and further strengthened while formulating plan of action for mainstreaming DRR in development.

7.6.1 Identify Development-induced Disasters

It is a well-known fact that inappropriate development processes are contributing to risk accumulation. There are many examples demonstrating how economic growth and social improvement lead to increase in disaster risk. Rapid urbanization is an example. The growth of informal settlements and inner city slums, whether fueled by international migration or internal migration from smaller urban settlements or the countryside, has led to the growth of unstable living environments. These settlements are often located in ravines, on steep slopes, along flood plains, or adjacent to noxious or dangerous industrial or transport facilities. One such development has led to an increase in risk due to landslides in urban areas of Chittagong city in Bangladesh. This is true in other megacities as well and in rapidly expanding small- and medium-sized urban centers. When the population expands faster than the capacity of the urban authorities or the private sector to supply housing or basic infrastructure, risk in informal settlements can accumulate quickly. In cities with transient or migrant populations, social and economic networks tend to be loose. Many people, especially minorities or groups of low social status, can become socially excluded and politically marginalized, leading to a lack of access to resources and increased vulnerability.

7.6.2 Develop Guidelines on Mainstreaming

All development projects should have mandatory guideline to address how exactly it is going to implement DRR in terms of social and physical vulnerability. Risk can be reduced by making efforts wherein either the vulnerability or the exposure is reduced. Risk can also be reduced by reducing the hazard probability, for example, while undertaking a road construction in hilly area, the slope stability measures can be built in such a way that the hazard probability can decrease, thus, reducing the overall risk. Similarly, the poverty alleviation or education program can also reduce the social vulnerability, thus, reducing overall disaster risk. Similarly, limiting development is high-risk area, it is possible to reduce exposure and, thus, overall risk is reduced.

7.6.3 Develop Sector-specific Guidelines on Mainstreaming

It is necessary that an appropriate strategy is developed to mainstream DRR into the following specific sectors with clear-cut guidelines and objectives. Some of the suggestive sectoral guidelines could be as under:

Infrastructure: Public Works, Roads, and Construction

- Promote use of hazard risk information in land-use planning and zoning regulations.
- Conduct disaster risk impact assessments as part of the planning process before the construction of new roads or bridges.
- Encourage use of hazard-resilient designs (e.g., flood proofing or seismic safety) in rural housing programs in hazard-prone areas.
- Promote utilization of national building codes that have special provisions for enhanced design standards for buildings in areas affected by natural disasters.
- Ensure compliance and enforcement of local building laws requiring prescribed standards under natural building codes in urban, hazard-prone areas.

Health

- To promote programs to identify hospitals and health facilities that are located in hazard-prone areas, analyze their internal and external vulnerability during emergencies, and increase the hazard resilience of these hospitals through "Safe Hospital" programs.
- To prepare and implement a hospital preparedness plan for all such health facilities.

Agriculture

- To promote effective programs of contingency crop planning to deal with year-to-year climate variations.
- To promote effective programs of crop diversification, including the use of hazard resistant crops, to deal with shifts in climate patterns.
- To ensure sustainable livelihoods in areas of recurrent climate risks (i.e., arid and semi-arid zones, and flood and cyclone-prone areas) by promoting supplementary income generation from off-farm (e.g., animal husbandry) and nonfarm activities (e.g., handicrafts).

- To promote effective insurance and credit schemes to compensate for agricultural-related damage and losses to livelihoods due to natural hazards.

Education

- To incorporate DRR modules into the school curriculum.
- To construct all new schools located in hazard-prone areas to higher standards of hazard resilience as has been attempted in Kashmir and Bhuj region under the "Safe School" program.
- To add features in schools in hazard-prone areas for use as emergency shelters such as facilities for water, sanitation, and cooking as envisaged in coastal areas as possible cyclone shelters.

Financial Services

- To incorporate provisions in microfinancing schemes to have flexible repayments schedules that can be activated in the event of recipients being affected by natural disasters.
- To encourage the financial services sectors and local capital markets to develop schemes for financing DRR measures.

7.6.4 Carry out Cross-sectoral Risk Analysis

Cross-sectoral risk analysis needs to be carried out at national, local, and regional levels. Ongoing schemes across the sectors should be critically revisited and, wherever possible, the development aspects of these schemes should be integrated for a better result. This should be done in a futuristic mode with immediate medium- and long-term planning. For example, if a hydroelectric project is being implemented, attempts must be made to assess the change in the hydrological regime and its impact on soil erosion and landsliding. This would require a multidisciplinary approach across sectors

7.6.5 Develop Area-specific Guidelines on Mainstreaming

Area-specific guidelines for mainstreaming DRR in development should be formulated with particular reference to coastal and hilly areas that are prone to disasters.

7.6.5.1 COASTAL ZONE MANAGEMENT

Coastal zone management would be critical for environment, natural resources, climate change adaptation, and DRR as well. It would then lead to a holistic development of the coastal zones in the region, which cater to a significant population of South Asia, the majority of whom are poor and vulnerable to any type of disaster. Therefore, in any coastal zone management effort, DRR with respect to multiple hazards must be considered.

7.6.5.2 HILLY AREA DEVELOPMENT

As South Asia encompasses large tracts of hilly area, it is important to use all developmental initiatives specific to hilly areas to implement the DRR strategy, which is very critical for environmental protection and sustainable development. It would then lead to a holistic development of the hilly area and its population, the majority of whom is poor and vulnerable to disaster and are often isolated from the mainland of development.

7.6.6 Create Techno-legal Regime for Mainstreaming

It is necessary that appropriate technolegal mechanism is developed to implement the regulations made with respect to the DRR strategy. There may be a statutory organization responsible for undertaking assessment on compliance and implementation on the ground. For example, the hydroprojects have a mandatory provision of afforestation, and it is imperative that it is implemented on the ground and proper assessment is done with respect to its positive impact.

7.6.7 Conduct Disaster Impact Assessment

The assessment of the potential risks to any place (village, city, nation, etc.) or element (infrastructure/land use, etc.) is a major part of DIA related to any developmental activity. Therefore, it is necessary to consider all possible impacts of various hazards that may arise due to the implementation of a project. This entire exercise could be very complex and may require comprehensive assessment of data related to natural as well as social sectors. Some elements of the DIA are similar to well-known practice of EIA and, therefore, it must be pursued under similar guidelines.

7.6.8 Private–Public Partnership

. In the present scenario, it is visualized that more and more unorganized and organized private sectors would play a major role in developmental activities. It is important to foster collaboration with private sector in a public–private partnership to address the implementation of DRR in development initiatives. This partnership could play a key role in communication, infrastructure, market, health, and many others areas. Recently, a leading software industry in Hyderabad, India, has demonstrated a disaster response system for the citizens of the city, which is operational 24/7 and is fully endorsed by the government.

7.6.9 Research and Development

It is one of the major elements of mainstreaming disaster mitigation/reduction into development. R&D capacity in earthquakes, floods, droughts, climate change, industrial, nuclear disasters, and many other fields must identify areas and strategies how to identify risk at early stage in a holistic manner and minimize it by suitably integrating mitigation measures in to development model. Various professional scientific organizations must re-orient their program to support the safe developmental needs. For example, the road development agencies must take into account the present requirement of mass transport and suggest a suitable infrastructure which is viable and environmentally sustainable.

7.6.10 Awareness Generation, Training, and Capacity-Building

It is important to make all stakeholders aware about the coupling of disaster and development. It must be understood and communicated that there exists a mechanism by which development can be implemented with DRR provisions. This awareness will lead to public demand for disaster audit and, in turn, will ensure sustainable development. It is important to note that awareness development must be initiated at all levels starting from school curricula to basic training in safe construction to advance project management. Capacity-building through education, training, and mid-career intervention using on-campus as well as off-campus model must be implemented for quickly covering large manpower base. Building on capacities that deal with existing disaster risk is an effective way to generate capacity to deal with future risk arising out of new context, which is often not visualized.

7.6.11 Recognition of Best Efforts

Recognition of efforts is one of the best incentives that promotes and attracts many to emulate good practices in implementing DRR in development. It also acts as stimulant for the recipients to carry on the good work and innovate ways in which the efforts will have far reaching results across the society. Numerous such examples can be cited from drought management and poverty alleviation programs that are being implemented in western part of India and have received international accolades.

South Asian countries have just made a beginning toward mainstreaming DRR in development. Pursing the HFA in the respective countries has led to some "foundation"-level initiatives, which would facilitate more specific national- and local-level activities on mainstreaming DRR. These include:

- development of a legislative framework and institutional capacity to prevent, mitigate, prepare, and manage hazards and disasters;
- undertaking hazard, risk, and vulnerability assessments;
- development of education, training, and information exchange programs;
- raising awareness of the community;
- development of partnerships with the stakeholders at each level; and
- the utilization of cooperative and information-sharing mechanisms and institutions across the region.

While there are efforts in South Asian countries to implement DRR in development through National Adaptation Plans of Action (NAPAs), their integration to DRR need specific priority. In order to address adaptation concerns as part of their national development plans, the explicit focus on disaster risk is seen only in few cases. For example, the Safe Island program of Maldives is an integrated effort on addressing vulnerability through strategic planning for climate change adaptation. Similarly, Bhutan has initiated plan of action in this direction through NAPA. It is expected that all member countries develop respective NAPAs with an aim to mainstream DRR in development.

Specific entry point activities for mainstreaming DRR in development have been taken up in the multihazard-prone regions.

7.6.12 Entry Points for Mainstreaming Disaster Risk Reduction in Development

Education: The building of appropriate school structures that not only adhere to safety measures but may also be useful as disaster shelters and the development of curricula and institutionalization of safety drills that provide information on DRR, particularly targeting women and children.

Health: Ensuring suitability of health infrastructure, compliance with building codes, availability of and accessibility to goods and services especially in times of emergency, and increased capacity to prepare for disaster events and the outbreak of infectious diseases.

Environment: Integrating disaster risk concerns into existing environmental assessment tools and planning mechanisms (environmental impact assessments and strategic environmental assessments), promoting greater compliance to existing environmental and risk management regulations, promoting integrated approaches to spatial planning, strengthening capacities to protect ecosystem services that reduce disaster risk (wetlands, coastal forests, watersheds, coral reefs, etc.), identifying potential sources of hazardous materials that can trigger acute environmental emergencies, and strategically assessing the environmental impacts of proposed post-disaster recovery plans.

Governance: Efficiency and accountability of governance structures at central and local levels should be strengthened, encouraging more inclusive and participatory decision-making processes. Local and national governments design and apply regulatory frameworks that ensure a safer environment, reduce structural vulnerabilities, and guide social behavior and economic decisions towards risk reduction and disaster prevention.

Employment and Livelihoods (including informal sector): Considering the possible impacts of disasters on livelihoods and jobs, particularly those affecting the informal sector and youth. Promote innovative mechanism to reduce underlying risk such as microfinance and risk-transfer schemes, especially targeting women. Promote greater compliance to existing workplace safety regulations and environmental standards, and raise awareness of DRR measures in relevant sectors (e.g., engineers/construction sector, chemical industry, etc.).

Agriculture: Increasing agricultural productivity through investments in soil health, water management, extension services, and research increases food availability for subsistence farmers. However, special focus is needed to mitigate the impact of hydrometeorological fluctuations through multiple cropping, water conservation, and biological control measures, with contingency cropping strategies linked to weather monitoring and early warning systems.

Gender: Improved women's participation in decision-making processes and productive activities should specifically include awareness of disaster risks, preparedness, and preventive measures that reinforce traditional coping measures undertaken by women and increase disaster resilience of communities. Research on the degree to which women suffer the negative impact of disasters could be undertaken, to better understand and address their specific vulnerabilities and needs.

Information and communication technologies: Steps to strengthen science advisory mechanisms, invest in higher education and research, promote private-sector development, and improve access to communications technologies can also be linked to better hydrometeorological monitoring, seismic risks monitoring, and the possibility of feeding into better early warning systems to save both lives and livelihoods.

7.7 Road Ahead for Disaster Management in Asia-Pacific Countries

The occurrence of natural disasters could not be stopped, but mitigation and preparedness will certainly reduce the fury of disaster. The following parameters need to be addressed for the futuristic approach on preventive and effective disaster management:

- New hazards and aggravated old hazards.
- Greater hazard frequency and intensity.
- Escalating costs of relief and rehabilitation during and after disasters.

- Growing societal vulnerability.
- Paradigm shift in the government approach from reactive management to proactive management.
- More importance from response management to hazard, risk, and vulnerability assessment and mitigation.
- Focus on adjustment of major conceptual revolution instead of making minor corrections in the acts and laws.

7.7.1 How Disaster Can Be Managed?

The chronological events or phases of managing a disaster after a disaster strikes are called a disaster cycle. The events or phases are

1. emergency response;
2. rehabilitation;
3. reconstruction;
4. prevention measures;
5. development;
6. disaster mitigation measures while developing;
7. disaster preparedness runs concurrently; and
8. warning.

Disaster strikes again, and we have to go back to step 1 (i.e., emergency response) and follow the cycle again. Now the question is, if this disaster cycle never ends, then how do we manage the situations?

As a concept, a disaster cycle is a never-ending cycle and we need to learn to live with it. The more we reduce the disaster impact, the easier and less costlier it will be to work on the eight components of disaster management.

In this context the points to be remembered are:

- Next disaster is on the way.
- No time is left for rest and complacency.
- Preparedness has to be round the clock and round the year.

Chapter 8

Response, Relief, Reconstruction, Rehabilitation, and Resilience

8.1 Disaster Response Mechanism and Challenges

Traditionally, people think of disaster management only in terms of the emergency relief period and the post-disaster rehabilitation. Successful disaster management planning must encompass the situation that occurs before, during, and after disasters. It has been experienced the world over that a prompt, well-coordinated, and effective response mounted in the aftermath of disasters not only minimizes loss of life and property but also facilitates early recovery. The important ingredients of an effective response system are integrated institutional arrangements, state-of-the-art forecasting and early warning systems, failsafe communication system, rapid evacuation of threatened communities, quick deployment of specialized response forces, and coordination and synergy among various agencies at various levels in dealing with any disaster. Most importantly, all the agencies and their functionaries must clearly understand their roles and responsibilities and the specific actions they have to take for responding to disasters or threatening disaster situations.

> Life on Earth is at the ever-increasing risk of being wiped out by a disaster, such as sudden global nuclear war, a genetically engineered virus or other dangers we have not yet thought of.
>
> —*Stephen Hawking*

There is a need to develop Standard Operating Procedure (SOP) as an institutional framework to deal with natural disasters like flood, cyclone, earthquake, landslides, avalanche, etc. at different levels of ministries concerned with these events. SOP should lay down, in a comprehensive manner, the specific actions required to be taken by various ministries and departments of national, provincial, and local government for responding to natural disasters of any magnitude and dimension.

The objectives of the SOP are to cater to the need of disaster response at various levels and to provide a concise and convenient form, a list of major executive actions involved in responding to natural disasters, and the necessary measures for preparedness, response, and relief required to be taken. The SOP also ensure that all concerned ministries, departments and organizations of the national, provincial (state), and local (district/subdistrict/municipality) Administrations know the precise measures required of them at each stage of the process, and all actions are closely and continuously coordinated. The instructions contained in this SOP should not be regarded as exhaustive of all the actions that might be considered necessary.

The five phases of disaster management for effective and efficient response to natural disasters are:

1. *Preparedness phase:* This phase will include taking all necessary measures for planning, capacity-building, and other preparedness so as to be in a state of readiness to respond in the event of a natural disaster. This stage will also include development of SAR teams, mobilization of resources, and taking measures in terms of equipping, providing training, conducting mock drills/exercises, etc.

2. *Early warning phase:* This phase will include all necessary measures to provide timely, qualitative, and quantitative warnings to the disaster managers to enable them to take preemptive measures for preventing loss of life and reducing loss/damage to the property. On the occurrence of a natural disaster or imminent threat thereof, all the concerned agencies will be informed/notified for initiating immediate necessary follow-up actions.

3. *Response phase:* This phase will include all necessary measures to provide immediate succor to the affected people by undertaking SAR, and evacuation measures.

4. *Relief phase:* This phase will include all necessary measures to provide immediate relief and succor to the affected people in terms of their essential needs of food, drinking water, health and hygiene, clothing, and shelter.

5. *Restoration stage:* This phase will include all necessary measures to stabilize the situation and restore the utilities.

8.2 Emergency Support Functions

Disaster response is a multi-agency function. The term was primarily coined and used by the Government of India. As per the emergency support functions (ESFs), there will be one lead or primary agency which will be responsible for managing and coordinating the response while other agencies will support and provide assistance in managing the incident. Each ESF will be headed by a lead ministry/organization responsible for coordinating the delivery of goods and services to the disaster area, and is supported by numerous other organizations. These ESFs will form integral part of the emergency operation centers (EOCs), and each ESF should coordinate its activities from the allocated EOC. Extension teams and workers of each ESF will be required to coordinate the response procedures at the disaster-affected site.

8.2.1 ESF Plan at National Level

An ESF plan will be prepared at the national level clearly indicating the areas of responsibility of each of the agency concerned that will provide mutual assistance in terms of resources, equipment, manpower, etc. during disasters. The name, address, and telephone numbers of NOs of the department/agencies will also be included in the plan. The ESF plans will be reviewed from time to time.

In the immediate aftermath of a major natural disaster requiring Central Government assistance to states/provinces, the responsible national agency (Ministry of Home Affairs in case of India, Ministry of Interior in case of Thailand) will activate the ESF plan to identify requirements and mobilize and deploy resources to the affected area to assist the state/province in its response actions.

Each ESF is headed by a primary agency, which has been selected based on its authorities, resources, and capabilities to support the functional area. The designated primary agency, acting as the lead agency, will be assisted by one or more support agencies (secondary agencies) and will be responsible for managing the activities of the ESF and assisting the state/provinces in the rescue and relief activities in the state/province.

8.3 Incident Command System

The incident command system (ICS) was originally developed through a cooperative effort among a number of federal, state, and local governmental agencies in response to the harmful disorder that occurred among various organizations (e.g., municipal and county fire departments, the California Department of Forestry, and state and federal governments) attempting to suppress massive wild fires in California during the 1970s. A task force investigating these incidents identified a number of recurring problems that suggested responding organizations lacked sufficient means to effectively coordinate activity in large, complex, and dynamic emergency situations. Examples of major deficiencies included a basic inability to adjust (e.g., expand or contract) organizations to accommodate shifting situational demands, nonstandard terminology, and communication procedures among responding agencies, and problematic action planning protocols at emergency scenes. To address these types of issues, the ICS approach turned out to be a major departure from previous large-scale emergency management methods. Although ICS was initially developed in response to problems associated with wildland fire fighting, it evolved into an all-risk system supposedly suitable for almost any type of emergency (e.g., natural disasters, riots, and terrorist attacks) and for emergencies of nearly any size (ranging from a minor incident involving a single unit, such as a fire engine company, to a major event involving numerous agencies). Consequently, the use of fundamental ICS principles has expanded rapidly. The ICS is also a cornerstone of the Federal Emergency Management Agency's Integrated Emergency Management System (IEMS). The IEMS has the objective of developing and maintaining a credible, nationwide emergency management capability involving all levels of the government and all types of hazards.

8.3.1 Genesis of ICS in India

The Government of India carried out a review of the emergency management/disaster management system in the country and found the strength of our system lies in the fact that there are designated coordinators at the district (district magistrate/collector), division (divisional commissioner), and the state (chief secretary) levels. These coordinators play a crucial role in disaster response management, and it is desirable that their response capability be continuously strengthened. Therefore, the home ministry at the center proposed a remedy to the existing weakness in the system by ensuring that the designated controlling/responsible authorities at different levels are backed by trained incident management teams, with the team members trained in different facets of emergency/disaster management.

In order to address the requirements of a professional on-scene response management team, it was proposed to institutionalize the training initiatives related to the incident management system in the country. A program for this has been taken up in conjunction with the US Forest Services (USFS). The program was aimed to build teams numbering 2–3 at the district level and 3–4 at the state level for responding to complex/critical disasters. Therefore, it was proposed to professionalize the emergency/disaster response management system in the country by the adoption of the flexible emergency response management system modeled along the line of the ICS presently being used in the USA. The system was discussed in the meeting of the relief commissioners and has been modified to suit Indian conditions.

8.3.2 ICS Structure

The ICS is an on-scene, all-risk, flexible modular system adaptable to any scale of natural as well as man-made emergency/incidents. The ICS seeks to strengthen the existing disaster response management system by ensuring that the designated controlling/responsible authorities at different levels are backed by trained incident command teams (ICTs) whose members have been trained in the different facets of emergency/disaster response management. The ICS will not put in place any new hierarchy or supplant the existing system but will only reinforce it. The members of the ICT will be jointly trained for deployment as a team. When an ICT is deployed for an incident, all concerned agencies of the government will respond as per the assessment of the team. This system, therefore, enables proper coordination amongst the different agencies of the government. The five command functions in the ICS are as follows:

1. *Incident command:* Has overall responsibility at the incident. Determines objectives and establishes priorities based on the nature of the incident, available resources, and agency policy.
2. *Operation:* Develops tactical organization and directs all resources to carry out the incident action plan.
3. *Planning:* Develops the incident action plan to accomplish the objectives. Collects and evaluates information, and maintains status of assigned resources.

4. *Logistics:* Provides resources and all other services needed to support the organization.
5. *Finance/administration:* Monitors costs related to the incident, provides accounting, procurement, time recording, cost analysis, and overall fiscal guidance.

The basic structure of a fully elaborated ICS may be appropriate for large-scale emergencies or disasters, such as the 1999 Orissa Super Cyclone, 2001 Bhuj Earthquake, and 2005 Tsunami shown in Figure 8.1 that defines major ICS elements. Jobs within the ICS are specialized, based on standardized routines, and require specific training. Positions are arranged hierarchically and are related to one another on the basis of formal authority. Basic system objectives and plans are established at or near the top of the hierarchy and are used as bases for decisions and behaviors at lower levels. In general, the ICS is constructed around five major functions: command, planning, operations, logistics, and finance/administration. These building blocks apply in both routine and nonroutine situations and for ICS structures of all sizes. According to ICS logic, even when the system is very small, involving as few as two individuals, all five elements are likely to be relevant to some extent. When the system is small, however, one person may be able to manage the five elements together. Large-scale incidents usually require that components be relegated to their own "sections" where different individuals can manage each one separately. In addition, the sections may be divided into smaller functions as needed.

The *incident commander* is the highest-ranking position within the ICS. The person occupying this position is ultimately responsible for all activities that take place at an incident, including the development and implementation of strategic decisions and the ordering and releasing of resources. The incident commander is the one functional position that is always filled. Up to four sections report directly to the incident commander. At the section level, the person in charge is designated as chief. The *operations section* is responsible for the development and execution of all tactical operations directly related to the primary goals and objectives of the ICS. The operations chief activates and supervises resources in accordance with the action plans developed by the planning section. The *planning section* develops the action plan to accomplish the system's objectives. It collects, evaluates, and disseminates information about the development of the incident and status of the resources. Information is needed to understand the situation, predict probable courses of events, prepare alternative strategies, and control operations. The *logistics section* provides facilities and services to support

Figure 8.1:
Elaborated ICS Organization Structure

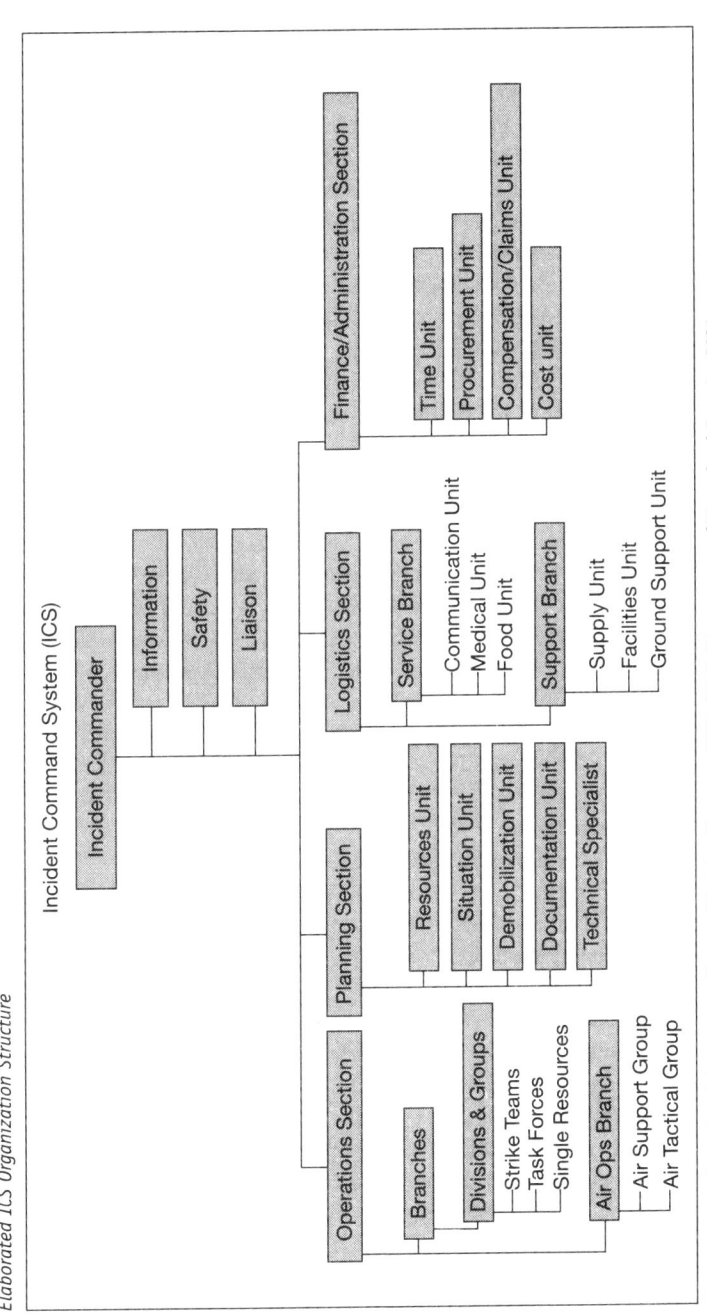

Source: National Incident Management System, Training Program. 2011. U. S. Department of Homeland Security, USA.

ICS personnel. The *finance/administration section* monitors costs related to the incident and provides accounting, procurement, and cost analysis.

In addition to the primary incident response activities of operations, planning, logistics, and finance/administration, the incident commander has responsibility for several other important services. Again, depending on the size and type of event, the incident commander may find it necessary to delegate authority for performing these functions to other individuals. These people constitute the command staff and are called officers. The *information officer* is responsible for the formulation and release of information about the incident to the news media and other agencies seeking information about the incident. The *safety officer* monitors and assesses safety conditions and develops measures to insure the safety of all personnel. The *liaison officer* is the point of contact for the assisting and cooperating agencies.

The ICS will comprise of two broad components, namely, *incident response* and *system institutionalization*. Incident response will involve three elements: (a) coordination, (b) ICT, and (c) specialized tactical resources. The coordination element will have the headquarters teams at the national, state, and district levels. The ICTs will be responsible for on-scene management and are formed at the district and state levels. As shown in the organizational chart, the following will be the eight core positions in the ICT:

1. incident commander;
2. operations section;
3. planning section;
4. logistics section;
5. FINANCE/administration section;
6. safety officer;
7. liaison officer; and
8. information officer.

The specialized tactical resources are being formed at the national or state levels having technical expertise in areas such as handling hazardous material, carrying out SAR, running field hospitals, etc. Such resources will be integrated into the incident.

ICS provides the management backbone with a solution to most of the problems being faced by the existing disaster response system in India. Safety, effectiveness, and efficiency may be achieved where a seamless integration of agencies is possible at an emergency situation. ICS provides the model for command, control, and coordination of an emergency response toward the common goal of stabilizing an incident and protecting life, property, and environment. The Government of India has already taken several initiatives to reinforce the existing emergency response mechanism with the adaptation of ICS and capacity-building. It is expected that the adaptation and assimilation of ICS in the response mechanism system of the country will reduce the risk of agency overlap and potential confusion at an emergency situation because of inadequate understanding and coordination.

8.4 Mock Drills

SAR teams at the national and state levels should carry out mock drills on various disasters situation annually. For floods and cyclones, these should be carried before the monsoon and cyclone period, tentatively in March and September for cyclones and June for floods. For earthquakes, landslides, etc., such drills can be done in the month of March itself.

At the subnational and state levels, mock exercises will be carried out for testing the effectiveness of all the preparedness machinery including manpower and equipment.

8.5 Resource Inventory: IDRN

Government of India has launched India Disaster Resource Network (IDRN), which is a web-enabled resource inventory for disaster management. The state governments will ensure that necessary entries have been made in the web portal and updated at least once in a month by the designated district authorities.

8.6 Implementation Challenges at the National and State Level in Indian Context

As most of the action takes place in the districts, preparation of comprehensive disaster management plans at the district level is given top priority. Pre-disaster planning is crucial for ensuring an efficient response at the time of a disaster. A well-planned and well-rehearsed response system can deal with the exigencies of calamities and also put up a resilient coping mechanism. Optimal utilization of scarce resources for rescue, relief, and rehabilitation during times of crisis is possible only with detailed planning and preparation. The District Magistrate/District Collector is the NO of emergency response in the event of a disaster at the district level. Preparation of district-level disaster management plans (DDMP) is imperative in order to provide a framework for disaster managers and district administration to prepare for and ultimately respond to a disaster event. Each DDMP needs to be prepared on the basis of the vulnerability of the district to various disasters and the resources available. Moreover, while preparing the DDMP, the structure identified should permit easy and quick retrieval of relevant information on which the authority/individual may have to act upon.

The DDMP is a multihazard response plan that assists and equips the district administration to organize its emergency preparedness, response, and mitigation functions in a timely and efficient manner within the district, and extend the necessary support to the state and Central Government. It is a plan that focuses on operations and defines the characterization of responder agencies within the district, from within and outside the government.

The DDMP establishes a structure for a systematic, coordinated, and effective response at the district level. The purpose of the plan is to:

- Define a system of coordination at the district level.
- Identify all the responder agencies at the district level and assign functional responsibilities to each of them.
- Establish a central facility in the district which enables all the responder agencies to interact and coordinate their efforts.
- Suggest hazard specific preparedness, response, and mitigation measures.
- Plan resource requirements and coordinate with the state government for requisitioning more resources.
- Provide an inventory of resources, key facilities, and addresses for deployment and assistance toward preparedness and mitigation.

2015 Nepal Earthquake

History of Earthquakes in Nepal

The history of earthquakes in Nepal was first recorded in 1255 AD with a magnitude (M) of 7.7 on the Richter scale, which took the life of then King Abhaya

Malla in Kathmandu. Since then, many earthquakes have been recorded. Recently, Udayapur district was severely affected by a 6.6M earthquake in 1988, killing 721 people and destroyed 64,467 houses. All of these earthquakes had a devastating effect on the lives and properties of people living in Nepal (DPNET 2015). Nepal faces two deadly earthquakes of which the magnitude 7.5–8.0 occur every 40 years and magnitude above 8.0 in Richter scale occur every 80 years approximately (DPNET 2015). According to the National Seismological Center (NSC) of Nepal 2012, a seismic gap lies in the western part of Nepal where the occurrence of 8.0M or higher was predicted. NSC further explained that no major earthquakes greater than 8.0 M has occurred between the Kangra earthquake of 1905 and the Bihar earthquake of 1934 AD.

Gorkha Earthquake 2015

The Gorkha or Nepal earthquake of 7.8M that occurred on April 25, 2015, at 11:56 NST, having an epicenter 77 km northwest of Kathmandu and a depth of 15 km, caused lot of devastation in Nepal. The earthquake of April 25, 2015 had its epicenter in Gorkha district in nearby Barpark village which left the self-sustaining village completely destroyed, with aftershocks felt about 130 km away (Table 8.1 and Figure 8.2).

Table 8.1:
Details of Gorkha Earthquake

Gorkha Earthquake	
Date of occurrence and epicenter	April 25, 2015—7.8M (Gorkha district)
Big Aftershocks	April 26, 2015—6.9M (Sindhupalchowk district); May 12, 2015—7.2M (Dolakha district)
Total deaths	8,844: 3,944 male; 4,894 female
Total injured	22,309
Total building damage	
Completely destroyed government houses	2,673
Partially destroyed government houses	3,757
Completely destroyed public houses	602,257
Partially destroyed public houses	285,099
Most Affected districts (14)	Gorkha, Kathmandu, Lalitpur, Bhaktapur, Sindhupalchowk, Rasuwa, Ramechhap, Nuwakot, Dolakha, Dhading, Kavre, Makawanpur, Sindhuli, and Okhaldhunga
Total aftershocks	376 (till June 2015)

Source: GON (2015).

(*Continued*)

(Continued)

Till June 24, 2015, about 8,832 people lost their lives and 22,309 people were injured. It was a powerful earthquake after 1934's Nepal–Bihar earthquake of 8.3M and felt as far as India, Tibet, and Bangladesh. The recent earthquake affected about 2.8 million population within the country (OCHA 2015).

Figure 8.2:
Gorkha Earthquake and Most Affected Districts

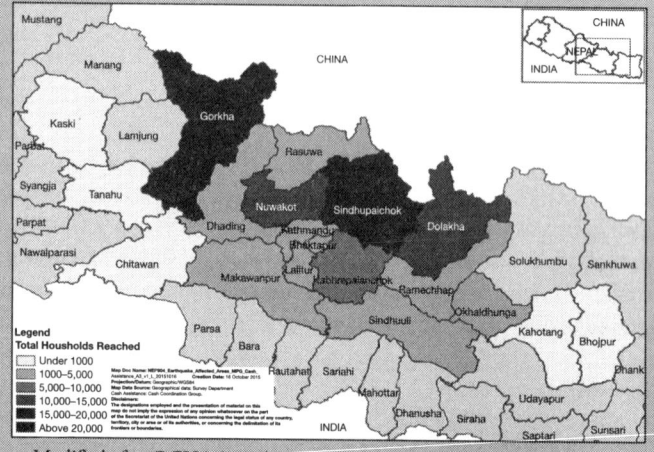

Source: Modified after OCHA (2015).

Immediately after the April 25 earthquake, the government declared an emergency and requested for international assistance. About 300 different agencies reached Nepal to provide immediate response. Out of the 75 districts, 31 were reported to be affected by the earthquake in central and western region of Nepal, whereas 14 out of 75 districts were severely affected by the earthquake in terms of rescue and relief priority (GON 2015). After some survey, the government announced that 500,717 houses were completely destroyed and 269,190 houses partially destroyed till May 25, 2015. Over 860,000 people needed immediate assistance to shelter before the monsoon period. On May 12, another aftershock of 7.2M occurred with the epicenter at Dolakha district which affected the eastern part of the country. About 150 people were killed, and further collapse of buildings occurred. Aftershocks were continuously felt even after 3 months of the disaster and it is still going on. Different agencies have been looking after their respective areas to provide much-needed assistance and relief in the affected areas. On May 4, about US$423 million was appealed for relief operation in Nepal (OCHA 2015).

The recent earthquake has befallen about 200 km west of the 1934 earthquake. Huge structural and nonstructural damages to buildings has been caused with loss of cultural heritage monuments and other secondary hazards such as landslides, flash floods, and avalanches in mountainous areas. About 17 people have been reported to be killed by an avalanche in Mount Everest. The mass movement deformed the land in different parts of the rupture zone, including Kathmandu. Within Kathmandu Valley, temples and old structures along with some engineered and nonengineered structures were completely flattened to ground, leaving huge piles of debris and dust. The intensity of the earthquake recorded on MMI scale in Kathmandu and Gorkha district was VIII and IX-X respectively, whereas in the epicentre it was X (Okamura, 2015).

8.7 Relief

In the aftermath of disaster, the affected people must be looked after for their safety, security, and the well-being and provided food, water, shelter, clothing, medical care, etc. so as to ensure that the affected people live with dignity. State/provincial governments shall be responsible for providing prompt and adequate relief assistance to the victims of disasters. The minimum standards of relief need to followed as laid down by the national and provincial level responsible agencies (in case of India its NDMA at the national level and SDMA at the state level. Some areas of concern need to be addressed in disaster-affected areas to ensure the well-being and help community to come back to normalcy.

8.7.1 Food and Nutrition

People affected by disasters may be deprived of food and, therefore, food aid shall be provided to sustain life. The following measures shall be taken:

- Where necessary, free distributions of food shall be made to those who need the food the most.
- The food distribution will be discontinued as soon as possible.
- Wherever possible, dry rations shall be provided for home cooking.
- Community kitchens for mass feeding shall be organized only for an initial short period following a major disaster, particularly where affected people do not have the means to cook.
- While providing food assistance, local food practices shall be kept in mind, and commodities being provided must be carefully chosen in consultation with the affected population.
- Foods must be of good quality, safe to consume, and appropriate and acceptable to its recipients.

- Rations for general food distributions shall be adopted to bridge the gap between the affected population's requirements and their own food resources.
- Food distributed should be of appropriate quality and fit for human consumption.
- Food should be stored, prepared, and consumed in a safe and appropriate manner at both household and community levels.
- Food should be distributed in a responsive, transparent, and equitable manner.
- NGOs, community-based organizations (CBOs), and other social organizations should be involved for supplementing the efforts of the government.
- The nutritional needs of the population should be met, and malnutrition and micronutrient deficiencies of identified at risk groups should be addressed.

8.7.2 Water

The water supply is invariably affected in natural disasters. Safe drinking water might not be available particularly in hydrometeorological disasters. The following measures shall be taken by the state governments/local administration:

- The local governments shall identify alternative sources of water and make necessary arrangements for supply to the affected population.
- The state governments shall ensure that affected people have adequate facilities and supplies to collect, store and use sufficient quantities of water for drinking, cooking and personal hygiene.
- It shall be ensured that drinking water supplied conforms to the prescribed quality standards
- It shall be ensured that water made available for personal and domestic hygiene should not cause any risk to health.

8.7.3 Health

During post-disaster phase, many factors increase the risk of diseases and epidemics. These include poverty, insecurity, overcrowding, inadequate quantity and quality of water, poor environmental and sanitary conditions, inadequate shelter, and food supply.

8.7.3.1 MEDICAL RESPONSE

Medical response has to be quick and effective. The execution of medical response plans and deployment of medical resources warrant special attention at the state and district levels in most of the situations. The following measures shall be taken by the states/districts:

- A mechanism for quick identification of factors affecting the health of the affected people shall be established for surveillance and reporting.
- An assessment of the health and nutritional status of the affected population shall be done by experts with experience of emergencies and, if possible, local knowledge.
- The voluntary deployment of the nearest medical resources to the disaster site, irrespective of the administrative boundaries, will be warranted.
- Mobile medical hospitals and other resources available with the national government shall be provided to the states/provinces.
- Adequate supply of medicines, disinfectants, etc., shall be made.
- Where necessary, inoculation shall be done.
- Vaccination of the children and pregnant women shall be undertaken.
- Vector-borne diseases are a major cause of sickness and death in many disaster situations. Vector control measures shall be undertaken.
- Water-borne diseases may cause sickness and deaths and therefore adequate measures shall be taken to prevent such outbreaks.

8.7.3.2 MENTAL HEALTH SERVICES

Disasters cause tremendous mental trauma to the survivors. Psychosocial support and mental health services should be made available immediately in the aftermath of disaster so as to reduce the stress and trauma of the affected community and facilitate speedy recovery. The following measures shall be undertaken by states/provinces:

- A nodal mental health officer shall be designated for each affected district.
- Rapid needs assessment of psychosocial support shall be carried out by the NO/health department.
- Trained manpower for psychosocial and mental health services shall be mobilized and deputed for psychosocial first aid and transfer of critically ill persons to referral hospitals.

- Psychosocial first aid shall be given to the affected community/population by the trained community-level workers and relief and rescue workers.
- Psychosocial first aid givers shall be sensitized to local, cultural, traditional, and ethical values and practices.
- Psychosocial support and mental health services shall be arranged in relief camps setup in the post-disaster phase.
- Where large number of disaster victims have to be provided psychosocial support, a referral system for long-term treatment shall be followed.
- The services of NGOs and CBOs may be requisitioned for providing psychosocial support and mental health services to the survivors of the disasters.
- Community practices such as mass prayers, religious discourse, etc. should be organized with four preventive and promotive mental health services.

8.7.4 Clothing and Utensils

During disasters, people lose their clothing and utensils. The following measures shall be taken by state/local authorities:

- The people affected by the disaster shall be provided with sufficient clothing, blankets, etc. to ensure their dignity, safety, and well-being.
- Each disaster-affected household shall be provided with cooking and eating utensils.

8.7.5 Shelter

In a major disaster, a large number of people are rendered homeless. In such situations, shelter becomes a critical factor for the survival of the affected people in the initial stages of a disaster. Further, shelter becomes essential for safety and security, and for protection from the adverse climatic conditions. Shelter is also important for human dignity and for sustaining family and community life in difficult circumstances. The following measures shall be taken by state/local (district) authorities for providing shelter to the affected people:

- Disaster-affected people who have lost their dwelling units or where such units have been rendered damaged/useless shall be provided sufficient covered space for shelter.

- Disaster-affected households shall be provided access to appropriate means of artificial lighting to ensure personal security.
- Disaster-affected households shall be provided with necessary tools, equipment, and materials for repair, reconstruction, and maintenance for safe use of their shelter.

8.7.6 Relief Camp

The following steps shall be taken for setting up relief camps in the affected areas:

- An adequate number of buildings or open spaces shall be identified where relief camps can be set up during emergency.
- The use of premises of educational institutions for setting up relief camps shall be discouraged.
- One member of the ICT of the district/local administration trained in the running and management of relief camps will be deputed for management of relief camps.
- The requirements for operation of relief camps shall be worked out in detail in advance.
- Agencies to supply the necessary stores will be identified in the pre-disaster phase.
- The temporary relief camps will have adequate provision of drinking water and of bathing, sanitation, and essential healthcare facilities.
- Adequate security arrangements shall be made by the local police.
- Adequate lighting arrangements shall be made in the camp area including at water points, toilets and other common areas.
- Wherever feasible, special task forces from amongst the disaster-affected families will be set up to explore the possibility of provision of food through community kitchens, provision of education through the restoration of schools, etc.
- Efficient governance systems like entitlement cards, identification cards, bank accounts for cash transfers, etc., shall be developed.

8.7.7 Sanitation and Hygiene

Sanitation services are crucial to prevent an outbreak of epidemics in post-disaster phase. Therefore, a constant monitoring of any such possibilities will be necessary. It should be ensured that disaster-affected households have access to sufficient hygiene measures. Soap, detergents, sanitary napkins, and other sanitary items should be made available to ensure personal hygiene, health, dignity, and well-being. In the relief camps, toilets should be sited, designed, constructed, and maintained in such a way as to be comfortable, hygienic, and safe to use.

8.7.8 Provision of Intermediate Shelters

In the case of devastating disasters, where extreme weather conditions can be life-threatening or when the period of stay in temporary shelters is likely to be long and uncertain, the construction of intermediate shelters with suitable sanitary facilities will be undertaken to ensure a reasonable quality of life to the affected people. Such shelters shall be designed to be cost-effective and as per local needs.

8.7.9 Management of Relief Supplies

Speedy supplies of relief materials shall be ensured in relief operations. A supply chain management system shall be developed. Standard protocols shall be put in place for ensuring the procurement, packaging, transportation, storage, and distribution of relief items. A mechanism shall be developed for receiving donations in cash or kind and their distribution.

8.9 Disaster Recovery

Disaster recovery is viewed by some people as a fight against Mother Nature to restore order in a community. However, the disaster recovery *process* is *not* a set of orderly actions triggered by the impact of a disaster upon a community. Rather, disaster recovery is a set of loosely related activities that occur before, during, and after a disastrous event. These activities can include:

- warning and ongoing public information;
- evacuation and sheltering;

- search and rescue;
- damage assessments;
- debris clearance, removal and disposal;
- utilities and communications restoration;
- re-establishment of major transport linkages;
- temporary housing;
- financial management;
- economic impact analyses;
- detailed building inspections;
- Re-development planning;
- environmental assessments;
- demolition;
- reconstruction; and
- hazard mitigation and preparation for the next disaster.

When disaster strikes, response activities and recovery activities are often uncoordinated, occur concurrently and, on occasion, overlap or conflict with one another. Often, management responsibility for these activities will be assigned to people unfamiliar with them. Decisions affecting community welfare—some of which may have long-lasting impacts—will have to be made under intense pressure and scrutiny, and it will be impossible to take into account the views of all the pertinent stakeholders. One consequence is that the community may miss opportunities to improve its infrastructure, economy, environment, or quality of life. The ideal disaster recovery *process* recognizes the possibilities of the situation, and manages the necessary activities so that they are solutions, not additional problems. A community should strive to fully coordinate available assistance and funding while seeking ways to accomplish other community goals and priorities, using the disaster recovery process as the catalyst. This *ideal* disaster recovery process is one where the community proactively manages:

- recovery and redevelopment decisions to balance competing interests so constituents are treated equitably and long-term community benefits are not sacrificed for short-term individual gains;
- multiple financial resources to achieve broad-based community support for holistic recovery activities;
- reconstruction and redevelopment opportunities to enhance economic and community vitality; and
- environmental and natural resource opportunities to enhance natural functions and maximize community benefits.

Exposure to risk to a level that is less than what it was before the disaster. This ideal disaster recovery process is consensus-based and compatible with long-term community goals, and it takes into account all the principles of sustainability. It will have both immediate and lasting impacts that are self-supporting, and will make a community better off than before. It is a *holistic disaster recovery*. Holistic disaster recovery is becoming the next step in a logical progression. If we include sustainability within the multi-objective mitigation we already incorporate during disaster recovery, it can become equally accepted and equally successful. Holistic disaster recovery *does not differ* from "normal" disaster recovery—it is *part* of what should be "normal" disaster recovery. A "good" recovery *is* a holistic recovery—one that considers the community's best interests overall, by including the principles of sustainability in every decision. The question is, "How does one make a holistic disaster recovery happen?" How can a decision-maker reshape a process that operates within an emotional, reactionary, time-sensitive, expensive, and politically charged atmosphere that is based upon incomplete information, disproportionate needs, and the worst working conditions imaginable? There are two important steps to get a community started. The first is identifying and understanding the obstacles that prevent a holistic disaster recovery from occurring. Second, a community needs to form and adopt new strategies, including the holistic disaster recovery framework and process that coordinate, lead, and manage post-disaster decisions in a way that starts to overcome these obstacles.

8.9.1 Role of Policy-makers in Disaster Recovery Process

Depending on the nature of the disaster, the development of a recovery framework can range from being relatively resource-light to being resource-intensive. Similarly, depending on the scale of recovery and reconstruction required, the framework can take from a few weeks to several months to develop.

8.9.2 Institutional Arrangements

The preferable arrangement for post-disaster recovery is to have a pre-existing entity for the core recovery planning and oversight functions required to meet recovery objectives. In the absence of such arrangements, it is critical to designate an agency to take the lead role in coordinating or planning recovery. A lead recovery agency should be designated early into the recovery process. This guide elaborates a multitude of possible institutional arrangements for recovery implementation. However, the key elements of an effective lead recovery agency are that it should have a clear mandate and that it should be backed by effective political and technical leadership.

8.9.3 Vision and Guiding Principles

Another important initial step in setting up a recovery framework is the articulation of the recovery vision and guiding principles. They enable the government to convey its recovery priorities to the public, stakeholders, and partners and to build national or subnational consensus around them. Early setting of the vision and guiding principles for recovery is important to ensure effective transition from the immediate humanitarian response to the medium- to long-term recovery. For example, guiding principles could encourage the use of shelter materials in the humanitarian response that could be reused to reconstruct homes.

Instruments such as the Post-disaster Needs Assessment (PDNA) provide a solid basis for quantifying recovery needs and formulating broad strategies. However, experience in the last several decades has shown that meeting recovery needs must go beyond simply conducting post-disaster assessments. Demand has been growing globally for a disaster recovery framework for resilient recovery that can build on the PDNA or other such assessments.

In response, the World Bank's Global Facility for Disaster Reduction and Recovery (GFDRR) collaborated with the European Union (EU) and the UN Development Programme (UNDP) to produce a guide to developing disaster recovery frameworks. The guide is intended as a practice-based, results-focused tool to assist governments and partners in planning for resilient post-disaster recovery.

8.9.4 Identification of Priority Sectors for Recovery

The next step entails the identification of priority sectors for recovery in line with the broader recovery vision and policy framework, and based on the detailed needs and damage assessment carried out at the PDNA stage. The typical breakdown of programmatic recovery includes the following sectors: rural/urban housing development, water and sanitation, governance, transport, power, communications infrastructure, environment, livelihoods, tourism, social protection, health, and education.

8.10 Planning for Disaster Recovery

Disaster recovery actually begins *before* a disaster occurs. Emergency managers refer to this as *preparedness*—that phase during which people get ready for the onslaught and aftermath of disaster with activities such as warning, evacuation, and sheltering. In disaster-prone regions, it is even common for debris removal, utility restoration, and the management of donations and volunteers to be preplanned. These pre-disaster activities have a dramatic impact upon a community's ability to respond to, and recover from, a disaster. *A community's response to a disaster lays the groundwork* for both short-term and long-term recovery. For example, to re-establish power quickly, downed lines are often immediately restrung on the poles, rapidly re-established the pre-existing risk with little or no thought as to why the power lines came down (quite often because trees fell across them) or why the poles themselves failed (were they blown down, broken by wind, or undermined by erosion?). An opportunity has thus been missed to "underground" the power lines to protect from future similar events or improve aesthetics. By studying some of the mitigation options before disaster strikes a community is better prepared for recovery. Decision-making could take place in a less-fettered environment, with appropriate funding, public input, and cost-benefit analysis.

If a community fails to adequately respond to a disaster, its credibility suffers. This loss of credibility can become a barrier to implementing a holistic disaster recovery. If a local government cannot re-establish power quickly, or clear the roads of debris from an event that they "should have known" would occur sooner or later, then how can that same government expect its constituents to believe in its ability to manage more complex long-term recovery issues?

Communities that are serious about disaster recovery tend to focus first on improving response activities (warning, evacuation, power restoration, debris management) before the more advanced concepts of holistic recovery. In the immediate post-disaster period, people often think that mitigation activities may not work, or that coupling community improvements with repairs may be too expensive, too disruptive, or take too long. Unfortunately, it is within this same timeframe that decisions affecting repairs and restoration are made, and thus the opportunity to integrate the principles of sustainability into the recovery process is lost.

Holistic disaster recovery is about change. Because the disaster recovery process begins before the disaster, the best chance to foster post-disaster change is to include sustainability issues in local pre-disaster planning. The six principles of sustainability can be integrated into post-disaster plans, but there is a better chance for implementation—because of timing and a less-pressured decision-making environment—if they are addressed beforehand. This concept has been called pre-event planning for post-event recovery (PEPPER) first advanced in the 1980s (Spangle, 1987).

In communities that endure repeated disasters, *after* one disaster is the same as *before* the next. Thus, the increased awareness created by the last disaster can provide impetus for pre-disaster planning for the next one, including the opportunity to incorporate sustainability in the next recovery.

8.10.1 Programmatic Approach

The lead recovery agency may help the government develop a framework that takes a programmatic approach to identify priority sectors that are critical for restoring livelihoods. Such a framework would enable the use of holistic recovery management. In it, the activities of government agencies, communities, and nongovernmental entities complement one another under a government-led framework. Because a significant portion of recovery activities are undertaken by the nongovernmental entities, an inclusive recovery process would help avoid duplications and gaps. For example, certain geographic areas or sectors may be allocated to particular donors, NGOs, and implementation partners.

The lead agency would oversee the development of the recovery framework and would play a critical role in its implementation. The lead agency also could play a central role in the coordination, oversight, and M&E of the progress of the recovery.

8.10.2 Financing for Recovery

Recovery implementation needs to be supported by the mobilization of funds and coordination mechanisms that channel funds to the implementation entities in a timely manner. Recovery may be funded through government funds, international aid, private-sector financing, and community contribution. To manage recovery in a holistic manner, it is recommended that the government have an effective funds tracking mechanism for both on-budget and off-budget funds. A good fund-tracking mechanism along with a strong public financial management system enhance donor confidence and help in mobilizing additional funds for recovery.

8.10.3 Simplified Procurement

The increase in the volume of transactions and the urgency with which they need to be completed often overwhelm existing government systems. Simplified procurement procedures can provide a robust mechanism for the timely purchase of goods and services. Experience suggests that responsible officers are reluctant at times to use the simplified procurement procedures even if they exist within the government systems. The mandate given to the lead recovery agency under the recovery framework is important in invoking and promoting the use of simplified procedures.

8.10.4 Communication

Recovery is often a multisectoral activity that encompasses a broad range of actors and affected communities. For this reason, it is crucial for the government to have a consolidated communications system that conveys the progress of the recovery and addresses the expectations of the affected communities. A coherent communications platform also is useful in communicating with donors and beneficiaries.

8.11 Policy and Strategy Setting for Recovery

8.11.1 Development of a Central Vision for Recovery

The articulation of a recovery vision enables the government to convey its recovery priorities, and build national or subnational consensus around them. The vision is the starting point around which the entire recovery process will be formulated. The core elements to be included in a recovery vision are as follows.

- *Ensuring that the vision is developed at the highest level of government is critical for building consensus among the range of stakeholders.* The government can invite groups of internal and external stakeholders to sessions in which it communicates and seeks input for its vision of recovery. Seeking agreement from stakeholders will smooth the way for unified planning. These consultations at the beginning of the recovery process can guide to meet up the expectations of the affected communities and reconstruction partners.

- *Ensuring coherence with development programs.* The recovery vision is intended to be coherent with the government's broader, longer term development goals and growth and poverty reduction strategies. The vision can provide a strategic continuum between pre- and post-disaster development planning by bridging both pre-existing development gaps and new gaps triggered by the disaster.

- *Incorporating resilience and BBB in recovery vision.* Resilient recovery is not well understood by most development practitioners. Countries are beginning to develop their own standards and definitions for resilient recovery. To support resilient recovery, this Disaster Risk Financing (DRF) guide recommends that countries pay particular attention to seven issues: BBB, gender concerns, equity, vulnerability reduction, natural resource conservation, environmental protection, and climate change adaptation.

- *Optimizing recovery across sectors.* Whenever possible, the recovery vision should encompass public and private sectors, and promote norms for nondiscriminatory and equitable asset disbursement among individuals and communities. In the past, infrastructure reconstruction often has dominated post-disaster recovery. However, equally important is the priority given to the recovery of the lives and livelihoods in disaster-affected communities. People-focused recovery can be facilitated by reconstructing private assets through direct subsidies, where affordable, or through other enabling policy measures, where appropriate. Showing sensitivity to the needs of the affected population also is important in meeting and managing public expectations (Figure 8.3).

Figure 8.3:
Recovery Planning Processes

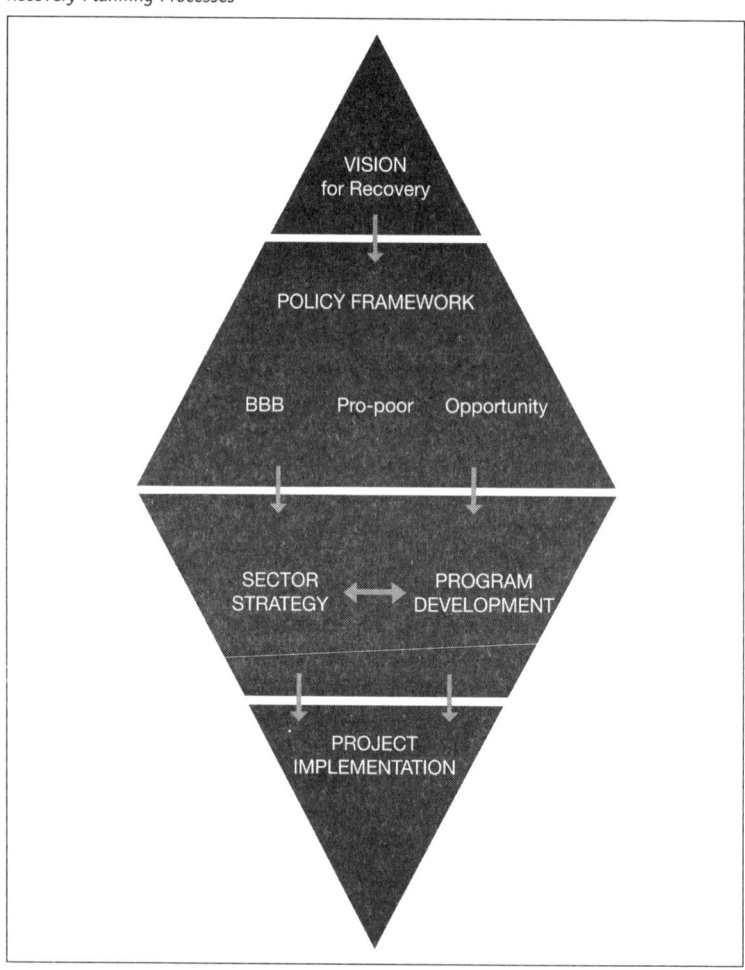

Source: Modified after Disaster Recovery Frameworks (2015).

8.12 Key Policy Imperatives for Recovery

Successful disaster recovery experiences from around the world have the adoption of at least three crucial policies in common: (a) BBB, (b) converting adversity into opportunity, and (c) prioritizing pro-poor recovery.

1. *Building back better:* BBB is the reconstruction approach that aims to reduce vulnerability and improve living conditions, while promoting more effective reconstruction. BBB addresses the importance of improving community resilience following disasters and identifies what is considered successful recovery. Recovery policy-makers and practitioners lack consensus on what BBB should include or not. However, at a minimum, BBB signifies policy commitment to right-sizing, right-siting, and improving the resilience of critical infrastructure.

2. *Converting adversity into opportunity:* Disaster recovery can be an opportunity to replace old infrastructure and update the service delivery systems with affordable, resilient improvements.

3. *Pro-poor recovery:* Prioritizing reconstruction planning to address the needs of socioeconomically vulnerable individuals and groups contributes to a more equitable society. If their needs are ignored, the poor and vulnerable are more susceptible to future hazards and shocks. Many disaster recovery programs include the provision of direct livelihood support, income generation opportunities, improved access to finance and microcredit, and new skills training. Governments also subsidize or facilitate the reconstruction of private assets, such as housing and local business enterprises. However, governments cannot substitute for private insurance to pay recovery costs.

8.13 Programmatic Framework for Recovery

8.13.1 Programmatic Approach to Recovery

- The achievement of the recovery goals, targets, and priorities, as defined by the vision and policy framework, requires the development and maintenance of a strategic and programmatic framework. This strategic framework is the central planning tool and oversight mechanism for cross-sectoral and integrated disaster recovery.

- Where the center of such recovery planning is located is not important. In cases of interprovincial recovery programs, it could be within a central government authority. In cases of subnational or local programs, the planning center could be located within the subnational recovery planning and oversight entities. What is important is that large-scale recovery should have a central meeting point, or node. At this point, the recovery plans and projects of the national, subnational, and local entities converge to provide the complete programmatic picture of recovery for policy-makers at all relevant levels.

• In many countries, the major portion of recovery interventions are undertaken by nongovernmental organizations (NGOs). A programmatic approach also provides an opportunity to exercise holistic management of recovery. In this approach, the activities of government agencies, NGOs, communities, and the private sector complement each other within a government-led framework.

8.13.2 Intersectoral Prioritization

• Prioritization across sectors can help ensure equitable and demand-responsive recovery across the affected jurisdictions and communities. Prioritization also can promote conflict-sensitive, pro-poor, pro-vulnerable, and gender-sensitive recovery agendas.

• The areas considered sectors include the social sectors (housing, land and settlements, education, health, and nutrition), production sectors (employment and livelihoods, agriculture, commerce and trade, and industry), and infrastructure sectors (community infrastructure, water, sanitation and hygiene, transport and telecommunications, and energy and electricity). Cross-cutting sectors include DRR, environment, gender, and governance.

• The first step in prioritizing is to identify the sectors targeted for reconstruction. Second, a criteria-based prioritization of recovery needs across competing intersectoral priorities should be made. Such prioritization broadens the resource allocation and the annual on- and off-budgetary flows for recovery throughout the expected recovery period.

• The rule of thumb for prioritization is, first, to determine the sectors and sectoral priorities that help leverage direct humanitarian impact in the shortest time. The case studies in this DRF guide show that housing and livelihood often take precedence over other sectors. These two sectors are given precedence because they directly impact disaster-affected populations. The interventions in these two sectors take place simultaneously with restoration of critical public infrastructure and service delivery. The next phase is medium- to long-term reconstruction and generation of sustainable livelihoods.

8.13.3 Developing Principles for Intersectoral Prioritization

The government needs to establish principles to determine the criteria for intersectoral prioritization to help ensure equitable and demand-responsive recovery across affected jurisdictions and communities. Prioritization grounded in principles ensures that sectoral program development remains consistent with the overall objectives of the reconstruction program.

Certain criteria used to prioritize recovery actions arise consistently in countries' experiences. These criteria include:

- potential for direct and widest humanitarian impact;
- pro-poor, pro-vulnerable, and gender-sensitive agendas;
- potential to generate sustainable livelihoods;
- balance between public sector and private sector recovery;
- balance between physical infrastructure reconstruction and less visible recovery (such as capacity-building and governance); and
- restoration and rebuilding of critical infrastructure and services.

8.14 Setting up Sector-level Recovery

Establishing sectoral strategies early can ensure that they are in line with the government's overarching principles for the recovery.

8.14.1 Developing Sector-specific Recovery Programs

After the policy framework and intersectoral strategies are established, the lead recovery agency typically undertakes a program-by-program approach to define sector-specific recovery programs. These programs translate the policy priorities into programs and projects that can be financed and implemented. Sector-specific recovery programs and projects are expected to reflect the policy framework and intersectoral strategies. The programs would draw on information from assessments and surveys to plan individual sectoral projects. This consultative process broadens ownership of the recovery program. Consultation ensures the implementation of the guiding principles at the project level.

8.14.2 Preparatory Actions and Information Collection for Sector Program Development

By developing broad sectoral strategies early in the recovery process, sectoral policies and reconstruction objectives can be aligned to ensure synergy between reconstruction activities and development goals.

The PDNA or a similar initial assessment is an important reference for the development of sector-specific recovery plans. These plans can be overseen by the lead implementation agency. Technical agencies would assist with the conceptualization and development of assessment frameworks, objectives, and instruments. The lead implementation agency may also engage other public-sector agencies, private-sector enterprises, or civil society and community organizations for these purposes.

To inform the development of sectoral recovery programs and projects, the following surveys/assessments may be carried out:

1. *Land risk survey/assessment:* An essential input for determining whether any relocation of communities is necessary.
2. *Land tenure survey/assessment:* Analyzes the issue of land and tenure records. Any disputes over ownership may delay, or even stop, the implementation of the sector planning recommendations.
3. *Land availability assessment:* Primary means to identify available and suitable land that may prove socially and economically viable for displaced populations.
4. *Governance and implementation capacity assessment:* Measures the government's capacity to implement programs.
5. *Social risks and vulnerability survey/assessment:* Assists in identifying vulnerable disaster-affected persons.
6. *Infrastructure and service delivery survey/assessment:* Provides results that may help design program components for rehabilitating infrastructure and resuming essential services.
7. *Economic and livelihood survey/assessment:* Assists in the adequate resumption of economic activities and livelihoods for beneficiaries of the land use and physical plans.
8. *Environmental assessment:* An essential input for the program to safeguard environmental objectives.

8.15 Who Will Manage Recovery?

• *Geographic and political spread of the disaster should inform the assessment.* Following a disaster, an immediate step is for the government to assess its existing capacity to conduct post-disaster recovery. The profiles of the lead manager and the lead agency for post-disaster reconstruction will depend on the magnitude and nature of the disaster. Nevertheless, the lead agency needs to be identified at the start of the recovery. Factors that may influence the selection of the lead manager and lead agency are the geographic impact of the disaster (such as cutting across jurisdictional lines) and whether existing government capacity is adequate for the estimated duration of the reconstruction.

• *Skills and logistical capacities need to be assessed.* The two main criteria to measure the capacity of an entity to manage recovery are *human resource capacity* and *skillsets.* Capacity assessments examine sector-specific requirements. Sufficient (perhaps even excess) expertise to successfully conduct recovery may reside in one sector. Another sector may be under skilled and under staffed. The lead agency's prior involvement in disaster recovery is not required. More important is its proven ability to produce results under tight deadlines; multitask; collaborate with other agencies, local authorities, and civil society; and be flexible about working within quickly evolving circumstances.

• *Capacity to manage contracts and procurements are critical.* Consideration of an agency's capacity to manage contracts is important for the procurement of reconstruction equipment and material, evaluation of tenders, and oversight of recovery projects. These processes require dedicated time and human resources as well as specific technical knowledge. In some recovery operations, third-party contractors form a substantial bulk of the implementers. In these cases, the skill and logistical capacity of the lead agency to manage contracts is critical to the successful implementation of the recovery.

8.16 Selecting an Effective Lead Agency

8.16.1 Lead Agency Coordinates Disparate Recovery Efforts

In developing the recovery program, the lead agency pays special attention to harmonizing strategies across sectors. Harmonizing strategies means distributing resources to avoid discrimination against minorities and inequities in spending and quality of delivery. The lead agency must also maintain the urgency to deliver results by keeping its focus on deliverables and targets.

8.16.2 Five Criteria Exist for Choosing the Lead Agency

Globally, post-disaster recovery experience reveals a range of potential institutional setups. The selection of the lead agency usually depends on five criteria.

1. characteristics of the disaster;
2. current governance structure;
3. the agency's prior disaster recovery experience;
4. the agency's ability to reach out and include communities in defining and implementing their recovery process, and the capacity to work with local authorities and nongovernmental organizations; and
5. the overarching coordination, monitoring, oversight, and control frameworks in operation among a country's agencies, line ministries, local governments, and civil society. The government may choose a lead agency after having necessary consultations with key stakeholders and future implementers of programs, both within and outside the government. Nevertheless, the decision must be made urgently.

8.16.3 Three Options for Structure of Lead Agency

The three most typical compositions of lead agencies are given further.

8.16.3.1 STRENGTHEN AND COORDINATE EXISTING SECTORAL LINE MINISTRIES TO LEAD THE RECONSTRUCTION BY SECTOR

This option depends on establishing recovery frameworks under which individual line ministries work independently to manage recovery, and to supervise and implement projects, in their sectors. This option usually begins with the line ministries jointly preparing an action plan for recovery that identifies the respective roles and activities of the line ministries to support reconstruction. In this option, the existing capacities of government line ministries must be adequate to deal with additional urgent responsibilities. Possible difficulties include:

- Rapid recruitment of temporary human resources may not adequately supplement the capacities.
- Recovery coordination may be difficult if the line ministry staff lack sufficient experience.
- Line ministries may struggle to focus on recovery programs at the expense of longer term goals.

8.16.3.2 CREATE A NEW INSTITUTION TO MANAGE RECOVERY

This option creates a single, lead-implementing agency. This agency envisions, strategizes, plans, implements, and controls the overall multisectoral reconstruction program.

- This option has several advantages. They are the agency's autonomy, the clear line of responsibilities, effective internal and external communication, and the capacity to handle complicated financial and M&E arrangements.
- Potential disadvantages of Option 2 include the lead agency's lack of authority to achieve results, possible lack of ownership by line ministries, and the line ministries' potential institutional resentment due to compromised authority and duplicated mandates at various levels of government. Another risk could be the insufficient inclusion of civil society and communities affected by the disaster in recovery planning. Moreover, start-ups will incur high administrative costs,

may inadequately represent local needs, and struggle to meet urgent planning and implementation demands.

8.16.3.3 HYBRID ARRANGEMENT

A third option increasingly being used by governments is a hybrid institutional model. It combines the advantages of the previously mentioned options while offsetting their risks. Under this arrangement, existing government structures are strengthened through the creation of a temporary agency with a built-in end-date. The agency will provide overarching central guidance, management, and support services to keep the reconstruction program on its planned course.

- The creation of a new institution may be desirable in situations in which the existing government agencies are unlikely to be able to coordinate and implement a high number of additional projects at an increased speed while sustaining their routine public services. The hybrid option ensures a relatively speedy delivery of reconstruction deliverables and meeting targets. It consolidates recovery into a single agency that will oversee the process. This agency will be the single point of coordination of national and international stakeholders. It will be responsible for ensuring the inclusion of line ministries, local authorities, the private sector, and civil society in all phases of the recovery. This agency will work with local governments and nongovernmental organizations to delegate implementation responsibilities.
- One drawback of the hybrid is that, as the recovery transitions to development and the temporary agency's mandate expires, its accumulated capacity, knowledge, and experience may be lost.

Hybrid Model in Indonesia

The hybrid model was used in Indonesia following the 2004 Indian Ocean tsunami. The sunset clause existed from the outset. The four-year mandate of the Agency for the Rehabilitation and Reconstruction of Aceh and Nias maintained the urgency for reconstruction and enforced a handover strategy to the existing administration in Indonesia.

Establishing Mandates and Operational Modalities in Pakistan

Pakistan's Earthquake Reconstruction and Rehabilitation Authority (ERRA) was set up following the 2005 earthquake as a time-bound central authority under the prime minister's office. The authority's purpose was to tackle early recovery, and long-term reconstruction and rehabilitation. Long-term efforts make up the overwhelming bulk of its mandate.

ERRA's scope of work included strategic planning, resource mobilization, coordination with all stakeholders, and monitoring reconstruction and rehabilitation activities in earthquake-affected areas. ERRA was established because of a recognized need for a central oversight body to coordinate the activities of the broad spectrum of actors participating in the reconstruction. These actors included multilateral and bilateral donors, international NGOs, civil society, and government agencies. It was anticipated that having multiple agencies overseeing the reconstruction would likely become unmanageable. Centralizing some functions within a single, dedicated body was seen as essential.

Setting up Legal Frameworks

Indonesia, Tsunami, 2004

Following the 2004 Indian Ocean tsunami, the Indonesian government established a regulatory framework for post-disaster responses. It identified the responsibilities of the central and local governments as well as the functions and duties of the national and regional disaster management agencies. The regulations outlined the disaster risk-financing framework, which is a shared responsibility between the central and local governments. The framework stipulated the three phases of a disaster: emergency, recovery, and reconstruction.

8.17 Government Coordination and Local Implementation

It is necessary to define the recovery vision and policy at the highest levels of government to ensure acceptance and coherent application across the many simultaneous ongoing reconstruction projects. A tiered implementation is recommended within the DRF process that balances national government policy setting with decentralized implementation. Program implementation is recommended to take place at the local level, closest to the affected communities and individuals.

It is the role of the lead agency to establish and oversee the coordination mechanisms that guarantee coherent policy application and effective implementation at the regional and local levels. The work of the implementing agency is overseen by the lead agency within the context of a coordination mechanism.

Multiple-Level Coordination and Implementation Structure

Following Pakistan's 2005 earthquake, the lead reconstruction agency, ERRA, combined central coordination alongside local implementation by creating a tiered coordination and implementation structure.

The ERRA Council acted as the leadership, which provided strategic direction for policy formulation and ensured adequate funding. The council was coupled with the ERRA Board, which ensured implementation of approved policy decisions. The board also developed and implemented annual plans, programs, and projects.

Similarly, at the provincial and state levels, the Provincial Steering Committee was partnered with the Provincial Earthquake Reconstruction and Rehabilitation Authority (PERRA). The State Steering Committee was coupled with the State Earthquake Reconstruction and Rehabilitation Authority (SERRA). At the district level, the district reconstruction advisory committees provided work-plan oversight to the district reconstruction units (DRUs) within designed programmatic interventions.

National government enabled local implementation by allocating independent budgets to PERRA, SERRA, and the DRUs. The independent budgets enabled the implementing organizations to create and manage their own work plans. Transferring ownership to the local levels helped ensure that projects were locally planned.

8.18 Financing for Recovery

In post-disaster recovery, there are four major financing challenges. They are to quickly quantify the economic costs of the disaster, develop recovery and recovery budgets, identify sources of financing, and set up the mechanisms to manage and track funds. Good financial practice across post-disaster experience shares the common characteristics of rapid disbursement, coordination of resources, and flexible sources of funding.

1. *Rapid disbursement:* Meeting the recovery objectives demands quick response. Actions must occur under significant time pressures and must be completed within the set timeframes. Compared to normal projects, the necessity for speed mandates short timelines for project preparation, approval, and procurement. Special dispensations or accelerated processes may be applied to disburse the funds available for recovery as quickly (yet transparently) as possible.
2. *Coordination of resources:* Often, numerous government and non-governmental actors engage in the recovery efforts. Their number poses significant coordination challenges for the lead agency. Having a variety of stakeholders and donors contributing to the same objectives requires the use of different types of coordination mechanisms to marry policy to funding and implementation. A range of such mechanisms is especially necessary when many funds will be managed not by the government (on-budget) but by the funding sources (off-budget).
3. *Flexible funding sources:* In post-disaster environments, conditions change so rapidly that unacceptable delays may occur if budgeting revisions have to wait until the normal budget cycle. The government may have established a *contingency fund* to respond to the immediacy of a disaster. Such funds are characterized by flexibility to respond appropriately, especially in the immediate aftermath of the disaster. Pooled funds from donors that are administered by a trustee also are characterized by their flexibility to finance recovery needs that may be unattractive to the bilateral donors or do not fit within the government's budget. Financial considerations of recovery start with budgeting within the pre-disaster and macroeconomic context. Depending on the scale of the disaster and the capacity of a national economy, the government may either rely largely on national resources, or appeal to external sources for funding. The latter option is useful, particularly when the government already has cooperation agreements with donors and/or multilateral agencies. Figure 8.4 details the elements of recovery financing from the variety of funding source possibilities—both domestic and external. The lead agency should ensure that all of these funds are allocated in accordance with the national recovery priorities, whether or not the funds are channeled on or off the national budgetary system.

Figure 8.4:
Global View of Post-disaster Financing

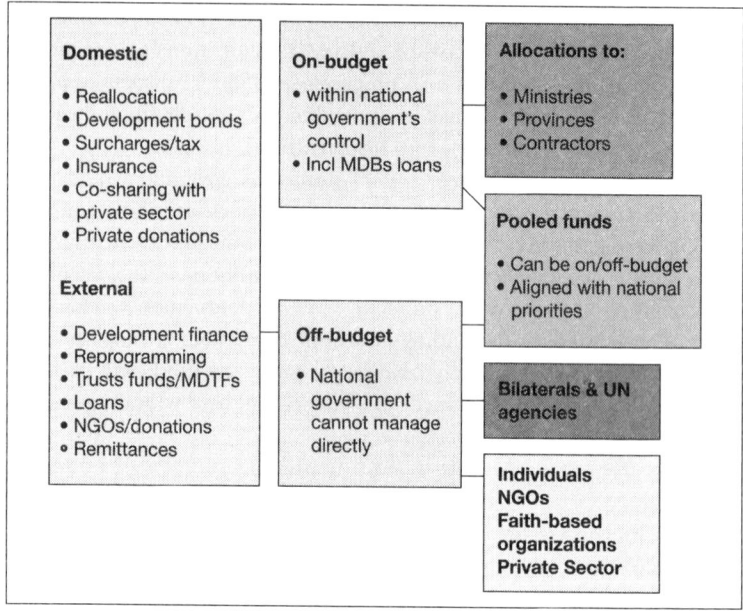

Domestic	On-budget	Allocations to:
• Reallocation • Development bonds • Surcharges/tax • Insurance • Co-sharing with private sector • Private donations	• within national government's control • Incl MDBs loans	• Ministries • Provinces • Contractors

Source: Modified after Disaster Recovery Frameworks (2015).

8.19 Implementation Arrangements and Recovery Management

The recovery policy framework, institutions, and financing are elements of paramount importance in the disaster management process. However, the issues and options related to them are of little relevance unless recovery programs are implemented quickly, they and visibly improve the lives of disaster-affected populations.

8.19.1 Coordination Mechanisms

In the context of implementing a recovery program, coordination refers to applying various tools to ensure coherent support for recovery policies and programs. Coordination also includes assigning different areas of recovery to the governmental or nongovernmental agencies in their areas of expertise. Coordination brings together a larger number of partners and stakeholders to support the recovery program. The growth in numbers results in resource pooling, new initiatives and innovations, and improvement in quality and speed of implementation. Involvement of a variety of actors in the implementation process makes recovery more transparent, and participatory. One coordination approach involves harnessing the ongoing cluster groups of humanitarian organizations. Their convening power can continue the coordination into the recovery phase of monitoring achievements and ongoing projects.

Coordination can take place both vertically and horizontally. When the implementing agency interacts with the national government and local administration, it is a case of vertical coordination. When the agency starts working with the private sector, nongovernmental organizations (NGOs), and civil society organizations (CSOs) to allocate areas of responsibilities and maximize the use of resources in the course of implementation, it is a case of horizontal coordination.

Several types of coordination mechanisms can be set up, depending on the type of coordination and stakeholders. A coordination mechanism at each level of policy-making, planning, and implementation is helpful in developing consensus and resolving conflicts and disputes. Some coordination mechanisms that are functioning or can be set up to support recovery implementation are:

- *Task force/empowered committee:* Consisting of senior politicians, administrators, and professional experts, the task force can be set up at a high level in the government to develop a recovery policy/program.
- *Donor coordination:* Coordination can be accomplished by the lead agency assigning a donor lead responsibility for specific sectors or projects.
- *NGOs' coordination committee:* The government can set up the committee at the subnational level to assist the NGOs with their participation in the recovery program. In this forum, the NGOs meet the government officials and resolve all the program issues. The committee also provides NGOs with the necessary support and authorization to implement.
- *Local-level project management committee:* This committee can consist of local government officials, NGOs, and representatives of affected communities. A number of local issues related to recovery can be discussed at this level.

8.19.2 Standard Implementation Procedures

Existing project approval and procurement, reporting, and staffing procedures in the country may need to be simplified to meet the pressing demands of the recovery process. Often recovery projects are stalled due to lengthy bureaucratic procedures for project approval and procurement. Even if fast-track approval processes exist, at times responsible officers are reluctant to use them [why?]. The authority given to the lead recovery agency by the government can play a critical role in promoting the use of simplified procedures and processes across all sector and entities for more rapid implementation.

8.19.3 Establish Reconstruction Standards

The key policy imperatives for disaster recovery are:

* building back better;
* converting adversity into opportunity; and
* pro-poor recovery.

As the government drafts its recovery framework, it needs to formulate the guiding principles for recovery based on these three key policy imperatives. As the recovery moves into the implementation phase, the guiding principles need to be translated into practical recovery and reconstruction standards. Local stakeholders from both the government and civil society, including NGOs and the private sector, can work together to detail these standards.

8.19.4 Reconstruction Standards Can Cover Recovery Sectors and Implementation Mechanisms

Reconstruction standards are specific to the sector and the type of natural disaster. Moreover, these standards must be detailed well ahead of actual implementation. For example, after an earthquake, the reconstruction must conform to appropriate seismic safety, quality, technological, and environmental standards. In another example, reconstruction of schools could include the standard that all schools must be rebuilt to function as shelters during a disaster. Reconstruction standards also could ensure that first consideration is given to local resourcing of materials and technical expertise.

Ensuring compliance with reconstruction standards during the implementation phase is key to resilient recovery. To ensure compliance, construction monitoring teams could be established by the lead agency to monitor technical aspects of both the inputs and outputs of reconstruction. In addition to alerting the relevant authorities of any missteps or lack of adherence to standards by the implementers, the lead agency also should support implementation entities to correct their procedures.

8.19.5 Monitoring and Evaluation

Effective M&E systems enable the progress of recovery to be assessed, ensure compliance with sectoral recovery policies and strategies, and provide early warning for corrective action. Ongoing M&E is critical to identify mid-course corrections in the implementation and adjust the strategy, particularly in response to community feedback about project design and results.

M&E reviews provide:

- a holistic assessment of recovery framework implementation;
- a fresh view of the recovery framework implementation;
- potentials for improvement;
- actionable, realistic, result-oriented, and concrete recommendations; and
- a learning opportunity for all involved. In addition, M&E provides substantive inputs into the periodic evaluations that donors require to continue funding projects.

8.19.6 Results Framework Implemented Best through Results Monitoring System

The results framework should be implemented through a systematic results monitoring system (RMS). The RMS specifies the M&E plans, data collection instruments, and indicator value-determination methodologies for all outputs and outcomes. Once fully developed, the RMS also will provide an overall medium-term M&E plan. This plan specifies the frequency, requirements, and means for monitoring, evaluating, and reporting, both at the broader level and for each of the selected outcomes. Following are the commonly used 10 steps for results-based M&E systems that can be applied to post-disaster recovery programs to create effective M&E systems.

1. Conduct a readiness assessment.
2. Agree on outcomes to monitor and evaluate.
3. Select key indicators to monitor outcomes.
4. Identify baseline data on indicators: ask "Where are we today?"
5. Plan for improvements: select results targets.
6. Monitor results.
7. Conduct evaluations.
8. Report findings.
9. Use findings.
10. Sustain the M&E system within the organization.
11. Dedicated management information systems (MIS) are required to build a results-based M&E system. MIS is the digital system to store all M&E information and collate results based on the different inputs.

8.19.7 Procurement

8.19.7.1 Rapid Procurement Systems

Rapid procurement of goods and services can be a crucial element for an efficient and successful recovery. However, procurement in post-disaster settings can be haphazard, leading to gaps in implementation and potential abuse of procedures. Several types of procurement systems will facilitate the purchase of goods and services during recovery. Two are pre-arranged procurement and fast-track procurement.

8.19.7.2 Pre-arranged Procurement

Pre-arranged procurement pre-establishes a list of qualified contractors. This list can be categorized by type of expertise and competencies. Having a pre-qualifying system in place expedites issuing contracts and evaluating tender responses. A pre-qualifying system also eliminates inexperienced contractors, who can significantly underbid more experienced competition, but who lack the expertise required to successfully implement the reconstruction project.

8.19.7.3 FAST-TRACK PROCUREMENT

Fast-track procurement means using simplified agreed tender and purchasing processes to quickly get goods and services to the areas in which they are needed. To further expedite procurement, a single source for the purchase of specific goods and services could be pre-determined. Fast-tracked procurement systems can be used by both the private sector and nongovernmental entities. To facilitate oversight and monitoring, it is helpful that all stakeholders that procure goods and services share some of the same procedures.

8.20 Recognizing Short-term and Long-term Disaster Recovery

Usually, communities think of preparing for a disaster *before* its onset, and response and recovery as activities for *after* the disaster. However, sometimes communities do respond before disaster happens. For example, in predictable events, such as slow-rise riverine flooding or most hurricanes, there is time to notify people of the impending danger, take some protective measures, and evacuate safely. Response actions are taken before anything happens. Doing so lessens the need to respond further, and lessens some of the elements of short-term recovery that might otherwise be necessary.

Traditional, post-event disaster recovery occurs in phases—short-term and long-term. SAR, damage assessments, public information, temporary housing, utility restoration, and debris clearance are essential elements of short-term recovery. How they occur will affect how some longer-term decisions are made (or not made).

Long-term recovery begins when a community starts to repair or replace roads, bridges, homes, and stores. It is also the period where improvement and changes for the better such as strengthening building codes, changing land use and zoning designations, improving transportation corridors, and replacing "affordable housing" stock are considered. Whether they are considered during pre-disaster planning or short-term post-disaster recovery, it is during the long-term recovery period that most changes in pre-existing conditions can and do occur. Changes that include sustained efforts to reduce loss of life and property from the next disaster, such as changes to building codes and land-use designations are examples of mitigation.

8.21 Different Perspectives on Disaster Recovery

It is important to recognize that not everyone within a community will have the same perspective or understanding of disaster recovery. The issues discussed thus far are presented from a "community recovery" point of view, that is, the activities that need to be managed in order for a local government to recover to an equal or improved state. However, there are also perspectives of the individual and of community economics that need to be taken into account.

The individual perspective is important, because as a community starts its recovery, most people are recovering emotionally, and this takes place at a slower pace than the external community recovery. Communities respond quickly, and with increasing resolve to re-establish utilities, provide access, and create reconstruction policies. Individuals experience a short period of cohesion during which people come together to help and comfort each other, followed by a longer period of disillusionment as personal, family, job, insurance, and disaster assistance issues begin to take their toll. The result is that constituents and stakeholders that are subject to the decisions being made on their behalf are in "a different place." This creates a "disconnect" between community recovery and individual recovery that leads to frustration, misunderstanding, and disillusionment.

Similarly, there is an economic perspective that differs from both that of the community and the individual. It is this economic perspective that highlights the interrelationship and interdependency between local governments and the business community. Businesses, from small "mom-and-pop" to "big box" national chains, are primarily concerned with minimizing theirdown time. The businesses often reach out to their employees to help them recover as individuals, because they need them as employees to help manage the business recovery. People forced to stand in line for water and ice, insurance appointments, and disaster assistance find it difficult to return to work to help their "other family" at the same time.

There is also an increased reliance of business upon local government. Without access to their facility, or power and water to run equipment and bathrooms, their recovery is hindered. Conversely, the longer it takes for businesses to recover, the greater the problems for local government (unemployment, loss of sales taxes, loss of business services, etc.).

Everyone in the community has a stake in disaster recovery, and the differing perspectives and interdependencies of individuals, the government, and businesses can create conflicts over priorities and timing. Local politics can also become a barrier to the holistic recovery. It is important to recognize the differing perspectives and agendas in order to tailor recovery actions that address those needs as much as possible.

8.22 Community-based Recovery

Even if the community is not preparing a formal recovery plan, the 10-step process is a useful guide to action. Holistic disaster recovery can be incorporated into this process as follows.

8.22.1 The 10-Step Process for Local Planning and Action

1. Get organized.
2. Involve the public.
3. Coordinate with other agencies, departments, and groups.
4. Identify the problem situation.
5. Evaluate the problem and identify opportunities.
6. Set goals.
7. Explore all alternative strategies.
8. Plan for action.
9. Get agreement on the action plan.
10. Implement, evaluate, and revise.

In *Step 1—Get organized*, the community can demonstrate its commitment to the process through there sources it provides for the planning process. This is where the holistic disaster recovery concept can be introduced, by encouraging appropriate staff and citizen input that reflects the principles of sustainability: environment, economic development, and disaster resilience.

In *Step 2—Involve the public,* the sustainability principle of using a participatory process is readily addressed by including the stakeholders directly.

In *Step 3—Coordinate with other agencies, departments, and groups,* a community can expand representation on the central recovery committee or task force to include those who can contribute expertise on each of the principles of sustainability. This could include, for example, state or local parks or wildlife departments, economic development directors, the business community, or social services personnel.

In *Step 4 —Identification of problems,* the community is facing

In *Step 5—Evaluate the problems* that conditions cause are described. Recovery team members should consider how the potential impacts might affect economic activities, natural resources, the overall quality of life, and people of different ages, races, and economic statuses. The team should also adopt a long-term viewpoint so that intergenerational equity is considered.

In *Step 6—Goals and objectives* are developed. The recovery team can use the matrix of opportunities to identify and incorporate short- and long-term recovery issues into the evolving plan. Coordination with other community plans and programs at this point can combine disaster recovery issues with existing comprehensive development, capital improvement, drainage, transportation, housing, and recreation plans. Multiple-objective opportunities give the community the opportunity to establish a coordinated recovery that maximizes available technical and financial recovery resources with pre-planned community goals and objectives.

In *Step 7—Explore all alternatives,* the recovery team reviews the options and tools available to achieve the selected goals and objectives. As part of this review, the six principles of sustainability are included among the criteria that assist the team in deciding which actions to take and in which order. The criteria should clearly identify the proposed actions that support sustainability as having high community value. The recovery team needs to be sure that the actions agreed upon do not undermine any of the aspects of sustainability. This step becomes the true litmus test for choosing activities that will help integrate sustainability into the community during its recovery.

Finally, in *Steps 8, 9, and 10, a plan is written, adopted by the elected governing board, and implemented.* Attention to sustainability details in these final steps will set the stage for managing the recovery and ensuring that the community maximizes the opportunities that are created by disaster.

This process does not guarantee that every sustainability principle ends up being addressed in the recovery, but including the principles as decision-making criteria ensures that they will at least be considered.

8.23 Making Sustainability Permanent

Disaster recovery provides the opportunity to introduce sustainability into a community. There are other ways to be sure, but the dramatic nature of disasters, and the frequent need to rebuild what has been destroyed, provides an opportunity to substantially improve the character of the community in a manner that rarely presents itself otherwise. However, the principles of sustainability may provide solutions to other problems that exist or that the community may soon be facing. Why should a community wait for a disaster before it pursues sensible objectives?

Sustainability goes far beyond just being an innovative disaster recovery strategy. It can inject the rejuvenating lifeblood that so many communities desperately need today. Communities that need this kind of help should consider incorporating sustainability into all development decisions—not just post-disaster redevelopment.

The most effective way to incorporate sustainability into acommunity is through adopting a "natural hazards element" within a local comprehensive plan. Following this concept, and the framework for smart growth from which it was derived would ensure that every development/redevelopment decision made, after a disaster or not, would be subject to the principles of sustainability. Holistic disaster recovery is really "sustainable redevelopment," which is a subset of a larger issue, sustainable development. As such, communities need to recognize that holistic disaster recovery is not the "end all," but rather one piece of the pie. Disaster recovery provides an opportunity to correct the unsustainable mistakes of the past. Disaster recovery is not, however, the driving force behind implementation of sustain ability, nor should it be. Disasters are simply catalysts for change. The post-disaster "window of opportunity" is a time when past mistakes can be assessed, and drawing upon experience, try to demonstrate the way for the future.

Chapter 9

Natural Disaster Risk Governance

9.1 Risk Governance Theories

National governments are expected to play a pivotal role in disaster risk management (DRM). The governments in the Asia and the Pacific region have developed a wide range of innovative solutions at the national level.

Governance, as defined by the UNDP (2004), is the exercise of political, economic, and administrative authority in the management of a country's affairs at all levels. It comprises mechanisms, processes, and institutions through which citizens and groups articulate their interests, exercise their legal rights, meet their obligations, and mediate their differences. Governance encompasses and also transcends government.

Governance refers generally to the set of instruments through which people living in a state, believing in common core values, govern themselves by the means of laws, rules, and regulations enforced by the state apparatus. It denotes a system of values, policies, and institutions by which society manages its economic, political, and social affairs through interaction among the state, civil society, and private sector.

It also denotes those processes and institutions through which citizens and groups articulate their interests, exercise their legal rights, meet their obligations, and mediate their differences. Governance has three components: economic, political, and administrative.

- Economic governance includes the decision-making processes that affect a country's economic activities and its relationship with other economies. This has major implications for equity, poverty, and quality of life.
- Political governance is the process of decision-making to formulate policies, including national disaster reduction and planning. The nature of this process and the way it brings together the state, on-state, and private-sector actors determines the quality of the policy outcomes.
- Administrative governance is a system of policy implementation and it requires the existence of well-functioning organizations at the

central and local levels. In the case of DRR, it requires functioning enforcement of building codes, land-use planning, environmental risk, and human vulnerability monitoring and safety standards.

Governance is widely regarded as the key to reducing disaster risks (Ahrens and Rudolph 2006; Castanos and Lomnitz 2009; UNISDR 2011; Wisner et al. 2004). Many developing countries need responsive, accountable, transparent, and efficient governance structures in DRM (Davis 2011; UNDP 2010). Governance is defined as an exercise of political, economic, and administrative authority in the management of a country's affairs. Governance influences how income and assets are distributed to the people, and it determines how the people protect themselves from hazards and how they access support in recovery (Turnbull and Pirson 2011). Since many developing countries lack the administrative, organizational, financial, and political capacity to effectively cope with disasters, the poor become particularly vulnerable. Since 1980, the economic costs of disasters in developing countries amounted to $1.2 trillion, about a third of all official development aid. Over that same period, low-income countries have accounted for only 9 percent of the total number of disasters, but 48 percent of the fatalities (World Bank 2012).

Governance is about the processes by which public policy decisions are made and implemented. It is the result of interactions, relationships, and networks between the different sectors (government, public sector, private sector, and civil society) and involves decisions, negotiations, and different power relations between stakeholders to determine who gets what, when, and how. The relationships between the government and different sectors of society determine how things are done and how services are provided. Governance is, therefore, much more than government or "good government" and shapes the way a service or set of services are planned, managed, and regulated within a set of political social and economic systems.

Governance is the exercise of political, economic, and administrative authority in the management of a country's affairs at all levels. It comprises formal and informal mechanisms, processes, and institutions through which citizens and groups articulate their interests, exercise their legal rights, meet their obligations, and mediate their differences. While governance encompasses government, it also includes all relevant groups in society, including private sector and CSOs, from household and local levels, to provincial, national, and international levels.

The concept of risk governance comprises a broad picture of risk. Not only does it include what has been termed "risk management" or "risk analysis," it also looks at how risk-related decision-making unfolds when a range of actors is involved, requiring coordination and possibly reconciliation between a profusion of roles, perspectives, goals, and activities. Indeed, the problem-solving capacities of individual actors, be they government, the scientific community, business players, NGOs, or civil society as a whole, are limited and often unequal to the major challenges facing society today. Risks such as those related to increasingly violent natural disasters, food safety, or critical infrastructures call for coordinated effort amongst a variety of players beyond the frontiers of countries, sectors, hierarchical levels, disciplines, and risk fields. Finally, risk governance also illuminates a risk's context by taking account of such factors as the historical and legal background, guiding principles, value systems, and perceptions as well as organizational imperatives.

Risk is an uncertain consequence of any event or an activity with respect to something that humans value (definition originally in Kates et al. 1985: 21). Risks always refer to a combination of two components: the likelihood or chance of potential consequences and the severity of consequences of human activities, natural events, or a combination of both. Such consequences can be positive or negative, depending on the values that people associate with them. International Risk Governance Council (IRGC) does not cover all risk areas but confines its efforts to (predominantly negatively evaluated) risks that lead to physical consequences in terms of human life, health, and the natural and built environment. It also addresses impacts on financial assets, economic investments, social institutions, cultural heritage, or psychological well-being as long as these impacts are associated with the physical consequences. In addition to the strength and likelihood of these consequences, the framework emphasizes the distribution of risks over time, space, and populations. In particular, the timescale of appearance of adverse effects is very important and links risk governance to sustainable development (delayed effects).

The focus on risk should be seen as a segment of a larger and wider perspective on how humans transform the natural environment into a cultural environment with the aims of improving living conditions and serving human wants and needs (Turner et al. 1990). These transformations are performed with a purpose in mind (normally a benefit to those who initiate them). When implementing these changes, intended (or tolerated) and unintended consequences may occur that meet or violate other dimensions of what humans value. Risks are not taken for their own sake; rather they are, actively or passively, incurred because of their being an integral factor in the very activity that is geared toward achieving the particular human need or purpose. In this context, it is the major task of risk assessment to identify and explore, preferably in quantitative terms, the types, intensities, and likelihood of the (normally undesired) consequences related to a risk.

In addition, these consequences are associated with special concerns that individuals, social groups, or different cultures may attribute to these risks. They also need to be assessed for making a prudent judgement about the tolerability or acceptability of risks. Once that judgement is made, it is the task of risk management to prevent, reduce, or alter these consequences by choosing appropriate actions. As obvious as this distinction between risk and concern assessment (as a tool of gaining knowledge about risks) and risk management (as a tool for handling risks) appears at first glance, the distinction becomes blurred in the actual risk governance process.

The most complex questions emerge, however, when one looks at how society and its various actors actually handle risk. In addition to knowledge gained through risk assessments and/or option generation and evaluation through risk management, the decision-making structure of a society is itself highly complicated and often fragmented. Apart from the structure itself—the people and organizations that share responsibility for assessing and managing risk—one must also consider the need for sufficient organizational capacity to create the necessary knowledge and implement the required actions, political and cultural norms, rules and values within a particular societal context, and the subjective perceptions of individuals and groups. These factors leave their marks on the way risks are treated in different domains and sociopolitical cultures. To place risk within a context of—sometimes closely interwoven—decision-making structures such as those prevalent in governments and related authorities, in the corporate sector and industry, in the scientific community, and in other stakeholder groups is of central concern.

9.2 Disaster Risk Governance and Decision-making

In the last decade, the term "governance" has experienced tremendous popularity in the literature on international relations, comparative political science, policy studies, sociology of environment, and technology as well as risk research. On a national scale, governance describes structures and processes for collective decision-making involving governmental and nongovernmental actors (Nye and Donahue 2000). Governing choices in modern societies is seen as an interplay between governmental institutions, economic forces, and civil society actors (such as NGOs). At the global level, governance embodies a horizontally organized structure of functional self-regulation encompassing state and nonstate actors bringing about collectively binding decisions without superior authority (c.f. Rosenau 1992; Wolf 2002). In this perspective, nonstate actors play an increasingly relevant role and become more important since they have decisive advantages of information and resources compared to single states. It is useful to differentiate between horizontal and vertical governance (Benz and Eberlein 1999; Lyall and Tait 2004). The horizontal level includes the relevant actors in decision-making processes within a defined geographical or functional segment (such as all relevant actors within a community, region, nation, or continent); the vertical level describes the links between these segments (such as the institutional relationships between the local, regional, and state levels). "Risk governance" involves the "translation" of the substance and core principles of governance to the context of risk and risk-related decision-making. The risk governance includes the totality of actors, rules, conventions, processes, and mechanisms concerned with how relevant risk information is collected, analyzed, and communicated and how management decisions are taken. Encompassing the combined risk-relevant decisions and actions of both governmental and private actors, risk governance is of particular importance in, but not restricted to, situations where there is no single authority to take a binding risk management decision but where, instead, the nature of the risk requires the collaboration of, and coordination between, a range of different stakeholders. Risk governance, however, not only includes a multifaceted, multi-actor risk process but also calls for the consideration of contextual factors such as institutional arrangements (e.g., the regulatory and legal framework that determines the relationship, roles and responsibilities of the actors and coordination mechanisms such as markets, incentives or self-imposed norms) and political cultures, including different perceptions of risk. When looking at risk governance structures, there is no possibility of including all the variables that may influence the decision-making process; there are too many. Therefore, it is necessary to limit one's efforts to those factors and actors that, by theoretical reasoning and/or empirical analysis, are demonstrably of particular importance with respect to the outcome of risk governance. The IRGC has highlighted the following aspects of risk governance that extend beyond risk assessment and risk management (Table 9.1).

- The structure and function of various actor groups in initiating, influencing, criticizing, and/or implementing risk policies and decisions
- Risk perceptions of individuals and groups
- Individual, social, and cultural concerns associated with the consequences of risk

Table 9.1:
Components of Pre-assessment in Handling Risks

Pre-assessment Components	Definition	Indicators
1. Problem framing	Different perspectives of how to conceptualize the issue	• Dissent or consent on goals of selection rule • Dissent or consent on relevance of evidence • Choice of frame (risk, opportunity, fate)
2. Early warning	Systematic search for new hazards	• Unusual events or phenomena • Systematic comparison between modeled ad observed phenomena • Novel activities or events

3. Screening (risk assessment	Establishing a procedure for screening hazards and risks, and determining assessment and management route	• Screening in place?
		• Criteria for screening:
		  o Hazard potential
		  o Persistence
		  o Ubiquity, etc.
		• Criteria for selecting risk assessment procedures for:
		  o Known risks
		  o Emergencies, etc.
		• Criteria for identifying and measuring social concerns

4. Scientific conventions for risk assessment and concern assessment	Determining the assumptions and parameters of scientific modeling and evaluating methods and procedures for assessing risks and concerns	• Definition of no adverse effect levels (NOAEL) • Validity of methods and techniques for risk assessments • Methodological rules for assessing concerns

Source: IRGC (2006).

- The regulatory and decision-making style (political culture)
- The requirements with respect to organizational and institutional capabilities for assessing, monitoring and managing risks (including emergency management)

In addition to these analytical categories, this document also addresses best practice and normative aspects of what is needed to improve governance structures and processes (EU 2001a). With respect to best practice, it is interesting to note that often risk creators, in particular when directly affected by the risk they generate, engage in risk reduction and avoidance out of self-interest or on a voluntary basis (e.g., industry "gentleman's agreements," self-restrictions, and industry standards). Other stakeholders' efforts in risk governance, therefore, have to be coordinated with what is tacitly in place already. The emphasis here is on cooperative models of public-private partnerships forming a governance system that aims at effective, efficient, and fair risk management solutions.

9.3 Disaster Risk Governance and Decision Framework

The framework's risk process, or risk handling chain, is illustrated in Figure 9.1. It breaks down into three main phases: "pre-assessment," "appraisal," and "management." A further phase, comprising the

Figure 9.1:
Risk Governance Framework

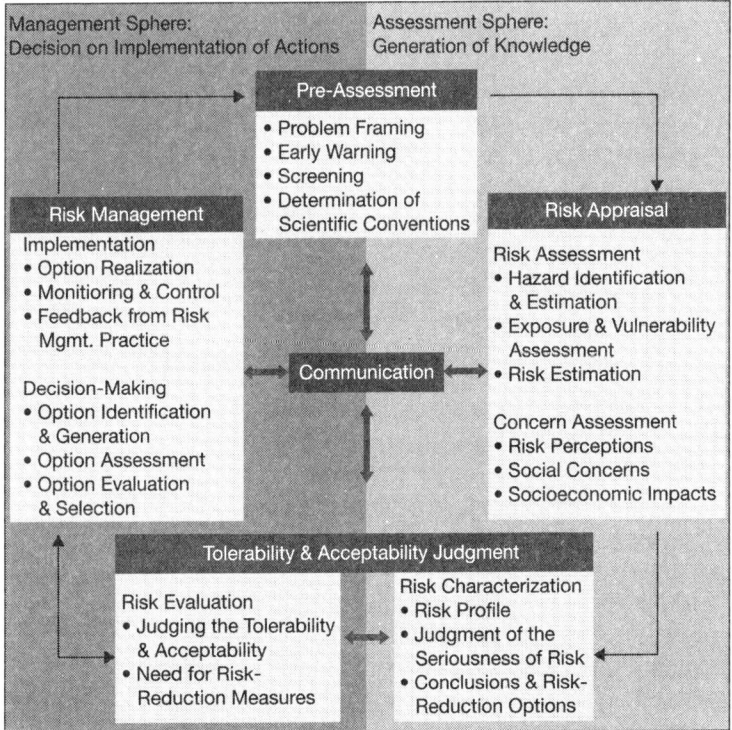

Source: Modified after IRGC (2006).

"characterization" and "evaluation" of risk, is placed between the appraisal and management phases and, depending on whether those charged with the assessment or those responsible for management, are better equipped to perform the associated tasks and can be assigned to either of them— thus concluding the appraisal phase or marking the start of the management phase. The risk process has "communication" as a companion to all phases of addressing and handling risk, and it is itself of a cyclical nature. However, the clear sequence of phases and steps offered by this process is primarily a logical and functional one and will not always correspond to reality.

The purpose of the pre-assessment phase is to capture both the variety of issues that stakeholders and society may associate with a certain risk as well as the existing indicators, routines, and conventions that may prematurely narrow down, or act as a filter for, what is going to be addressed as risk. What counts as a risk may be different for different groups of actors. The first step of pre-assessment, risk framing, therefore, places particular importance on the need for all interested parties to share a common understanding of the risk issue(s) being addressed or, otherwise, to raise awareness amongst those parties of the differences in what is perceived as a risk. For a common understanding to be achieved, the actors need both to agree with the underlying goal of the activity or event generating the risk and be willing to accept the risk's foreseeable implications on that very goal. A second step of the pre-assessment phase, early warning and monitoring, establishes whether signals of the risk exist that would indicate its realization. This step also investigates the institutional means in place for monitoring the environment for such early warning signals. The third step, pre-screening, takes up and looks into the widespread practice of conducting preliminary probes into hazards or risks and, based on prioritization schemes and existing models for dealing with risk, of assigning a risk to pre-defined assessment and management "routes." The fourth and final step of pre-assessment selects major assumptions, conventions, and procedural rules for assessing the risk as well as the emotions associated with it.

The objective of the risk-appraisal phase is to provide the knowledge base for the societal decision on whether or not a risk should be taken and, if so, how the risk can possibly be reduced or contained. Risk appraisal thus, comprises a scientific assessment of both the risk and the questions that stakeholders may have concerning its social and economic implications.

The first component of risk appraisal, risk assessment, seeks to link a potential source of harm, a hazard, with likely consequences, specifying probabilities of occurrence for the latter. Depending on the source of a risk and the organizational culture of the community dealing with it, many different ways exist for structuring risk assessment. Despite such diversity, three core steps can be identified. These are the identification and, if possible, estimation of the hazard; an assessment of related exposure and/or vulnerability; and an estimation of the consequent risk. The latter step—risk estimation—aggregates the results of the first two steps and states, for each conceivable degree of severity of the consequence(s), a probability of occurrence. Confirming the results of risk assessments can be extremely difficult, in particular when cause–effect relationships are hard to establish, when they are instable due to variations in both causes and effects, and when effects are both scarce and difficult to understand. Depending on the achievable state and quality of knowledge, risk assessment is, thus, confronted with three major challenges that can best be summarized using the risk categories outlined previously—"complexity," "uncertainty," and "ambiguity." For a successful outcome to the risk process and, indeed, overall risk governance, it is crucial that the implications of these challenges are made transparent at the conclusion of risk assessment and throughout all subsequent phases.

Equally important to understanding the physical attributes of the risk is the detailed knowledge of stakeholders' concerns and questions—emotions, hopes, fears, and apprehensions—about the risk as well as the likely social consequences, economic implications, and political responses. The second component of risk appraisal, concern assessment, thus, complements the results from risk assessment with insights from risk perception studies and interdisciplinary analyses of the risk's (secondary) social and economic implications.

The most controversial phase of handling risk, risk characterization and evaluation aims at judging a risk's acceptability and/or tolerability. A risk deemed "acceptable" is usually limited in terms of its negative consequences so that it is taken on without risk-reduction or mitigation measures being envisaged. A risk deemed "tolerable" links undertaking an activity—which is considered worthwhile for the value addition or benefit it provides—with specific measures to diminish and limit the likely adverse consequences. This judgment is informed by two distinct, but closely related efforts to gather and compile the necessary knowledge that, in the case of tolerability, must additionally support an initial understanding of required risk-reduction and mitigation measures. While risk characterization compiles scientific evidence based on the results from the risk appraisal phase, risk evaluation assesses broader, value-based issues that also influence the judgement. Such issues, which include questions such as the choice of technology, societal needs requiring a given risk agent to be present, and the potential for substitution as well as for compensation, reach beyond the risk itself and into the realm of policy-making and societal balancing of risks and benefits.

The risk management phase designs and implements the actions and remedies required to tackle risks with an aim to avoid, reduce, transfer, or retain them. Risk management, thereby, relies on a sequence of six steps that facilitate systematic decision-making. To start with, and based on a reconsideration of the knowledge gained in the risk-appraisal phase and while judging the acceptability and/or tolerability of a given risk, a range of potential risk management options are identified. The options are then assessed with regard to such criteria such as effectiveness, efficiency, minimization of external side effects, sustainability, etc. These assessment results are next complemented by a value judgment on the relative weight of each of the assessment criteria, allowing an evaluation of the risk-management options. This evaluation supports the next step in which one (or more) of the of risk management options is selected, normally after consideration of possible trade-offs that need to be made between a number of second-best options. The final two steps include the implementation of the selected options and the periodic monitoring and review of their performance.

Based on the dominant characteristic of each of the four risk categories ("simple," "complexity," "uncertainty," and "ambiguity"), it is possible to identify specific safety principles and, consequently, design a targeted risk-management strategy (Table 9.1). "Simple" risk problems can be managed using a "routine-based" strategy that draws on traditional decision-making instruments and best practice as well as time-tested trial-and-error. For "complex" and "uncertain" risk problems, it is helpful to distinguish the strategies required to deal with a risk agent from those directed at the risk-absorbing system. Complex risks are thus usefully addressed on the basis of "risk-informed" and "robustness-focused" strategies, while uncertain risks are better managed using "precaution-based" and "resilience-focused" strategies. Whereas the former strategies aim at accessing and acting on the best available scientific expertise and at reducing a system's vulnerability to known hazards and threats by improving its buffer capacity, the latter strategies pursue the goal of applying a precautionary approach in order to ensure the reversibility of critical decisions and of increasing a system's coping capacity to the point where it can withstand surprises. Finally, for "ambiguous" risk problems, the appropriate strategy consists of a "discourse-based" strategy that seeks to create tolerance and mutual understanding of conflicting views and values with a view to eventually reconciling them.

9.4 Good Governance and Disaster Risk Reduction

Good governance that drives the achievement of development results must also now rise to the challenge of achieving the equitable and sustainable development will secure our common future. Good governance for building disaster and climate resilience takes place when capable, accountable, transparent, inclusive, and responsive governments work together with the civil society, private-sector, and at-risk populations to create an enabling environment to improve society's ability to prepare and respond to disasters and their capacity to adapt to changes in the climate.

The very concept of "good governance" at local levels denotes quality, effectiveness, and efficiency of local administration and public service delivery; the quality of local public policy and decision-making procedures, their inclusiveness, their transparency, and their accountability; and the manner in which power and authority are exercised at the local level. While local government is the essential institutional building block for local governance, the wider governance sphere comprises a set of state and nonstate institutions, mechanisms, and processes through which public goods and services are delivered to citizens and through which citizens can articulate their interests and needs, mediate their differences, and exercise their rights and obligations.

The concepts of local governance and decentralization, at times used interchangeably, are related but different. Decentralization is primarily a national, political, legislative, institutional, and fiscal process. While local governance can be affected by the decentralization processes, UNDP has a holistic approach to defining the field of local governance and decentralization by using the concept of *decentralized governance for development*. Decentralized governance is not a panacea or a quick fix. The key to human development-friendly, decentralized governance is to ensure that the voices and concerns of the poor, especially the women, help guide its design, implementation and monitoring.

Decentralized governance for development is considered to be a key area of democratic governance, which in turn is crucial to attaining human development and the SDGs. For development and governance to be fully responsive and representational, people and institutions must be empowered at every level of society—national, provincial, district, city, town, and village.

There are four main *objectives* to undertaking an assessment of governance at the local level.

1. *Diagnostic:* An assessment will be done to identify a problem and its scope.
2. *Monitoring:* An assessment will be done at regular intervals to keep a check on the success or failure of an initiative, policy, or program.
3. *Evaluation:* An assessment will be done to assess whether an initiative, policy, or program has achieved its pre-defined results and outcomes.
4. *Dialogue:* An assessment will also serve to engage citizens and communities in informed discussions about shared goals and priorities.

Common stakeholders in most assessments of local governance will include:

1. *Local government representatives:* Local political and administrative leaders are crucial for launching, implementing, and using governance assessments. In many cases, local government will be in the driving seat of the assessment process. Local government representatives are especially active in the partnership, promoting, development, action planning/dissemination, and policy implementation phases.
2. *Central government representatives:* The central government (e.g., the ministry/department responsible for local government) is important in assessments as it has a significant role to play in capacity development of local authorities, including in the setting and maintaining of standards of performance, monitoring, ensuring the establishment of mechanisms of accountability, and the formulation and approval of local government policy frameworks. The central government may be especially active in the policy implementation phase, integrating the assessment results into its local government-monitoring mandate.
3. *Local government associations:* These are comprised of local councils and express their collective voice in the national arena. Some illustrative examples of their roles and objectives include: shaping public debate on local government issues, influencing policy at the national level, supporting capacity development that enables councils and their partnerships to deliver services, and working to enhance democratic accountability and transparency in local government institutions. Where these exist, they represent an important stakeholder in assessments, especially in the partnership promotion, development, and policy implementation phases.
4. *Civil society organizations:* The existence of a vibrant and diverse civil society is an important indicator of good local governance. CSOs also need to play a role in the assessment process, including

identifying and drawing attention to local governance deficits as well as using their expertise in data collection and analysis in the implementation of the assessment. CSOs are potentially active in every phase of the assessment process.

5. *Community-based organizations:* These are a form of organized citizens and have a role in mobilizing local people around community development actions and to act as a watchdog. CBOs are also important for reflecting the views, rights, and interests of vulnerable or marginalized groups in communities.

9.5 Disaster and Emergency

In any situation where the community becomes annihilated by an incident that is beyond their scope of control and cannot handle the resources effectively. requiring external assistance is considered a disaster (Madry 2015). On the other hand, emergency is "an event that can be responded by using resources available at hand with no need of external assistance" (UNOOSA 2015). Disaster management is "an applied science which seeks, by the systematic observation and analysis of disasters, to improve measures relating to prevention, mitigation, preparedness, emergency response and recovery" (Carter 1999, cited in Pettit and Beresford 2005). It can also be described as "the action or initiative taken to formulate policy, strategy, decision-making and build legal, financial and administrative interventions within a disaster cycle" (Bhandary et al. 2014). According to UNISDR, the terms "disaster management" and "emergency management" can be used interchangeably. Emergency management consists of four different phases: mitigation, preparedness, response, and recovery (Alexander 2002; Waugh and Streib 2006). Emergency response and rescue (ERR) is considered a key phase in emergency management (Subramaniam et al. 2010).

9.6 Emergency Response

Response is defined as "the provision of emergency services and public assistance during or immediately after a disaster in order to save lives, reduce health impacts, ensure public safety and meet the basic subsistence needs of the people affected" (UNISDR 2009). It is also known for its intricate process that involves extreme time pressures and high uncertainty. Emergency response in post-disaster of earthquake will be discussed focusing on the immediate aftermath of the disaster and first-hand response in post-disaster situation. The professionals involved in handling the emergency situation in a community or nation who are specialized in conducting emergency activities are emergency managers (Madry 2015). After a disaster strikes, the immediate response is the SAR of affected people in danger, emergency shelter, medical care, and food distribution by the community members, police officers, and other local agencies. For effective and smoother coordination, an emergency response plan is a must for any organization. Prioritization of information sharing and decision-making is critical for disaster response (Chen et al. 2011).

Earthquakes are catastrophic natural disasters that not only cause death but also cause injuries to victims trapped inside collapsed buildings (Bartels and VanRooyen 2012). Almost half of the survivors are rescued by the local people immediately following the earthquake with the help of security forces in the early hours of disaster, while the remaining victims tend to get trapped until professional rescuers rescue them (Mochalski et al. 2015). In extreme events like an earthquake, the first 72 hours are critical for SAR efforts (Ochoa and Santos 2015). In urban areas, advance equipment designed for scrapping the concrete structures, debris removal, and lifting are required to rescue trapped survivors. For instance, the Foreign Exchange Management Act (FEMA) uses a structural triage for collecting information where the priorities are probability of victims, occupancy, structural failure, access time, and other known conditions (Chen et al. 2011). Also, they have used methods of marking buildings already searched providing information to other SAR teams and avoiding redundancy. The example provides a systematic system of response following a disaster. In contrast, many developing countries and least developed countries lack advanced SAR equipment, and only a handful of people can be rescued in the early hours of disasters through local gear. These countries have to rely heavily on developed countries for advanced SAR in urban areas. SAR teams usually comprise fire fighters, security forces, structural engineers, medical experts, etc., depending on the location and rescue operation.

Rapid response is required in unpredictable disasters for minimization of fatalities and property damage. Sending out sufficient rescue teams with equipment and emergency resources on dot to affected citizens will benefit with speedy emergency activities; however, earthquake response and rescue is still a global challenge specifically in developing countries (Doyle 1996). Rescue and response after an earthquake occur in limited time and information, making decisions harder, which further validates the difficulty faced in emergency response and rescue operations (Kusumasari and Alam 2012). Many of the developing countries have poorer building structures causing devastation of massive life and properties as observed in the Sumatra earthquake in 2004, Haiti earthquake in 2010, and Pakistan earthquake in 2005. Furthermore, the inadequacy can be observed during rescue and response management such as absence of professional relief agencies, inadequate aid materials and storage space, and lack of appropriate rescue equipment, which leads to delay of rescue and response operation (Lu and Xu 2014). The incompetence of developing countries leaves them heavily dependent on international agencies for professional response.

In any emergency management, government performance can increase by integrating information from multiple agents (Schooley and Horan 2007). Marincioni (2007) concluded that interpersonal interaction and knowledge sharing were the key factors. Multi-agency collaboration, optimal resource management, and local knowledge are significant for good response (Gopalakrishnan and Okada 2007). Various research in emergency response has shown specific topics under response and rescue being studied. Most of them have reported time, interorganizational coordination, and relief mobilization as three critical rescue and response indicators (Lu and Xu 2014).

Prompt response from emergency responders following any disaster determines the degree of loss examined (Subramaniam et al. 2010). Victim survival rate rises with effective response and rescue, which has been classified as the action taken promptly to distribute relief and deploy rescue with minimum delay (Edrissi et al. 2013). A nation has a set of pre-planned actions available for emergency that are only valued if resources are available to mobilize them (Valcik and Tracy 2013). Participation of multiple organizations with collaboration help in effective response and rescue. Different emergency agencies such as firefighting teams, medical teams, and police have to come together and deliver a joint and supportive teamwork to produce an effective ERR (Granot 1997). When groups analyze and understand a scenario better prior to the disaster event, the ERR operations become more effective. The ERR includes multifaceted relations among government agencies, NGOs, individuals, and the private sector even at a local level (Kapucu 2008). Kapucu suggested that greater importance should be given to local, state, and federal resource cooperation and coordination than to intergovernmental and interorganizational response as they are difficult task. A very important step in emergency situation is an understanding and unity among various stakeholders to mobilize the resources.

Chen et al. (2008) recommended a coordination patterns of the emergency response life cycle that had five elementary elements, that is, task flow, resource, information, decision, and responder. Allocating the available resources in the optimal way is a difficult task in the ERR period (Fiedrich et al. 2000). Time demand and resource availability are important parameters of disaster relief distribution. Different research suggest plans to mobilize response. For example, O¨zdamar et al. (2004) suggested an urgent logistics planning model to response to natural hazards. After earthquake events, numerous case studies paid attention to specified ERR operations. The various themes under which the characteristics of the formal and emergent organizations' response to the 1985 Mexico earthquake were summarized include the emergency response coordination, the adequacy of organizational resources, the types of organizations that respond to disasters, and the performance of key organizational personnel (Quarantelli 1993). There were certain gaps not covered by the government programs in the Indian Ocean earthquake and the tsunami of December 26, 2004, and these gaps were being filled by an integrated network of twelve NGOs (Kilby 2008). The collective response by Japanese Central and local governments and other civil defense forces to handle the unpredicted scale of the 2011 Tohoku earthquake and the tsunami was also studied (Suzuki and Kanko 2013). After the 2008 Wenchuan earthquake, there was an upgrade in ERR of China with medical experts fully trained during Lushan earthquake in 2013 (Chen and Booth 2011; Yang et al. 2013).

9.7 Role of the Government in Disaster Response

Policies and regulation are important to respond to disasters but equally important is its implementation during disaster response. The role of district heads responding to a disaster according to the needs of community and the international agencies supporting the states differ widely in terms of time and space as well as the type of disaster. The international assistance started between 1970 and 1980, with collaboration from UN agencies, International Federation of Red Cross (IFRC), and other INGOs after facing the hassle of corruption within the government (Harvey 2010). While the arrangement of the international organizations are significant, the role of the government also plays an essential role in humanitarian assistance. In any response situation, it is critical to have actions taken by the government that give more meaning to the conduct and delivery of aids from an international organization (Harvey 2010). Government's consent is a crucial factor in emergency response for international organization to perform response actions in certain states, which marks the importance of policies and legislation of the affected country. The importance of the government in disaster field have been identified recently, such as in Sphere's standard that has mentioned about the responsibilities of government authorities involved in guidelines, policies, and legislation in national context (Sphere 1994). Although little has been mentioned on the structure and management of response in disaster for government, it still plays a vital role of coordination and communication affecting the efficiency of response in any kind of disaster.

9.8 Role of the Military/Army

The military has a crucial role to play in the context of natural disasters. But, existing guidelines such as the UN Military and Civil Defence Asset (MCDA) Register and the Oslo Guidelines has focused on international forces' deployment in emergencies rather than using the military of a nation (UN 2003). The Inter-Agency Standing Committee (IASC)'s reference paper, Civil–Military Relationship in Complex Emergencies, covers the role of national militaries (IASC 2004). In recent years, some of the countries' militaries have a portrayed prodigious role during disasters. For example, the Pakistan military had a positive and imperative role during response of the 2005 earthquake. IASC has mentioned about the military being the primary provider of food, medical services, livelihood, and shelter to the affected citizens (IASC 2006). Military has played a vital role in disaster response and recovery in India and Indonesia during the 2001 Gujarat earthquake, and the 2004 tsunami and earthquake respectively. Yet, the international organizations, realizing the importance of coordination, also putting aside concerns of neutrality and independence, find it difficult to participate together with military on different levels (Wilder 2008). A military having good coordination capacities, decision-making skills, logistics, and willingness to learn and listen can mobilize a most effective humanitarian response in a large-scale disaster (Harvey 2010; Wilder 2008). Any large-scale disaster will benefit from cooperating military and international agencies working together to assist in response to affected citizens for response, relief, and longer term recovery.

9.9 Role of Coordination in Disasters

Coordination has been defined as a planned response to tasks within an organization or among actors where they are dependent on one another for best results (Jahre and Jensen 2010; Tatham et al. 2012). It is an important aspect of humanitarian relief operation to meet the goals of an organization as well as create a better environment to cooperate among different organizations. Relief environments consist of governments, local and international organizations, militaries, and private companies who are involved with their mandates, capacity, logistics, and specific interests. Also, a single actor cannot respond effectively without the help of other actors in a large-scale disaster (Balcik et al. 2010; Bui et al. 2000). Coordinating and collaborating with numerous humanitarian actors at a chaotic time like a disaster is a challenge. If a country doesn't have good governance to take a lead during the post-disaster phase, the relief operation becomes unsuccessful. Other factors for coordination challenge are a large amount of humanitarian organizations at the ground and insufficient local resources and manpower to maintain coordination. Coordination has become a weakness faced by humanitarian organizations (Rey 2001). It has been difficult to coordinate, mostly during large-scale disasters such as the 2004 tsunami in the Indian Ocean and the Darfur crisis (Balcik et al. 2010; Bui et al. 2000). Factors such as number and diversity of actors, funding structure, media and competition for funding, resource security, coordination cost, and unpredictability have important roles during coordination. Two types of coordination system have been derived based upon the actors involved: horizontal and vertical coordination. Horizontal coordination occurs within a chain of actors by coordinating among organizations at same level, whereas vertical coordination consists of organizations coordinating at different levels (Balcik et al. 2010). To clarify it more, vertical coordination involves suppliers, humanitarian actors, and customers, whereas horizontal coordination consists of external and internal organizations along with their competitors. The cluster approach has been used many times among humanitarian actors to meet coordination and cooperation. Through clusters, organizations have profited from cooperation as well as competition for better provision of information, relief resources, and services (Jahre and Jensen 2010). However, lack of coordination among different clusters when supplying relief from international to local level has also been noted and is considered a key challenge for better relief operation (Adinolfi et al. 2005; Jahre and Jensen 2010; Stoddard et al. 2007). For the success of coordination, a strategy should be formed that should be tactical to deal with problems and operational at a crisis situation (Jahre and Jensen 2010). The review elucidates coordination as a broad term that incorporate different factors to operate smoothly considering all the actors cooperate together during a disaster.

9.10 Risk Transfer/Insurance Mechanism

Over the past two decades, there has been increase in disaster, costing human and economic losses. This is due to ever-increasing vulnerabilities of people to natural disasters. There is a need to reduce disaster risks by improving the capabilities of people by ensuring preparedness, mitigation, and response and planning processes at various levels. The objective is to look at the entire cycle of disaster management in reducing risk and linking it to developmental planning process. In the past, disasters were viewed as isolated events, responded to by the governments and various agencies without taking into account the social and economic causes and long-term implications of these events. In short, disasters were considered as on-off emergencies.

However, the devastation and losses can be prevented to a great extent. While the hazard events cannot be prevented from occurring, the impact can be reduced on the basis of pre-disaster activities for mitigating risks. One of the most crucial aspects of DRR, mitigation activities, forms a part of a coordinated strategy or plan. Natural hazard mitigation is an important policy issue because "monetary damages from natural disasters are reaching catastrophic proportions" and is only expected to increase over the years (Figure 9.2).

9.11 Risk Management Process

Risk management is a dynamic and well-established discipline practiced by many companies around the world. Traditional forms of risk management—loss control, loss financing, and risk reduction, arranged through mechanisms such as insurance and derivatives—have been actively used by companies for many decades, and they remain an essential element of most corporate strategies. But newer forms of risk protection—including

Figure 9.2:
Economic Impacts of (Left) and Exposure to (Right) River Floods in the Asia-Pacific Region

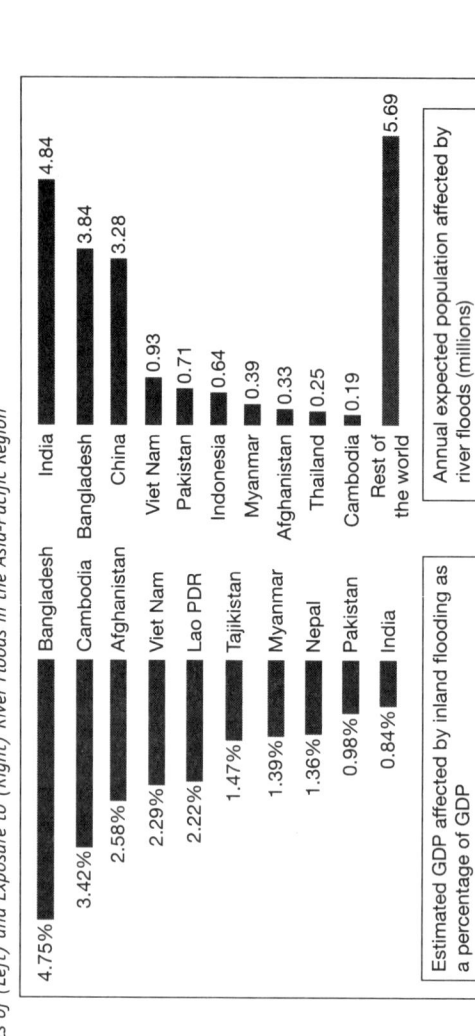

Source: ESCAP (2016).

those from the *alternative risk transfer* (ART) market, which we define as the combined marketplace for innovative insurance and capital market solutions—form an increasingly important element of overall risk management. Indeed, a firm seeking to develop an optimal risk management structure should consider all available risk techniques before deciding on a strategy. In this chapter, we explore issues related to risk and return, general risk management processes and techniques, and fundamental risk concepts and measures. We then consider the nature of the ART market, its background and origins, and the forces of market/product convergence.

Active risk management is increasingly important for companies interested in maximizing enterprise value. Creating a framework by which to properly evaluate financial and operating risks, which may be of a pure or speculative nature, demands proper focus on the expected losses, probability of ruin, risk aversion, and expected utility, as well as the techniques of loss control, loss financing, and risk reduction. Key drivers of the disciplined risk management process include analysis of costs and benefits, establishment of pre- and post-loss management goals, and definition of a risk philosophy.

Risk is a broad, complex, and vitally important topic that touches on virtually all aspects of modern corporate operation. We begin by defining *risk*, in its most general form, as an uncertainty associated with a future outcome or event. To apply this more specifically to corporate activities, we can say that risk is the expected variance in profits, losses, or cash flows arising from an uncertain event. Other terms commonly associated with risk such as *peril* and *hazard* are often encountered in the risk management industry.

A *peril*, for instance, is a cause of loss, while a *hazard* is an event that creates, or increases, peril. While both have a bearing on risk, risk itself is a broader concept. Companies are exposed to a wide range of risks that might, at any time, include such things as business interruption, catastrophic and noncatastrophic property damage, product recall/liability, directors and officers liability, credit default/loss, workers' compensation, and environmental liabilities. These risks must be managed if the market value of the company is to be increased or, at a minimum, if the probability of financial distress is to be lowered. Some of the risks can be retained as part of core business operations, while others are best transferred to others when it is cost-effective to do so. Though we shall consider risks in more detail later, we begin by classifying them broadly as operating risks and financial risks.

Operating risk: The risk of loss arising from the daily physical (nonfinancial) operating activities of a firm

Financial risk: The risk of loss arising from the financial activities of a firm. Operating and financial risks can be decomposed further. For example, within the general category of operating risks, we can consider subclasses such as personal liability and commercial property/casualty (P&C) liability. Within commercial P&C liability, we might differentiate between losses related to commercial property (direct/indirect), machinery, transportation (inland/marine), crime, commercial liability, commercial auto, workers' compensation, and employers' liability. Similar decomposition is possible within the broad category of financial risks, where we might first divide exposures into credit risk, market risk, liquidity risk, and model risk. A category such as market risk might then be segregated into directional risk, volatility risk, time decay risk, curve risk, basis risk, spread risk, correlation risk, and so forth. We can also categorize financial and operating risks as being pure or speculative (Figure 9.3).

Pure risk: Risk that has the prospect of loss/no loss, but no prospect of gain.

Speculative risk: Risk that has the possibility of loss, no loss, or gain.

The insurance market is premised on two fundamental mechanisms: the transfer of exposure from a single party to a broad group, and the sharing of losses by all those in the group. Risk transfer, as the name suggests, occurs when one party pays a second party a small, certain cost (e.g., a risk premium) in exchange for coverage of uncertain losses; this is equal to a shifting of exposures. The risk-averse firm, in defining its risk philosophy, may decide to shed an exposure by transferring it through one of several different mechanisms, including insurance/reinsurance, derivatives, or hybrid structures. The amount of risk that a firm transfers is a function of overall tolerance (i.e., its level of risk aversion), the specific benefits it hopes to derive, and the total cost; this is often determined in a cost–benefit analysis framework (Figure 9.4).

9.12 Catastrophic Events in South Asia and Economic Consequences

The Intergovernmental Panel on Climate Change (IPCC 2007, 2012) forecasted that the intensity and frequency of natural disasters would

Figure 9.3:
Generic Risk Management Process

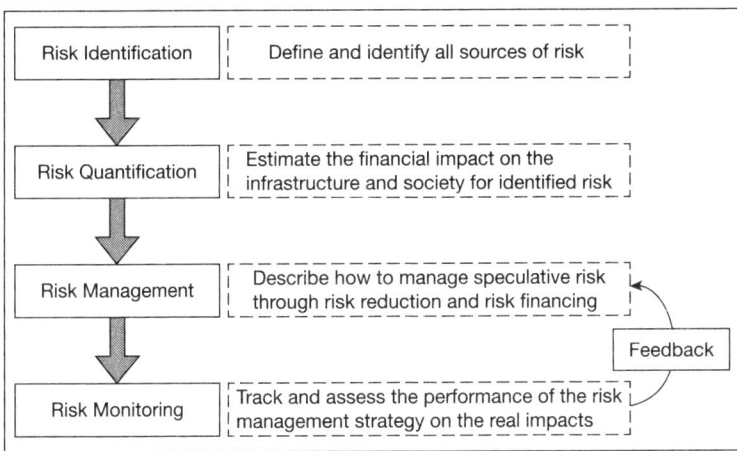

Source: Authors.

Figure 9.4:
Risk Management Techniques

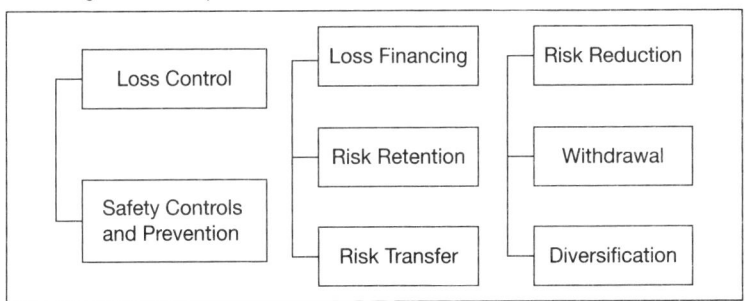

Source: Authors.

increase in the coming years. The number of natural catastrophes has increased in recent years affecting the lives and livelihoods of people especially in developing and less developed economies (IRIN 2005). Every year, some part of South Asia is affected by a natural disaster, such as cyclones, floods, droughts, or even heat and cold waves. The worst catastrophic events in the last decade affecting the economies in South Asia are presented in Table 9.2.

Table 9.2:
South Asia: 25 Worst Natural Disasters During 2006–15

Victims	Insured Loss[a]	Year	Event	Country
8,960	160	2015	Earthquake Mw 7.8, avalanche on Mt. Everest, aftershocks	Nepal, India, China, Bangladesh
5,748	500	2013	Floods caused by heavy rains	India
4,234	–	2007	Cyclone Sidr, floods	Bangladesh, India
2,248	–	2015	Heatwave	India
1,980	–	2010	Floods caused by heavy rains	Pakistan
1,500	–	2007	Floods caused by heavy rains	India, Bangladesh
1,270	–	2015	Heatwave	Pakistan
950	–	2008	Floods caused by monsoon rains	India
678	–	2007	Floods caused by heavy rains	Bangladesh
665	–	2014	Severe monsoon floods	India, Pakistan
605	–	2006	Floods, mudslides, heavy rains	Nepal
600	–	2006	Heavy storm causing floods	India, Bangladesh
531	–	2013	Heatwave	India
520	–	2009	Floods caused by monsoon rains	India
456	–	2011	Floods caused by heavy rains	Pakistan
450	–	2006	Floods caused by heavy rains	India
399	–	2013	Earthquake Mw 7.7, aftershocks	Pakistan
399	–	2015	Earthquake Mw 7.5	Afghanistan, Pakistan, India
388	–	2013	Coldwave	India, Bangladesh, Nepal
361	–	2012	Floods after monsoonal rains	Pakistan
350	407	2006	Floods caused by monsoon rains	India
340	–	2007	Cyclone Yemyin, heavy rains	Pakistan
300	–	2008	Earthquake Mw 6.4	Pakistan
300	51	2009	Floods caused by heavy rains	India
289	755	2015	Severe flash floods in Chennai	India

Source: Compiled from data released by Swiss Re (various issues).
Note: [a]In million US$.

9.13 The Insurance Industry in South Asia

In 2015, the emerging Asia share in the world insurance market was 11.51 percent (Swiss Re 2016). If we further consider only South Asia, the share falls below 2 percent in 2015, indicating that the insurance industry is relatively underdeveloped. For example, there is hardly any information on insurance outreach in Bhutan. The Royal Monetary Authority reports that finance and insurance sector contributed 5.4 percent to the country's nominal GDP in 2014 with two insurance companies providing life and nonlife insurance coverage to little over 100,000 policyholders in a country with less than a million population.

The insurance industry in Nepal is also underdeveloped. Nepal reported the lowest insurance penetration of 1.4 percent in 2011–12 (Beema Samiti 2012) compared to other South Asian economies. The total premium collected by life and nonlife insurers amounted to NRs. 244 million in 1990, which increased to NRs 30.28 billion in 2013–14.

Following liberalization of the insurance industries in Bangladesh (1990), Pakistan (1992), Nepal (1993), India (1999), and Sri Lanka (2001), the size and number of insurance companies have increased and the industry has become more stable. The indicators reported by *Swiss Re* (2016) suggest that during the last decade (2005–15), contribution of insurance to the GDP has fallen marginally. Although "insurance density" is criticized for not being the best measure to assess insurance outreach and spread, but because of absence of a suitable alternative, premium-per capita continues to be the parameter to assess insurance consumption level (Table 9.3).

The per capita consumption or expenditure on insurance continues to remain at lower levels in South Asia. The premium for life insurance is higher than that of nonlife insurance. For example, in Bangladesh, India, and Pakistan, the share of life insurance premium in total business for 2015 are 73.5 percent, 79 percent and 66.5 percent respectively. For Sri Lanka, the share of life insurance in total business in 2015 was 43 percent. Moreover, the nonlife insurance penetration is considerably lower than that of the life insurance penetration. Therefore, the insurance industry has scope for increasing nonlife insurance service delivery and raise general awareness for insuring property and assets in these economies. It is noteworthy that the nonlife insurance policies do not cover damages due to natural disasters until and otherwise mentioned in the policy statement. In India, the premium rates for property (residential or commercial dwellings) insurance policy with additional coverage for earthquake are higher

Table 9.3:
Insurance Density and Penetration

	Density		Penetration (%)	
	2005	*2015*	*2005*	*2015*
Bangladesh	2.9	9.1	0.77	0.67
India	25.6	54.7	3.58	3.44
Pakistan	4.7	11.5	0.71	0.82
Sri Lanka	19	43.1	1.52	1.15
Asia	196.9	311.7	6.82	5.34
World	520.0	621.2	7.47	6.23

Insurance Density: Total premiums per-capita in US$.
Insurance Penetration: Total premiums as % of GDP.

Source: Swiss Re.

compared to a traditional fire policy which does not cover indemnification for disaster-related losses. These exclusions are often due to demand-side problems attributed to adverse selection, undervaluation of risks, ignorance, etc. But, given the threats of natural disasters on the financial sustainability of the insurance companies, the supply side also seems reluctant in selling such insurance policies.

9.14 Disaster Risk Management Financial Strategy

Developing a disaster financial protection strategy requires effective leadership by a country's Ministry of Finance. As a first step, the ministry should prioritize its key policy objectives and, ideally, identify in advance its post-disaster spending priorities. Next, officials would have to consider possible solutions and decide on the country's own ideal combination of financial tools.

9.14.1 International Financial Institutions Contingency Funding

International financial institutions (IFIs) can contribute significantly, both technically and financially, toward creating contingency disaster risk funding mechanisms in less developed countries. In more developed or transitional economies, IFIs can help set up advanced risk-transfer mechanisms. The scope is huge for enhanced development cooperation and aid harmonization across IFIs and donors in the area of disaster risk financing.

9.14.2 Contingency Budgeting

As part of disaster risk finance strategies, when governments are identifying financial instruments to protect against losses, contingency funding mechanisms are an important one. A contingency instrument could be the establishment of a tax or surcharge to be placed into a fund that can be drawn on when a disaster occurs. Another example is to put unused funds at the end of a budgetary year into a special budget specifically for disaster recovery. While the sources of the funds could differ, the primary aim is to have them in place before the disaster for a more rapid recovery.

9.14.3 Financial Management Mechanisms

There are two levels of financial mechanisms that could be established before a disaster. The first is to build the lead agency's capacity to receive large donor contributions. This mechanism consists of establishing draft agreements with potential donor governments and setting up mechanisms to receive and manage contributions. The second financial mechanism is internal to the country. It is the aid-tracking mechanism that enables the lead agency to manage, disburse, and account for funds with local implementers.

9.14.4 Disaster Risk Insurance

Disaster risk insurance aims to increase the financial response capacity of national and subnational governments to secure cost-effective access to adequate funding for emergency response, reconstruction, and recovery.

9.15 Disaster Assessments for Resource Mobilization

First, the damages to infrastructure and assets are valued in physical terms (number, extension of area or surface, as applicable). Second, damages are assigned monetary value, expressed as the replacement costs, according to the market prices prevailing just before and after the disaster. These costs are the baseline cost. The reason is that the calculation of recovery costs would have to account for additional costs. They are post-disaster price alterations, improvements associated with risk reduction, and the concept of build back better introduced by the recovery framework. Additional economic losses calculated refer to changes in the economic flows arising from the disaster. Changes in flows continue until the achievement of full economic recovery, in some cases requiring several years: up to a decade or more.

9.15.1 Resource Mobilization

The challenge of post-disaster recovery is to mobilize additional resources. To the extent possible, recovery should not be at the expense of normal, ongoing development processes. Depending on the nature and scale of the disaster, recovery funding can come from domestic or external resources.

9.15.2 Domestic Funding

Domestic resources generated by disaster-affected governments are

- Reallocation among the budget items from "less" to "more" disaster-hit sectors.
- Issuing sovereign reconstruction or development bonds.
- Levying tax or surcharge for recovery.
- Introducing policy incentives for the private sector to share recovery costs.
- Voluntary civil society and private philanthropies' contributions.

• *Insurance.* Most important, a huge amount of recovery is supported by the people themselves. The public sector's share in recovery can vary widely. It depends on the nature and scale of disaster damage and the relative balance of public- and private-sector asset ownership in the affected areas. In most cases, the biggest contribution to recovery financing comes from the citizens within the country and abroad. These sources of funding, among them remittances, are becoming increasingly important in recovery programs.

9.15.3 External/International Sources of Funding

External resources for post-disaster recovery can be sourced from multilateral development banks, regional development banks, bilateral development partners, international NGOs, private philanthropies and charities, remittances and, most recently, social media. Frequent methods used access external or international funds are international appeals and donor conferences.

9.15.4 International Appeals

National, regional, and international relief systems are able to mobilize and respond to large-scale disasters that require a system-wide response to humanitarian crises by launching appeals. A renewed appeal is usually launched after the first appeal that covers the recovery needs in detail.

9.15.5 Donor Conferences

An international donors' conference may be organized as soon as possible by the government or international community, preferably within the first three months following a large-scale disaster or complex emergency. Holding a donor conference is an effective and coherent way of sourcing funding for the post-disaster activities from governments. Donors commit resources for humanitarian needs as well as long-term recovery and reconstruction in keeping with their own strategic priorities.

9.15.6 International Financial Institutions

IFIs, such as the World Bank and regional development banks (including the Inter-American Development Bank and the Asian Development Bank), have increasingly been engaged in providing lending and nonlending services to developing countries for post-disaster recovery. The financial assistance, generally provided as soft loans, is used to rebuild physical assets, including private housing. Nonlending assistance from IFIs includes damage and loss assessments, acting in an advisory role, and other forms of technical assistance. Joint assessments have become an important mechanism for engaging with other donors and ensuring that borrower needs are met without overlaps. In almost all major disasters in the recent past, IFIs have been one of the most important sources of financial assistance for recovery.

9.16 Private Sector and Disaster Management

The changing scope and scale of disasters, both natural and technological, have altered the ways in which disaster management and financing are addressed and the roles of private-sector organizations specifically. Businesses and nonprofit organizations are increasingly central to the process, offering critical support in immediate disaster response, but also contributing necessary redevelopment funding that supports community recovery. Although these new expectations position the private sector as a key leader in community resilience, these responsibilities have not been fully met with established guidance or clear metrics for how and when these organizations should participate in disaster recovery and financing.

The key issues confronting the private sector in disaster recovery financing are what roles private-sector entities have played and where there has been successful integration or leadership of these organizations. There are challenges the private-sector faces, with particular attention to issues of information use and application, coordination in response and recovery, and timing of funding. Given the continued data gaps in this field, the authors offer opportunities for research and policy analysis.

Private-sector entities also have a large untapped potential to help provide skilled services in the form of technical manpower or in-kind donations of goods or services for the preparedness and emergency response phase of disaster management.

Where does the role of the private sector lie in disaster management? In is obvious that this is a multifaceted and multidisciplinary one, including those listed below and more.

- In defining, assigning, and implementing clear and coherent institutional roles. In training, equipping and achieving proficiency for effective response capacity for high risk communities.
- In assessing institutional needs, and in developing and implementing programs to assist key organizations with sustainability issues and measures.
- In improving the disaster consciousness of the general population.
- In improving access to accurate information and basic communication, and energy and water systems for high-risk communities by facilitating appropriate technology alternatives on credit to qualifying communities.

At the same time, local authorities need to provide a more effective framework for unleashing the full potential of private-sector contribution. This framework can include policies that contribute directly to safer industrial operation, and incentives for business to become more involved in disaster management programs. Mechanisms whereby businesses and the authorities meet to discuss their respective roles and contributions to national disaster security need to be set up and operationalized on a regular basis.

**Disaster Risk Governance of 2013
Cyclone Phailin in INDIA (Pal et al. 2017)**

Around the past half century, Odisha has witnessed sea cyclones in 1971, 1973, 1977, 1981, 1983, 1984, 1985, 1987, 1989, 1999, and 2013 that have damaged life and property. Odisha, with its long coastline of 450 km, occupies the face of the attracting tunnel for cyclones from the Indian Ocean northwards, with indentations of Ganjam, Khordha, Puri, Jagatsinghpur, and Kendrapada to Bay of Bengal, usually these costal districts facing the direction of flow of the cyclonic storm (Patra et al. 2013).

On October 8, 2013, the India Meteorological Department (IMD) reported the formation of a depression over the North Andaman Sea. A warning was issued to the NDMA and Odisha State Disaster Management Authority (OSDMA), which was subsequently relayed to the District Disaster Management Authority (DDMA). The Emergency Operation Centre (EOC) was activated on October

---

(Continued)

(Continued)

8 at the OSDMA and DDMA. On the evening of October 12, 2013, a very severe tropical cyclone, Phailin, brought torrential downpours, damaging winds of more than 220 km per hour (km/h), and storm surges of up to 3.5 meters (m) to the eastern Indian states of Odisha and Andhra Pradesh (GoO 2013), especially the ravaged Ganjam district, which was the location of landfall of the cyclone Phailin. The cyclone was accompanied with torrential rains for 3 consecutive days, leading to floods in number of rivers.

The impacts of Phailin and the ensuing floods affected more than 13.2 million people, left five districts of Odisha under water, and caused hundreds of millions of dollars (GoO 2013) in damage to homes, schools, crops, and the fishing industry (Froberg 2013). However, early warning alerts, disseminated four days before Phailin struck land, allowed for the evacuation of approximately 400,000 people on or by October 11 (Senapati 2013). Ultimately, a total of nearly 1.2 million people were evacuated (GoO 2013), resulting in the largest evacuation operation in India in 23 years (IFRC 2013). Early warning also allowed for the relocation of more than 30,000 animals. A total of 21 lives were lost as a result of the cyclone and an additional 23 lives due to severe flash flooding in the aftermath of the cyclone (GoO 2013). About 3,212 villages, 22 townships, and a population of over 3.7 million was directly affected in Ganjam district. Nineteen fishermen lost their lives in deep sea during fishing due to cyclone philin. The standing crops in 195,674 hectares of land was damaged. The fishermen communities were severely affected due to damage to their boats and nets.

A comparable cyclone, Cyclone 05B, hit the same area in 1999 with winds of up to 260 km/h (IFRC 1999), but had a much more devastating outcome: more than 10,000 lives were lost (World Bank 2013). Government cooperation, preparedness at the community level, early warning communication, and lessons learned from Cyclone 05B contributed to the successful evacuation operation, effective preparation activities, and impact mitigation. This event exhibits the importance, benefits, and effectiveness of the use of early warning for a massive disaster.

On the conterary, during Super Cyclone 1999, Odisha, India (Cyclone 04 and 05), emergency response, the DAC in the Ministry of Agriculture functioned as the nodal agency for relief and rehabilitation measures in the wake of natural calamities. On receipt of the first information from the IMD on October 26, 1999, regarding the cyclonic storm, DAC requested the chief secretaries and relief commissioners of Orissa, West Bengal, and Andhra Pradesh to take all preparatory measures including evacuation of the vulnerable people. The cabinet secretary, secretary, DAC and other senior officers remained in constant touch with these state government authorities. National Crisis Management Committee (NCMC) under the chairmanship of the cabinet secretary had reviewed the status of preparedness in the meetings held at 2200 hrs (IST) on October 27, 1999, and 1600 hrs (IST) and 2300 hrs (IST) on October 28, 1999. The representatives of various ministries and departments of GOI, that is, Defence, Home Affairs, Power, Telecom, Shipping, Road Transport, Railways,

Petroleum and Natural Gas, Information and Broadcasting, the IMD and the resident commissioners of the concerned states were present in this meeting. Various preparatory measures taken at the central level were reviewed during the NCMC meetings. Most of the actions and decisions have been taken during the 1999 Odisha Super Cyclone was in ad hoc basis (World Bank 2013).

From 1970 to 2010, the Asia-Pacific population living in cyclone-prone areas increased from 71.8 million to 120.7 million, expanding the magnitude of vulnerability to disasters (ESCAP and UNISDR 2012). However, significant improvements in disaster management, preparedness, forecasting capabilities and early warning, such as the improvements exhibited by India during Cyclone Phailin in October 2013, have helped to mitigate some disaster-related impacts. Preparedness and early warning communications, cyclone shelters in Ganjam district, and activities had been much improved since the comparable Cyclone 05B 14 years earlier (Singh 2013). In response to Cyclone 05B in 1999, Odisha established the first state agency in India to address disaster management specifically (Odisha State Disaster Management Authority [OSDMA]); World Bank 2013). This initiative has led to the construction of 200 new cyclone shelters, operating in places such as schools and community centers to ensure regular maintenance (Singh 2013). Cyclone shelters have proven to be useful as 75 shelters operated by the Indian Red Cross provided safety to more than 100,000 people during Phailin (Mukherji and Agarwal 2013) with some shelters holding up to 500 people (Froberg 2013). Regarding forecasts, the IMD was able to predict wind velocity more accurately, contributing to better forecasts and more effective early warning communications (TOI 2013). Warnings from the IMD were disseminated as early as four days before Phailinmade landfall, as compared with two days of warning provided for Cyclone 05B in 1999 (Senapati 2013). In addition to early warning alerts that prompted evacuations, precautions to protect cattle were taken and reservoirs were lowered to mitigate anticipated flooding (GoO 2013). Several means of communication were exercised in the days prior to the landfall of Phailin to disperse early warning information. Different means and methods of communication were used to reach a large population includes Constant news coverage before and throughout the event (Jain, 2013). These include the use of email, fax, telephone and print media to communicate warnings and alerts (GoO, 2013), including text message broadcasting through mobile network, warnings and alerts delivered by the IMD through channels such as online news networks (Kumar, 2013). Loudspeakers were also used to warn residents on impending danger and to warn fishing boats that were out at sea (Senapati, 2013) in several districts. Distribution of satellite phones to representatives in the 14 most vulnerable districts ensured the warning communications remain continued during the storm (GoO, 2013).

The enactment of the National Disaster Management Act in 2005 by the Government of India has marked a paradigm shift in the mitigation and management of disaster in the country in line with the principles of HFA 2005–15.

(*Continued*)

(*Continued*)

The present research has attempted to show how the enactment of a new disaster management law and implementation of a policy provided the institutional means for achieving the intended aims, that is, achieving effective mitigation and response to a disaster thereby minimizing the casualties and losses to the community. Intended outcomes possibly achieved by the clear delineation of the aims, and identification and activation of the key mechanisms. Dynamic agenda-setting and the attribution of opportunity played a key role in overcoming the main obstacles to the compliance of the policy that had nationwide implications. Disaster response requires a multilayered and multisectoral interventions and combined/integrated strength of such a state-of-art institutional base, which facilitated effective and timely response to disasters within the country.

In the case of disaster management policy, a three-tier machine bureaucratic structure and vesting emergency powers in key actors theoretically provide strengths toward the flexible "complex-adaptive structures" (Comfort 2007) that are considered ideal for disaster management. However, the need for unity of command and span of control has resulted in are latively rigid structure. It must also be remembered that ideal Incident Command System (ICS)-type flexible system works better in the context where the personnel are adequately trained and systems have matured. In India, capacity-building in disaster management is still in a nascent stage, and it would be sometime before the institutions migrate to a more flexible and a dynamic mode.

It must be factored that Odisha, with a history of cyclones and tsunamis, had certain advantages that enabled better management of the situation. The experience and institutional memory about the risk appraisal and the cognition part of the disaster appear to have contributed decisively in government's successful response strategy. The vibrancy of the disaster management machinery and the overall feedback suggests the efficacy of the program. Disaster management being a rather experiential discipline would do well to benefit from the case study of the cyclone Phailin and iteratively draw from the learning to fine tune the response strategies for the future.

9.17 Private Sector and Disaster Recovery Financing

The private sector contributes to disaster recovery financing in a variety of ways.

- The private sector contributes to disaster recovery financing in a variety of ways, including playing a key role in early response and long-term recovery, collaborating with the public sector in public–private partnerships, driving innovation and facilitating technology use, helping smaller communities manage influxes of funds, and supplementing federal disbursement processes.

- Public–private partnerships are integral and they can help to increase efficiency and effectiveness in disaster management. The partnerships between private actors and public-sector partners and recipients can alter the strategic focus of disaster management agencies.
- Public–private partnerships can reduce the burdens placed on government to provide certain goods and services immediately and, over time, permitting the public sector to focus on other important strategic priorities.
- For-profit businesses can provide important templates informing the design of public sector programs.
- The private sector, particularly businesses, is key to disaster recovery.
- The private sector is key to developing and implementing flexible financing models. It is also critical in ongoing resilience development.

The private sector faces challenges when it comes to financing

- These challenges arise from information availability, tracking the flow and timing of funds, and basic challenges in what financial supports can be provided, particularly by business. Further, explaining the value of business investment in recovery can be difficult.

Recommendations

- Determine the full extent of private-sector contributions in disaster recovery.
- Extend private- and public-sector disaster recovery analyses to include comparative value of private-sector engagement and assessment of coordination.
- Assess initiatives in which private-sector organizations, particularly businesses, assist in supporting resilience.

Chapter 10

Tools and Techniques in Disaster Management

10.1 Social Survey and Disaster Research

In conventional understanding, a survey is any activity that collects information in an organized and methodical manner about characteristics of interest from some or all units of a population using well-defined concepts, methods, and procedures, and compiles such information into a useful summary form. A survey usually begins with the need for information where no data—or insufficient data—exist. Sometimes this need arises from within the statistical agency itself, and sometimes it results from a request from an external client, which could be another government agency or department, or a private organization. Typically, the statistical agency or the client wishes to study the characteristics of a population, build a database for analytical purposes, or test a hypothesis.

A survey can be thought to consist of several interconnected steps, which include defining the objectives, selecting a survey frame, determining the sample design, designing the questionnaire, collecting and processing the data, analyzing and disseminating the data, and documenting the survey. The life of a survey can be broken down into several phases. The first is the planning phase, which is followed by the design and development phase, and then the implementation phase. Finally, the entire survey process is reviewed and evaluated.

10.2 Steps of a Survey

It may appear that conducting a survey is a simple procedure of asking questions and then compiling the answers to produce statistics. However, a survey must be carried out step by step, following precise procedures and formulas, if the results are to yield accurate and meaningful information. In order to understand the entire survey process, it is necessary to understand and follow the individual tasks and their interlinkages.

The steps of a survey are

- formulation of the statement of objectives;
- selection of a survey frame;
- determination of the sample design;
- determination of the questionnaire design;
- data collection;
- DATA capture and coding;
- editing and imputation;
- estimation;
- data analysis;
- data dissemination; and
- documentation.

10.3 Types of Scientific Research

Depending on the purpose of research, scientific research projects can be grouped into three types: exploratory, descriptive, and explanatory. Exploratory research is often conducted in new areas of inquiry where the goals of the research are (a) to scope out the magnitude or extent of a particular phenomenon, problem, or behavior, (b) to generate some initial ideas (or "hunches") about that phenomenon, or (c) to test the feasibility of undertaking a more extensive study regarding that phenomenon. For instance, if the citizens of a country are generally dissatisfied with governmental policies regarding during an economic recession, exploratory research may be directed at measuring the extent of citizens' dissatisfaction, understanding how such dissatisfaction is manifested, such as the frequency of public protests, and the presumed causes of such dissatisfaction, such as ineffective government policies in dealing with inflation, interest rates, unemployment, or higher taxes. Such research may include the examination of publicly reported figures, such as estimates of economic indicators, such as GDP, unemployment, and consumer price index, as archived by third-party sources, obtained through interviews of experts, eminent economists, or key government officials, and/or derived from studying historical examples of dealing with similar problems. This research may not lead to a very accurate understanding of the target problem but may be worthwhile in scoping out the nature and extent of the problem and may serve as a useful precursor to more in-depth research.

10.3.1 Descriptive Research

Descriptive research is directed at making careful observations and detailed documentation of a phenomenon of interest. These observations must be based on the scientific method (i.e., must be replicable, precise, etc.), and therefore, these are more reliable than casual observations by untrained people. Examples of descriptive research are tabulation of demographic statistics by the United States Census Bureau or employment statistics by the Bureau of Labor, who use the same or similar instruments for estimating employment by sector or population growth by ethnicity over multiple employment surveys or censuses. If any changes are made to the measuring instruments, estimates are provided with and without the changed instrumentation to allow the readers to make a fair before-and-after comparison regarding population or employment trends. Other descriptive research may include chronicling ethnographic reports of gang activities among adolescent youth in urban populations, the persistence or evolution of religious, cultural, or ethnic practices in select communities, and the role of technologies such as Twitter and instant messaging in the spread of democracy movements in Middle Eastern countries.

10.3.2 Explanatory Research

Explanatory research seeks explanations of observed phenomena, problems, or behaviors. While descriptive research examines the what, where, and when of a phenomenon, explanatory research seeks answers to why and how types of questions. It attempts to "connect the dots" in research, by identifying causal factors and outcomes of the target phenomenon. Examples include understanding the reasons behind adolescent crime or gang violence, with the goal of prescribing strategies to overcome such societal ailments. Most academic or doctoral research belongs to the explanation category, though some amount of exploratory and/or descriptive research may also be needed during initial phases of academic research. Seeking explanations for observed events requires strong theoretical and interpretation skills, along with intuition, insights, and personal experience. Those who can do it well are also the most prized scientists in their disciplines.

10.4 Popular Research Designs and Data Collection Tools

Research designs can be classified into two categories—positivist and interpretive—depending on their goals in scientific research. Positivist designs are meant for theory testing, while interpretive designs are meant for theory building. Positivist designs seek generalized patterns based on an objective view of reality, while interpretive designs seek subjective interpretations of social phenomena from the perspectives of the subjects involved. Some popular examples of positivist designs include laboratory experiments, field experiments, field surveys, secondary data analysis, and case research, while examples of interpretive designs include case research, phenomenology, and ethnography. Note that case research can be used for theory building or theory testing, though not at the same time. Not all techniques are suited for all kinds of scientific research. Some techniques such as focus groups are best suited for exploratory research, others such as ethnography are best for descriptive research, and still others such as laboratory experiments are ideal for explanatory research. Following are brief descriptions of some of these designs.

10.4.1 Survey

A survey is a method of collecting information directly from people about their demographic and socioeconomic background. A survey can be a self-administered questionnaire, a structured and semi-structured interview, focused group discussions, etc.

10.4.2 Experimental Studies

These studies are intended to test the cause–effect relationships (hypotheses) in a tightly controlled setting by separating the cause from the effect in time, administering the cause to one group of subjects (the "treatment group") but not to another group ("control group"), and observing how the mean effects vary between subjects in these two groups. For instance, if we design a laboratory experiment to test the efficacy of a new drug in treating a certain ailment, we can get a random sample of people afflicted with that ailment, randomly assign them to one of two groups (treatment and control groups), and administer the drug to subjects in the treatment group, but only give a placebo (e.g., a sugar pill with no medicinal value). More complex designs may include multiple treatment groups, such as low versus high dosage of the drug, and multiple treatments, such as combining drug administration with dietary interventions. In a true experimental design, subjects must be randomly assigned between each group. If random assignment is not followed, then the design becomes quasi-experimental. Experiments can be conducted in an artificial or laboratory setting, such as at a university (laboratory experiments), or in field settings, such as in an organization, where the phenomenon of interest is actually occurring (field experiments). Laboratory experiments allow the researcher to isolate the variables of interest and control for extraneous variables, which may not be possible in field experiments. Hence, inferences drawn from laboratory experiments tend to be stronger in internal validity, but those from field experiments tend to be stronger in external validity. Experimental data are analyzed using quantitative statistical techniques. The primary strength of the experimental design is its strong internal validity due to its ability to isolate, control, and intensively examine a small number of variables, while its primary weakness is limited external generalizability since real life is often more complex (i.e., involves more extraneous variables) than contrived lab settings. Furthermore, if the research does not identify ex-ante relevant extraneous variables and control for such variables, such lack of controls may hurt internal validity and may lead to spurious correlations.

10.4.3 Field Surveys

Field-based surveys are one of the important elements in disaster management research. The field surveys are nonexperimental designs that do not control for or manipulate independent variables or treatments but that measure these variables and test their effects using statistical methods. Field surveys capture snapshots of practices, beliefs, or situations from a random sample of subjects in field settings through a survey questionnaire or less frequently, through a structured interview. In cross-sectional field surveys, independent and dependent variables are measured at the same point in time (e.g., using a single questionnaire), while in longitudinal field surveys, dependent variables are measured at a later point in time than the independent variables. The strengths of field surveys are their external validity (since data are collected in field settings), their ability to capture and control for a large number of variables, and their ability to study a problem from multiple perspectives or using multiple theories. However, because of their nontemporal nature, internal validity (cause–effect relationships) is difficult to infer, and surveys may be subject to respondent biases (e.g., subjects may provide a "socially desirable" response rather than their true response), which further hurts internal validity.

10.4.4 Secondary Data Analysis

It is an analysis of data that have previously been collected and tabulated by other sources. Such data may include data from government agencies such as employment statistics from the US Bureau of Labor Services or development statistics by country from the UNDP, data collected by other researchers (often used in meta-analytic studies), or publicly available third-party data, such as financial data from stock markets or real-time auction data from eBay. This is in contrast to most other research designs where collecting primary data for research is part of the researcher's job. Secondary data analysis may be an effective means of research where primary data collection is too costly or infeasible, and secondary data are available at a level of analysis suitable for answering the researcher's questions. The limitations of this design are that the data might not have been collected in a systematic or scientific manner and, hence, be unsuitable for scientific research; since the data were collected for a presumably different purpose, they may not adequately address the research questions of interest to the researcher; and the interval validity is problematic if the temporal precedence between cause and effect is unclear.

10.4.5 Case Study Research

Case study–based research is an in-depth investigation of a problem in one or more real-life settings (case sites) over an extended period of time. Data may be collected using a combination of interviews, personal observations, and internal or external documents. Case studies can be positivist in nature (for hypotheses testing) or interpretive (for theory building). The strength of this research method is its ability to discover a wide variety of social, cultural, and political factors potentially related to the phenomenon of interest that may not be known in advance. Analysis tends to be qualitative in nature, but heavily contextualized and nuanced. However, the interpretation of findings may depend on the observational and integrative ability of the researcher, lack of control may make it difficult to establish causalities and findings from a single case site may not be readily generalized to other case sites. Generalizability can be improved by replicating and comparing the analysis in other case sites in a multiple case design.

10.4.6 Focus Group Research

Focus group research involves a small group of subjects (typically 6–10 people) at one location and having them discuss a phenomenon of interest for a period of 1.5–2 hours. The discussion is moderated and led by a trained facilitator, who sets the agenda and poses an initial set of questions for participants, makes sure that ideas and experiences of all participants are represented, and attempts to build a holistic understanding of the problem situation based on the participants' comments and experiences. Internal validity cannot be established due to lack of controls and the findings may not be generalized to other settings because of small sample size. Hence, focus groups are not generally used for explanatory or descriptive research but are more suited for exploratory research.

10.4.7 Key Informants

Key informants involve survey of key stakeholders (e.g., disaster managers) of different institutions, coordinating agency, government personnel, community leaders, some NGOs and INGOs through conducting interview to understand their role in the respective disaster management cycle. Focus group research is being useful to conduct among local authorities, community and political leaders as required to get an insight into the local knowledge of the place and the primary concern following the disasters.

10.4.8 Action Research

Action research assumes that complex social phenomena are best understood by introducing "interventions" or "actions" into those phenomena and observing the effects of those actions. In this method, the researcher is usually a consultant or an organizational member embedded within a social context such as an organization who initiates an action such as new organizational procedures or new technologies in response to a real problem such as declining profitability or operational bottlenecks. The researcher's choice of actions must be based on theory, which should explain why and how such actions may cause the desired change. The researcher then observes the results of that action, modifying it as necessary, while simultaneously learning from the action and generating theoretical insights about the target problem and interventions. The initial theory is validated by the extent to which the chosen action successfully solves the target problem. Simultaneous problem-solving and insight generation is the central feature that distinguishes action research from all other research methods, and hence, action research is an excellent method for bridging research and practice. This method is also suited for studying unique social problems that cannot be replicated outside that context, but it is also subject to researcher bias and subjectivity, and the generalizability of findings is often restricted to the context where the study was conducted.

10.4.9 Field Observation

The real case scenario can be observed through field observation. When in the site, the actual disaster event will be evident along with the affected population's current situation. Also, progressive work carried out at the study area was clearly observed. The researcher will get an opportunity to interact with the local residents and observe their present condition. Field observation will help in depicting the numerous difficulties present within an affected community after the disaster.

10.4.10 Weighted Average Index

A weighted average index (WAI) is an accurate measurement of scores or investments that are of relative importance to each other. It combines choice weights and question weights to produce a single index for all responses. It is used for checklist score analysis and to compute a weighted average for questionnaire. The WAI is based on five-social scaling technique to analyze a respondent's answers on the post-disaster needs, response level toward a community, and their level of satisfaction gathered through a household survey. The scale used in the WAI for data collection and analysis in the field-based study is usually based on five ranges: no impact, low, moderate, high, and very high.

10.5 Statistical Terms in Survey Sampling

10.5.1 Probability Sampling/Random Sampling

A probability sampling method is any method of sampling that utilizes some form of random selection. In order to have a random selection method, you must set up a process or procedure that assures that the different units in your population have equal probabilities of being chosen. Humans have long practiced various forms of random selection, such as picking a name out of a hat or choosing the short straw. These days, we tend to use computers as the mechanism for generating random numbers as the basis for random selection.

10.5.1.1 SOME DEFINITIONS

Before we explain the various probability methods, we have to define some basic terms.

These are

- N = the number of cases in the sampling frame;
- n = the number of cases in the sample;
- $_NC_n$ = the number of combinations (subsets) of n from N;
- $f = n/N$ = the sampling fraction.

That's it. With these terms defined we can begin to define the different probability sampling methods.

10.5.2 Simple Random Sampling

The simplest form of random sampling is called simple random sampling. Pretty tricky, huh? Here is a quick description of simple random sampling.

- *Objective:* To select *n* units out of *N* such that each $_NC_n$ has an equal chance of being selected.
- *Procedure:* Use a table of random numbers, a computer random number generator, or a mechanical device to select the sample.

This is a somewhat stilted, if accurate, definition. Let us see if we can make it a little more real. How do we select a simple random sample? Let us assume that we are doing some research with a small service agency that wishes to assess clients' views of quality of service over the past year. First, we have to get the sampling frame organized. To accomplish this, we will go through the agency's records to identify every client over the past 12 months. If we are lucky, the agency has good, accurate computerized records and can quickly produce such a list. Then, we have to actually draw the sample. Decide on the number of clients you would like to have in the final sample. For the sake of the example, let us say you want to select 100 clients to survey and that there were 1,000 clients over the past 12 months. Then, the sampling fraction is $f = n/N = 100/1,000 = 0.10$ or 10 percent. Now, to actually draw the sample, you have several options. You could print off the list of 1,000 clients, tear then into separate strips, put the strips in a hat, mix them up real good, close your eyes and pull out the first 100. But this mechanical procedure would be tedious and the quality of the sample would depend on how thoroughly you mixed them up and how randomly you reached in. Perhaps, a better procedure would be to use the kind of ball machine that is popular with many of the state lotteries. You would need three sets of balls numbered 0–9, one set for each of the digits from 000 to 999 (if we select 000 we will call that 1,000). Number the list of names from 1 to 1,000 and then use the ball machine to select the three digits that selects each person. The obvious disadvantage here is that you need to get the ball machines. (Where do they make those things, anyway? Is there a ball machine industry?)

Neither of these mechanical procedures is very feasible and, with the development of inexpensive computers there is a much easier way. Here is a simple procedure that is especially useful if you have the names of the clients already on the computer. Many computer programs can generate a series of random numbers. Let us assume you can copy and paste the list of client names into a column in an EXCEL spreadsheet. Then, in the column right next to it paste the function =RAND(), which is EXCEL's way of putting a random number between 0 and 1 in the cells. Then, sort both columns— the list of names and the random number—by the random numbers. This rearranges the list in random order from the lowest to the highest random number. Finally, all you have to do is take the first 100 names in this sorted list. Pretty simple! You could probably accomplish the whole thing in under a minute.

Simple random sampling is simple to accomplish and is easy to explain to others. Because simple random sampling is a fair way to select a sample, it is reasonable to generalize the results from the sample back to the population. Simple random sampling is not the most statistically efficient method of sampling, and you may, just because of the luck of the draw, not get good representation of subgroups in a population. To deal with these issues, we have to turn to other sampling methods.

10.5.3 Stratified Random Sampling

Stratified random sampling, also sometimes called proportional or quota random sampling, involves dividing your population into homogeneous subgroups and then taking a simple random sample in each subgroup.

Objective: Divide the population into nonoverlapping groups (i.e., strata) $N_1, N_2, N_3,\ldots N_i$, such that $N_1 + N_2 + N_3 + \cdots + N_i = N$. Then do a simple random sample of $f = n/N$ in each strata.

There are several major reasons why you might prefer stratified sampling over simple random sampling. First, it assures that you will be able to represent not only the overall population but also key subgroups of the population, especially small minority groups. If you want to be able to talk about subgroups, this may be the only way to effectively assure you will be able to. If the subgroup is extremely small, you can use different sampling fractions (f) within the different strata to randomly oversample the small group (although you will then have to weigh the within-group estimates using the sampling fraction whenever you want overall population estimates). When we use the same sampling fraction within strata, we are conducting proportionate stratified random sampling. When we use different sampling fractions in the strata, we call this disproportionate stratified random sampling.

Second, stratified random sampling will generally have more statistical precision than simple random sampling. This will only be true if the strata or groups are homogeneous. If they are, we expect that the variability within groups is lower than the variability for the population as a whole. Stratified sampling capitalizes on that fact.

For example, let us say that the population of clients for our agency can be divided into three groups: Caucasian, African-American, and Hispanic-American. Furthermore, let us assume that both the African-Americans and Hispanic-Americans are relatively small minorities of the clientele (10 percent and 5 percent, respectively). If we just did a simple random sample of $n = 100$ with a sampling fraction of 10 percent, we would expect by chance alone that we would only get 10 and 5 persons from each of our two smaller groups. And, by chance, we could get fewer than that! If we stratify, we can do better. First, let us determine how many people we want to have in each group. Let us say we still want to take a sample of 100 from the population of 1,000 clients over the past year. But we think that in order to say anything about subgroups, we will need at least 25 cases in each group. So, let us sample 50 Caucasians, 25 African-Americans, and 25 Hispanic-Americans. We know that 10 percent of the population, or 100 clients, are African-Americans. If we randomly sample 25 of these, we have a within-stratum sampling fraction of $25/100 = 25$ percent. Similarly, we know that 5 percent or 50 clients are Hispanic-Americans. So our within-stratum sampling fraction will be $25/50 = 50$ percent. Finally, by subtraction we know that there are 850 Caucasian clients. Our within-stratum sampling fraction for them is $50/850 =$ about 5.88 percent. Because the groups are more homogeneous within group than across the population as a whole, we can expect greater statistical precision (less variance). And, because we stratified, we know we will have enough cases from each group to make meaningful subgroup inferences.

10.5.4 Systematic Random Sampling

Here are the steps you need to follow in order to achieve a systematic random sample.

- number the units in the population from 1 to N
- decide on the n (sample size) that you want or need
- $k = N/n =$ the interval size

- randomly select an integer between 1 and *k*
- then take every *k*th unit.

All of this will be much clearer with an example. Let us assume that we have a population that only has $N = 100$ people in it and that you want to take a sample of $n = 20$. To use systematic sampling, the population must be listed in a random order. The sampling fraction would be $f = 20/100 = 20$ percent. In this case, the interval size, *k*, is equal to $N/n = 100/20 = 5$. Now, select a random integer from 1 to 5. In our example, imagine that you chose 4. Now, to select the sample, start with the 4th unit in the list and take every *k*th unit (every 5th, because $k = 5$). You would be sampling units 4, 9, 14, 19, and so on to 100 and you would wind up with 20 units in your sample.

For this to work, it is essential that the units in the population are randomly ordered, at least with respect to the characteristics you are measuring. Why would you ever want to use systematic random sampling? For one thing, it is fairly easy to do. You only have to select a single random number to start things off. It may also be more precise than simple random sampling. Finally, in some situations there is simply no easier way to do random sampling. For instance, I once had to do a study that involved sampling from all the books in a library. Once selected, I would have to go to the shelf, locate the book, and record when it last circulated. I knew that I had a fairly good sampling frame in the form of the shelf list (which is a card catalog where the entries are arranged in the order they occur on the shelf). To do a simple random sample, I could have estimated the total number of books and generated random numbers to draw the sample; but how would I find book #74,329 easily if that is the number I selected? I could not very well count the cards until I came to 74,329! Stratifying would not solve that problem either. For instance, I could have stratified by card catalog drawer and drawn a simple random sample within each drawer. But I would still be stuck counting cards.

Instead, I did a systematic random sample. I estimated the number of books in the entire collection. Let us imagine it was 100,000. I decided that I wanted to take a sample of 1,000 for a sampling fraction of $1,000/100,000 = 1$ percent. To get the sampling interval *k*, I divided *N* by $n = 100,000/1,000 = 100$. Then I selected a random integer between 1 and 100. Let us say I got 57. Next I did a little side study to determine how thick a thousand cards are in the card catalog (taking into account the varying ages of the cards). Let us say that on average I found that two cards that were separated by 100 cards were about 0.75 inches apart in the catalog drawer. That information gave me everything I needed to draw the sample.

I counted to the 57th by hand and recorded the book information. Then, I took a compass. (Remember those from your high-school math class? They're the funny little metal instruments with a sharp pin on one end and a pencil on the other that you used to draw circles in geometry class.) Then I set the compass at 0.75," stuck the pin end in at the 57th card and pointed with the pencil end to the next card (approximately 100 books away). In this way, I approximated selecting the 157th, 257th, 357th, and so on. I was able to accomplish the entire selection procedure in very little time using this systematic random sampling approach. I would probably still be there counting cards if I had tried another random sampling method.

10.5.5 Cluster (Area) Random Sampling

The problem with random sampling methods when we have to sample a population that is disbursed across a wide geographic region is that you will have to cover a lot of ground geographically in order to get to each of the units you sampled. Imagine taking a simple random sample of all the residents of New York State in order to conduct personal interviews. By the luck of the draw you will wind up with respondents who come from all over the state. Your interviewers are going to have a lot of traveling to do. It is for precisely this problem that "cluster or area random sampling" was invented.

In cluster sampling, we follow the steps listed.

- Divide population into clusters (usually along geographic boundaries)
- Randomly sample clusters
- Measure *all* units within sampled clusters

10.5.6 Multistage Sampling

The four methods we have covered so far—simple, stratified, systematic, and cluster—are the simplest random sampling strategies. In most real applied social research, we would use sampling methods that are considerably more complex than these simple variations. The most important principle here is that we can combine the simple methods described earlier in a variety of useful ways that help us address our sampling needs in the most efficient and effective manner possible. When we combine sampling methods, we call this multistage sampling.

10.5.7 Nonprobability Sampling

The difference between nonprobability and probability sampling is that nonprobability sampling does not involve random selection and probability sampling does. Does that mean that nonprobability samples are not representative of the population? Not necessarily. But it does mean that nonprobability samples cannot depend upon the rationale of a probability theory. At least with a probabilistic sample, we know the odds or probability that we have represented the population well. We are able to estimate confidence intervals for the statistic. With nonprobability samples, we may or may not represent the population well, and it will often be hard for us to know how well we have done so. In general, researchers prefer probabilistic or random sampling methods over nonprobabilistic ones and consider them to be more accurate and rigorous. However, in applied social research there may be circumstances where it is not feasible, practical, or theoretically sensible to do random sampling. Here, we consider a wide range of nonprobabilistic alternatives.

We can divide nonprobability sampling methods into two broad types: accidental or purposive. Most sampling methods are purposive in nature because we usually approach the sampling problem with a specific plan in mind. The most important distinctions among these types of sampling methods are the ones between the different types of purposive sampling approaches.

10.5.8 Accidental, Haphazard, or Convenience Sampling

One of the most common methods of sampling goes under the various titles listed here. I would include in this category the traditional "man on the street" (of course, now it is probably the "person on the street") interviews conducted frequently by television news programs to get a quick (although nonrepresentative) reading of public opinion. I would also argue that the typical use of college students in much psychological research is primarily a matter of convenience. (You do not really believe that psychologists use college students because they believe they are representatives of the population at large, do you?) In clinical practice, we might use clients who are available to us as our sample. In many research contexts, we sample simply by asking for volunteers. Clearly, the problem with all of these types of samples is that we have no evidence that they are representative of the populations we are interested in generalizing to—and in many cases we would clearly suspect that they are not.

10.5.9 Purposive Sampling

In purposive sampling, we sample with a purpose in mind. We usually would have one or more specific predefined groups we are seeking. For instance, have you ever run into people in a mall or on the street who are carrying a clipboard and who are stopping various people and asking if they could interview them? Most likely they are conducting a purposive sample (and most likely they are engaged in market research). They might be looking for Caucasian females between 30 and 40 years old. They size up the people passing by and stop anyone who looks to be in that category to ask if they will participate. One of the first things they are likely to do is verify that the respondent does in fact meet the criteria for being in the sample. Purposive sampling can be very useful for situations where you need to reach a targeted sample quickly and where sampling for proportionality is not the primary concern. With a purposive sample, you are likely to get the opinions of your target population, but you are also likely to overweight subgroups in your population that are more readily accessible.

All of the methods that follow can be considered subcategories of purposive sampling methods. We might sample for specific groups or types of people as in modal instance, expert, or quota sampling. We might sample for diversity as in heterogeneity sampling. Or, we might capitalize on informal social networks to identify specific respondents who are hard to locate otherwise, as in snowball sampling. In all of these methods we know what we want—we are sampling with a purpose.

10.5.10 Modal Instance Sampling

In statistics, the mode is the most frequently occurring value in a distribution. In sampling, when we do a modal instance sample, we are sampling the "most frequent" case, or the "typical" case. In a lot of informal public opinion polls, for instance, they interview a "typical" voter. There are a number of problems with this sampling approach. First, how do we know what the "typical" or "modal" case is? We could say that the modal voter is a person who is of average age, educational level, and income in the population. But, it is not clear that using the averages of these is the fairest (consider the skewed distribution of income, for instance). And, how do you know that those three variables—age, education, and income—are the only or even the most relevant for classifying a typical voter? What if religion or ethnicity is an important discriminator? Clearly, modal instance sampling is only sensible for informal sampling contexts.

10.5.11 Expert Sampling

Expert sampling involves the assembling of a sample of persons with known or demonstrable experience and expertise in some area. Often, we convene such a sample under the auspices of a "panel of experts." There are actually two reasons you might do expert sampling. First, because it would be the best way to elicit the views of persons who have specific expertise. In this case, expert sampling is essentially just a specific subcase of purposive sampling. But the other reason you might use expert sampling is to provide evidence for the validity of another sampling approach you have chosen. For instance, let us say you do modal instance sampling and are concerned that the criteria you used for defining the modal instance are subject to criticism. You might convene an expert panel consisting of persons with acknowledged experience and insight into that field or topic and ask them to examine your modal definitions and comment on their appropriateness and validity. The advantage of doing this is that you are not out on your own trying to defend your decisions—You have some acknowledged experts to back you. The disadvantage is that even the experts can be, and often are, wrong.

10.5.12 Quota Sampling

In quota sampling, you select people nonrandomly according to some fixed quota. There are two types of quota sampling: proportional and nonproportional. In "proportional quota sampling," you want to represent the major characteristics of the population by sampling a proportional amount of each. For instance, if you know the population has 40 percent women and 60 percent men, and that you want a total sample size of 100, you will continue sampling until you get those percentages and then you will stop. So, if you have already got the 40 women for your sample, but not the 60 men, you will continue to sample men; but even if legitimate women respondents come along, you will not sample them because you have already "met your quota." The problem here (as in much purposive sampling) is that you have to decide the specific characteristics on which you will base the quota. Will it be by gender, age, education race, religion, etc.?

10.5.13 Nonproportional Quota Sampling

Nonproportional quota sampling is a bit less restrictive. In this method, you specify the minimum number of sampled units you want in each category. Here, you are not concerned with having numbers that match the proportions in the population. Instead, you simply want to have enough to assure that you will be able to talk about even small groups in the population. This method is the nonprobabilistic analogue of stratified random sampling in that it is typically used to assure that smaller groups are adequately represented in your sample.

10.5.14 Snowball Sampling

In snowball sampling, you begin by identifying someone who meets the criteria for inclusion in your study. You then ask them to recommend others who they may know who also meet the criteria. Although this method would hardly lead to representative samples, there are times when it may be the best method available. Snowball sampling is especially useful when you are trying to reach populations that are inaccessible or hard to find. For instance, if you are studying the homeless, you are not likely to be able to find good lists of homeless people within a specific geographical area. However, if you go to that area and identify one or two, you may find that they know very well who the other homeless people in their vicinity are and how you can find them.

10.6 Flood Forecasting Model

Rainfall-runoff and channel-routing models are the foundation of flood forecasting systems. Floods could be forecast using rainfall-runoff models (also called hydrological models), or routing models, or a combination of both. Hydrological modeling is the process of mathematically representing the response of a catchment system (runoff) to precipitation events during the time period under consideration. Hydrological modeling is a very effective tool in generating runoff forecast, based on weather forecast. Hydrological models use climatic variables (e.g., precipitation, temperature, and evapotranspiration), catchment topography, and land-use characteristics to simulate runoff. Precipitation is the activating signal of a hydrological process. Runoff or streamflow is the part of precipitation that appears in a stream and represents the total response of a basin. The total runoff consists of surface flow, subsurface flow, groundwater or base flow, and precipitation falling directly on the stream. Streamflow data are the most important data in hydrology, as it is required for planning, operation, and control of any water resource project. Hydrological cycle is the endless circulation of water between the earth and its atmosphere. It is the most fundamental principle of hydrology. Hydrological phenomena are extremely complex and highly nonlinear, and exhibit a high degree of spatial and temporal variability. It is not possible to measure everything that is required to be known about hydrological systems. Therefore, modelling of hydrological variables becomes one of the important aspects in the field of hydrology. The ultimate aim of prediction, using models, is to assist in decision-making in hydrological problems, such as flood protection, water resources planning, etc. Hydrological models are classified based on process, spatial representation, or randomness (Figure 10.1).

Figure 10.1:
Classification of Hydrological Models

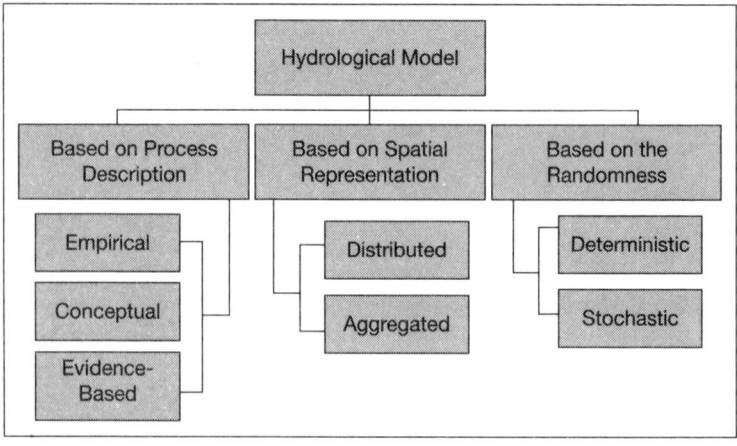

Source: Authors.

10.7 A Model for an Ideal Transboundary Flood Forecasting System

A model flood forecasting and early warning systems has been discussed in this section. It is now realized that ideal river flow and flood forecasting and early warning systems should be those that have effective and efficient regional (basin-wide) hydrometeorological monitoring, scientific data analysis, and forecasting models at an appropriate center producing timely warning and forecast products. The systems comprise reliable and rich data and information sources; the forecasting center and flood areas linked with real-time communication to enable operations for flood forecasting models save lives and protect property and infrastructure from destruction by floods. Essential features of comprehensive end-to-end flood early warning systems are given in succeeding paras. End-to-end flood forecasting and early warning system flood early warning, to be effective, should provide adequate lead time for institutions and communities at risk to undertake preparatory and mitigating actions. The chain that starts with monitoring extreme weather and climate events, leading up to community-level response, can be functionally disintegrated into steps wherein developmental interventions can contribute to preparedness and reduction in disaster risks at the community level. It is end to end when it involves a chain of activities that connect the technical and societal aspects of warning, from understanding and mapping the hazard and monitoring and forecasting/predicting impending/emerging harmful disasters to processing and disseminating understandable warnings to authorities and the population and undertaking appropriate and timely actions in response to the warnings by involvement and participation of all stakeholders. Stakeholder feedback is a key feature, allowing post-disaster assessment for learning lessons, identifying good practices, and providing recommendations for improving the early warning system. These components of an end-to-end flood early warning system are illustrated in Figure 10.2.

An operational end-to-end flood forecasting and warning system has the following basic elements:

- real-time monitoring system;
- forecasting system;
- numerical weather prediction system;
- data preprocessing system;

Figure 10.2:
End-to-end Flood Early Warning System

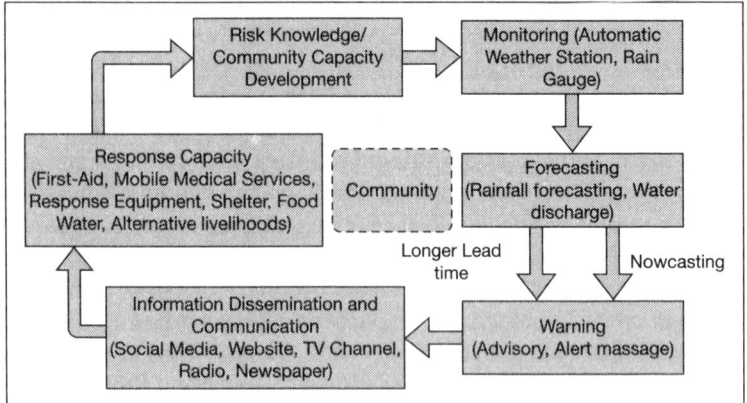

Source: Authors.

- hydrological modeling system;
- hydraulic modeling system;
- error correction system;
- warning system;
- decision support system;
- dissemination and communication system;
- preparedness and response system; and
- feedback system.

10.8 Post-disaster Needs Assessment

10.8.1 Objectives of a PDNA

The main goal of conducting a post-disaster needs assessment (PDNA) is to assist governments to assess the full extent of a disaster's impact on the country and, on the basis of these findings, to produce an actionable and sustainable recovery strategy for mobilizing financial and technical resources. And, if necessary, request additional external cooperation and assistance to implement it, given the affected country's capacities—financial, technical, and institutional. More specifically, a PDNA sets out the following objectives,

- Support country-led assessments and initiate recovery planning processes through a coordinated inter-institutional platform integrating the concerted efforts of the United Nations system, the European Union, the World Bank, other participating international donors, financial institutions, and NGOs.
- Evaluate the effect of the disaster on
  o infrastructure and assets;
  o service delivery and access to goods and services across all sectors, particularly the availability of basic services and the quality of service delivery;
  o governance and social processes; and
  o assessing needs to address underlying risks and vulnerabilities so as to reduce risk and build back better (BBB).
- Estimating the damage and loss caused by the disaster to physical infrastructures, productive sectors and the economy, including an assessment of its macroeconomic consequences;
- Identify all recovery and reconstruction needs.
- Develop the recovery strategy outlining priority needs, recovery interventions, expected outputs, and the cost of recovery and reconstruction that would form the basis for a comprehensive recovery framework.
- Provide the basis for mobilizing resources for recovery and reconstruction through local, national, and international sources.

The PDNA guide includes the main elements of the Damage and Loss Assessment (DaLA) method and the Human Recovery Needs Assessment (HRNA) approach and process for a comprehensive assessment of damages, losses, and needs, which would lead to the development of a recovery strategy. Since the assessment and recovery strategy developed during the PDNA is completed in a relatively short period, it requires more comprehensive recovery planning, particularly in the case of large-scale disasters. This is done through the development of national recovery frameworks (Figure 10.3).

Figure 10.3:
Typical Sectors Assessed in the PDNA

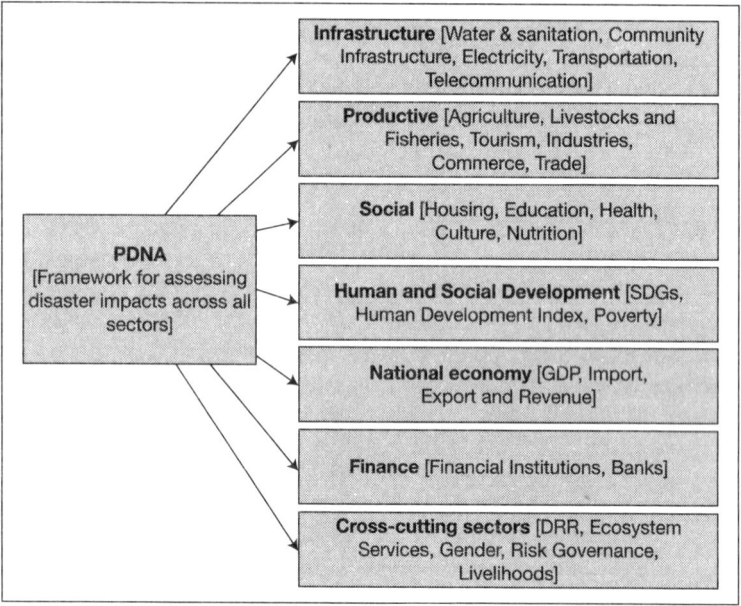

Source: Authors.

10.8.2 The Deliverables of PDNA

The PDNA produces the following four core deliverables:

- One consolidated assessment report, based on sector reports, presenting the overall effect and impact of the disaster on each sector, the recovery needs for each, as well as the explicit impact on cross-cutting themes, with a gender perspective, environmental considerations, risk reduction, and governance.

- A recovery strategy that defines the vision for national recovery; provides a strategy for recovery actions within each sector and affected region, armed with clear objectives and interventions; directs it toward expected results; and defines the timeframe as well as the cost for the recovery process.

- The basis for resource mobilization in support of the country's recovery, including a donor conference where required.
- An outline for a country-led implementation mechanism for recovery.

10.8.3 Core Principles of PDNA

All PDNAs would be guided by the following core principles:

- Adhere to the core principles of humanitarianism, impartiality, and neutrality.
- Acknowledge the national ownership of PDNA and ensure that it is a demand-driven and country-led process, with the fullest possible leadership and engagement of national authorities in assessment, recovery planning and implementation, from the highest political levels to local levels, and at the level of technical expertise.
- Support local ownership and the fullest possible engagement of local authorities and CBOs in the planning and execution of recovery, and building specific capacities where needed.
- Provide coordination at all stages of the process and at all levels, ensuring collaboration and partnership between the UN, the WB, and the EU, as well as with the national government, donors, NGOs, civil society, and other stakeholders engaged in the PDNA.
- Ensure one team, one process, and one output.
- Adhere to the principle of *Primum non nocere*—"first, do no harm"— ensuring that the process does not have a detrimental effect on life-saving relief to the affected population and on the country.
- Adopt a conflict-sensitive approach and ensure that the assessment does not exacerbate the existing tension and that the recovery strategy takes into account potential disaster-related conflicts.
- Support and strengthen national and local capacities to lead and manage recovery and reconstruction.
- Ensure transparency and accountability in the PDNA process as well as in post-disaster recovery and reconstruction.
- Integrate DRR measures in the recovery process to enhance the resilience of affected populations and countries with regard to future disasters.

- Develop a recovery plan that addresses the gap created by the disaster, which effectively helps people to BBB and reduce future risks without expanding recovery needs and priorities into a full-fledged development plan that goes beyond the disaster.
- Ensure the participation of the affected population in the assessment of needs and priorities and in the recovery process, at the same time providing support to their spontaneous recovery efforts.
- With a gender perspective, focus on the most vulnerable sections, including female-headed households, children, orphans, the landless, people with special needs, the youth and the aged.
- Promote equality to prevent discrimination of any kind on grounds of race, color, nationality, ideology, sex, ethnicity, age, language, religion, disability, property, and birth, among others.
- Mainstream cross-cutting issues such as gender, environment, governance, human rights, HIV/AIDS, among others.
- Ground recovery in the principles of sustainable development.
- Build on national development strategies as required.
- Monitor, evaluate, and learn from practice.
- Complete the assessment in a timely manner to capitalize on the limited window of opportunity to start recovery, resource mobilization, and resilience building initiatives.

Glossary

Adaptation: The adjustment in natural or human systems in response to actual or expected climatic or other stimuli or their effects, which moderates harm or exploits beneficial opportunities.

AEP: Annual exceedance probability is the estimated probability that an event of specified magnitude will be exceeded in any year.

ALOS: Advanced Land Observing Satellite is a land-observing satellite launched in January 2006 by the Japan Aerospace Exploration Agency (JAXA).

APRSAF: The Asia-Pacific Regional Space Agency Forum was established in 1993 to enhance space activities in the Asia-Pacific region.

Audit: An official examination and verification of accounts and records to analyze the legality and regularity of project expenditures and income, in accordance with laws, regulations, and contracts, such as loan contracts and accounting rules. It also may analyze the efficient and effective use of funds.

Avalanche: An avalanche describes a quantity of snow or ice that slides down a mountainside under the force of gravity. It occurs if the load on the upper snow layers exceeds the bonding forces of the entire mass of snow. It often gathers material that is underneath the snowpack such as soil, rock etc. (debris avalanche). Any kind of rapid snow/ice movement.

Baseline Data: Initial information collected during an assessment, including facts, numbers, and descriptions. This information will allow us to do the comparison between the situation that exists before the disaster and measurement of the impact of the project implemented.

Basic Needs: The items that people need to survive. They can include safe access to essential goods and services such as food, water, shelter, clothing, healthcare, sanitation, and education.

Biological Disasters: Disasters caused by the exposure of living organisms to germs and toxic substances.

Build Back Better (BBB): An approach to reconstruction to reduce vulnerability and improve living conditions, while promoting a more effective and sustainable reconstruction. BBB uses the opportunity of having to rebuild to examine the suitability of reconstructing in the same location and making a home warmer, drier, and cheaper to run.

Building Code: A set of ordinances or regulations and associated standards intended to control aspects of the design, constructions, materials, alteration, and occupancy of structures that are necessary to ensure human safety and welfare, including resistance to collapse and damage.

Capacity-Building: The process by which individuals, groups, and organizations build their knowledge, abilities, relationships, and values to solve problems and achieve development objectives. The impacts of capacity-building thus may be seen at different scales—individuals, households, communities, and governments.

Capacity: The combination of all physical, institutional, social, and/or economic strengths, attributes, and resources available within a community, society, or organization that can be used to achieve the agreed goals. It also includes collective attributes such as leadership and management.

Capacity: The physical, human, and social capital in any society.

Cash Transfers: Direct payments or vouchers to provide resources to affected populations.

Civil Society: Individuals and groups active between the governance and market, including the institutions within the interaction.

Climate Change Resilience: The ability to resist, absorb, adapt to, and recover from meteorological changes attributed directly or indirectly to human activities that alter the composition of the global atmosphere or the natural climate variability.

Climatological Disasters: Events caused by long-lived/meso to macroscale processes (in the spectrum from intraseasonal to multidecadal climate variability).

Cold Wave: A cold wave can be both a prolonged period of excessively cold weather and the sudden invasion of very cold air over a large area. Along with frost, it can cause damage to agriculture, infrastructure, and property. The damage caused by low temperatures.

Community Contracting: Procurement by or on behalf of a community. While there are many different models of community contracting, a common feature is that they seek to give the community degrees of control over investment and implementation, to encourage ownership and sustainability.

Community: A social group of any size whose members reside in a specific locality, share government, and often have a common cultural and historical heritage.

Community: Association of individuals, cutting across any geographical boundary, language, religion, ethnicity, etc.

Complementarities: Complementarities refer to a situation where two or more factors increase each other's effects on performance.

Consequences: Outcomes of an event, such as a landslide hazard. Depend on the exposure and vulnerability of the elements-at-risk, such as human beings, houses, and infrastructure.

Coping Capacity: The strength to face and overcome an adverse situation.

Corruption: Misuse of an entrusted position for private gain by using bribery, extortion, fraud, deception, collusion, and money laundering. Includes gains accruing to a person's family members, a political party, or an institution in which the person has an interest.

Direct Costs (or Damage): Reconstruction costs incurred by total or partial destruction of the physical assets existing in the affected area. The damage occurs during and immediately after the disaster and is measured in physical units. Its monetary value, it is expressed in terms of replacement costs according to prices prevailing just before the event.

Disaster Risk Management (DRM): Systematic process of using administrative directives, organizations, and operational skills and capacities to implement strategies, policies, and improved coping capacities to lessen the adverse impacts of hazards and the possibility of disaster.

Disaster Risk Reduction (DRR): Concept and practice of reducing disaster risks through systematic efforts to analyze and manage the causal factors of disasters. Results of DRR include reduced exposure to hazards, lessened vulnerability of people and property, wise management of land and the environment, and improved preparedness.

Disaster: A situation or event that overwhelms local capacity, necessitating a request to a national or an international level for external assistance; an unforeseen and often sudden event that causes great damage, destruction, and human suffering.

Drought: A long-lasting event triggered by a lack of precipitation. A drought is an extended period of time characterized by a deficiency in a region's water supply that is the result of constantly below average precipitation. A drought can lead to losses in agriculture, affect inland navigation and hydropower plants, and cause a lack of drinking water and famine.

Early Warning System: The set of capacities needed to generate and disseminate timely and meaningful warning information to enable individuals, communities, and organizations threatened by a hazard to prepare and to act appropriately and in sufficient time to reduce the possibility of harm to or loss of life or livelihoods, injury, damage to property, and damage to the environment. A people-centered early warning system comprises four key elements. They are (a) knowing the risks; (b) monitoring, analyzing, and forecasting the hazards; (c) communicating or disseminating alerts and warnings; and (d) developing the local capacities to respond to the warnings. The term "end-to-end warning systems" is used to emphasize that warning systems need to span all steps from detecting hazards to the community's response.

Earthquake: The shaking and displacement of ground due to seismic waves. This is the earthquake itself without secondary effects. An earthquake is the result of a sudden release of stored energy in the Earth's crust that creates seismic waves. They can be of tectonic or volcanic origin. At the Earth's surface, they are felt as a shaking or displacement of the ground. The energy released in the hypocenter can be measured in different frequency ranges. Therefore, there are different scales for measuring the magnitude of a quake according to a certain frequency range. These are (a) surface wave magnitude (Ms), (b) body wave magnitude (Mb), (c) local magnitude (ML); and (d) moment magnitude (Mw).

Effective Recovery: Achieving the intended outcomes of medium- to long-term recovery such as the rehabilitation and reconstruction of damaged infrastructure and the re-creation of sustainable livelihoods and income-generating opportunities.

Efficient Recovery: Stabilizing lives and livelihoods to return to normal and rapidly restoring critical social, physical, and productive infrastructure and service delivery.

Empowerment: The authority given to an institution, organization, or individual to determine policies and make decisions.

Enabling Environment: The rules and regulations, both national and local, which provide a supportive environment for a specific activity, such as a community participation or DRM, to take place.

Epidemic: Either an unusual increase in the number of cases of an infectious disease that already exists in the region or population concerned or the appearance of an infection disease previously absent from a region.

Equity: The quality of being impartial and "fair" in the distribution of development benefits and costs and the provision of access of opportunities for all.

EX-post Measures: Actions taken after a disaster has occurred to seek to mitigate or repair all damages caused by the disaster.

Exposure: People, property, systems, or other elements present in hazard zones that thereby are subject to potential losses.

Extensive Risk: Widespread risk associated with the exposure of dispersed populations to repeated or persistent hazard conditions of low or moderate intensity, often of a localized nature. Such persisting exposure can have debilitating cumulative disaster impacts. This type of risk is a characteristic primarily of rural areas and urban margins. See also "risk" and "intensive risk."

Extreme Winter Condition: Damage caused by snow and ice. Winter damage refers to the damage to buildings, infrastructure, and traffic (especially navigation) inflicted by snow and ice in the form of snow pressure, freezing rain, frozen waterways, etc.

Fault: A fracture inside the underground rock mass caused by shifting of the Earth's crust and energy released during the earthquake.

Flash Food: Rapid inland floods due to intense rainfall. A flash flood describes sudden flooding with short duration. In sloped terrains, the water flows rapidly with a high destruction potential. In flat terrains, the rainwater cannot infiltrate into the ground or runoff (due to small slope) as quickly as it falls. Flash floods typically are associated with thunderstorms. A flash flood can occur at virtually any place.

Flood Forecasting: Use of real-time precipitation and streamflow data in rainfall runoff and streamflow routing models to forecast flow rates and water levels from a few hours to a days ahead, depending on the size of the watershed or river basin.

Flood: A general and temporary condition of partial or complete inundation of normally dry land areas from (a) the overflow of inland or tidal waters, (b) the unusual and rapid accumulation or runoff of surface waters from any source, or (c) mudflows or the sudden collapse inland of shoreline.

Floodwall: A concrete or rigid wall on a levee, which protect low-lying lands from inundation by rising water.

Forecast: Definite statement or statistical estimate of the likely occurrence of a future event or conditions for a specific area.

Forest fire: Fires in forests that cover extensive damage. They may start by natural causes such as volcanic eruptions or lightning, or they may be caused by arsonists or careless smokers, by those burning wood, or by clearing a forest area.

Fungibility: The property of a good or a commodity whose individual units are capable of mutual substitution.

General Flood: Gradually rising inland floods (rivers, lakes, and groundwater) due to high total depth of rainfall or snowmelt. A general flood is caused when a body of water (river or lake) overflows its normal confines due to rising water levels. The term general flood additionally comprises the accumulation of water on the surface due to long-lasting rainfall (water logging) and the rise of the groundwater table above the surface. Furthermore, inundation by melting snow and ice, backwater effects, and special causes such as the outburst of a glacial lake or the breaching of a dam are subsumed under the term general flood. General floods can be expected at certain locations (e.g., along rivers) with a significantly higher probability than at others.

Geophysical Disasters: Events originating from solid earth.

Green Growth: Growth that is efficient in its use of natural resources; clean in that it minimizes pollution and environmental impacts and resilient in that it takes into account natural hazards and the role of environmental management and natural capital in preventing physical disasters.

Hazard: Natural process or phenomenon or human activity that has the potential to cause property damage, loss of livelihoods and services, social and economic disruption, and/or environmental degradation.

Heat Wave: A heat wave is a prolonged period of excessively hot and sometimes also humid weather relative to the normal climate patterns of a certain region. A period of unusual hot weather, with higher surface temperature than normal, that last for a few days.

Housing: The immediate physical environment, including the inside and outside of buildings, in which families and households live and so serves as a shelter.

Housing-sector Assessment: Assessment that collects data including demographic, housing types, housing tenure status, settlement patterns before and after the disaster, government interventions in the housing sector, infrastructure access, construction capacity, and market capacity to provide materials and labor for reconstruction.

Humanitarian Relief: A process that seeks to lead to sustainable development opportunities by generating self-sustaining processes for post-disaster recovery. Humanitarian relief encompasses livelihoods, shelter, governance, environment, and social dimensions, including the reintegration of displaced populations. It also addresses the underlying risks that contributed to the crisis.

Hydrological Disasters: Events caused by deviations in the normal water cycle and/or overflow of bodies of water caused by wind setup.

Infrastructure: Systems and networks by which public services are delivered. These services include water supply and sanitation, energy and other utility networks, and transportation networks for all forms of travel.

Insect infestation: Pervasive influx and development of insects or parasites affecting humans, animals, crops, and materials.

Intensive Risk: The risk associated with the exposure of large concentrations of people and economic activities to intense hazard events that can lead to potentially catastrophic disaster impacts involving high mortality and asset loss. A characteristic primarily of large cities or densely populated areas that not only are exposed to intense hazards but also have high levels of vulnerability to them. See also "risk" and "extensive risk."

Key Performance indicators (KPIS): Quantitative and qualitative measures of project outputs and outcomes used to evaluate the progress of success of the project.

Landslide: Any kind of moderate to rapid soil movement including mudslide and debris flow. A landslide is the movement of soil or rock controlled by gravity, and the speed of the movement usually ranges between slow and rapid. It can be superficial or deep, but the materials have to make up a mass that is a portion of the slope or the slope itself. The movement has to be downward and outward with a free face.

Levee: A high embankment or reinforced earth wall, which protects low-lying areas from inundation.

Livelihoods: The ways in which people earn access to the resources that they need, individually and communally, including food, water, clothing, and shelter.

Local Windstorm (Orographic Storm): A local windstorm refers to strong winds caused by regional atmospheric phenomena which are typical for a certain area. These can be katabatic winds, foehn winds, mistral, bora, etc.

Loss Assessment: An assessment that analyzes the changes in economic flows that occur after a disaster and over time, valued at current prices.

Losses: This includes the decline in output in productive sectors and the lower revenues and higher operational costs in the provision of services. Also considered losses are the unexpected expenditures to meet emergency needs. Losses are expressed in current values.

Meteorological Disasters: Events caused by short-lived/small to mesoscale atmospheric processes (in the spectrum from minutes to days).

Mitigate/Mitigation: The use of reasonable care and diligence to minimize damage; to take protective action to avoid additional injury or loss, and to lessen or limit the adverse impact of hazards and disasters.

Monitoring: The ongoing task of collecting and reviewing program-related information that pertains to the program's goals, objectives, and activities.

Needs Assessment: The process for estimating (usually based on a damage assessment) the financial, technical, and human resources needed to implement the agreed program of recovery, reconstruction, and risk management.

Node: The central location for staff and materials during a disaster event.

Nonstructural Measure: Any measure not involving physical construction that uses knowledge, practice, or agreement to reduce risks and impacts, particularly through policies and laws, public awareness-raising, training, and education. See also "structural measures."

Off-budget Financing: Could not be managed directly by the national government or is not comprised in its budget.

On-budget Financing: Within the national government's control, including own-source revenue (OSR) as well as external funding and loans.

Participatory: People's participation in decision-making and policy formulation.

Partners: A donor community or any group or individual taking part and sharing the responsibility of the reconstruction and recovery process. In contrast, see "stakeholders."

Physical Planning: A design exercise based on a land-use plan to propose optimal infrastructure for public services, transport, economic activities, recreation, and environmental protection for a settlement or area. A physical plan can have rural and urban components.

Policy: A principle or protocol to guide decisions and achieve rational outcomes.

Post-disaster Needs Assessment (PDNA): A multisectoral assessment that measures the impact of disasters on the society, economy, and environment of the disaster-affected area.

Preliminary Assessment: An assessment that provides immediate information on needs, possible interventions, and resource requirements. It may be conducted as a multisectoral assessment or in a single sector or location.

Preparedness: The knowledge and capacities developed by governments, professional response and recovery organizations, communities, and individuals to effectively anticipate, respond to, and recover from the impacts of likely, imminent, or current hazard events or conditions.

Prevention: To avoid and minimize the adverse impact of related environmental, technological, and biological disasters by raising public awareness and providing education related to disaster risk reduction, changing attitudes and behavior.

Prior Measures (ex ante): Actions taken in advance of a disaster in the expectation that they will either prevent or significantly reduce the impacts of a possible disaster.

Project Outputs: Results of a project that are measurable at the immediate point of project completion.

Reconstruction: It focuses primarily on the construction or replacement of damaged physical structures, and the restoration of local services and infrastructure.

Recovery Framework: A pragmatic, sequenced, prioritized, programmatic, yet living (and flexible) action plan that ensures resilient recovery after a disaster.

Recovery: Decisions and actions taken after a disaster to restore or improve the pre-disaster living conditions of the affected communities while encouraging and facilitating necessary adjustments to reduce disaster risk. It focuses not only on physical reconstruction but also on the revitalization of the economy and restoration of social and cultural life.

Relief: The provision of assistance or intervention immediately after a disaster to meet the life preservation and basic subsistence needs of the persons affected.

Relocation: A process whereby a community's housing assets and public infrastructure are rebuilt in another location.

Residual Risk: The risk that remains in unmanaged form, even when effective disaster risk-reduction measures are in place, and for which emergency response and recovery capacities must be maintained. The presence of residual risk implies a continuing need to develop and support effective capacities for emergency services, preparedness, response, and recovery together with socioeconomic policies such as safety nets and risk-transfer mechanisms.

Resilience: The ability of a system, community, or society exposed to hazards to resist, absorb, accommodate, and recover from the effects of a hazard in a timely and efficient manner, including through the preservation and restoration of its essential structures and functions. Resilience is determined by the degree to which the community has the necessary resources and is capable of organizing itself both prior to and during times of need.

Resilient Recovery: Builds resilience during recovery and promotes resilience in regular development. Resilient recovery is a means to sustainable development. See also "resilience," "recovery," "disaster risk management," and "disaster risk reduction."

Response: The provision of emergency services and public assistance during or immediately after a disaster to save lives, reduce health impacts, ensure public safety, and meet the basic subsistence needs of the people affected. See also "humanitarian relief."

Right-siting: Rebuilding facilities in areas that are less prone to disasters and accessible to the community.

Right-sizing: Rebuilding facilities such that they adequately respond to the existing demand; for example, if classes are crowded, more classes could be built.

Risk: The combination of the probability of an event and its negative consequences.

Risk Transfer: The process of formally or informally shifting the financial consequences of particular risks from one party to another. In this transaction, one party (household, community, enterprise, or state authority) will obtain post-disaster resources from another party in exchange for ongoing or compensatory social or financial benefits.

Rockfall: Quantities of rock or stone falling freely from a cliff face. It is caused by undercutting, weathering, or permafrost degradation.

Scoping: An investigation or discussion to determine the effect that a proposed policy or project would have on a community or the environment.

Stakeholders: Groups who have any direct or indirect interest in the recovery interventions, or who can affect or be affected by the implementation and outcomes. The term includes groups undertaking, managing, reporting on, affected by, promoting, and funding the interventions. Stakeholders include vulnerable segments of the population, local governments that are in direct dialogue with communities.

Storm Surge: An abnormal rise in the sea along a shore primarily due to high winds. Coastal flood on coasts and lake shores induced by wind. A storm surge is the rise of the water level in the sea, an estuary, or a lake as result of strong wind driving the seawater towards the coast. This so-called wind setup is superimposed on the normal astronomical tide. The mean high water level can be exceeded by five and more meters. The areas threatened by storm surges are coastal lowlands.

Structural Measure: Any physical construction to reduce or avoid possible impacts of hazards, or application of engineering techniques to achieve hazard-resistance and resilience in structures or systems. See also "nonstructural measures."

Subsidence: The downward motion of the Earth's surface relative to a datum (e.g., the sea level). Dry subsidence can be the result of geological faulting, isostatic rebound, or human impact (e.g., mining, extraction of natural gas). Wet subsidence can be the result of karst, changes in soil water saturation, permafrost degradation (thermokarst), etc.

Subsidiarity: The principle by which matters ought to be handled by the smallest, lowest, or least centralized competent authority.

Sustainable Development: Development that meets the needs of the present without compromising the ability of future generations to meet their own needs. This 1987 Brundtland Commission definition does not address questions regarding the meaning of the word "development" and the social, economic, and environmental processes involved. Disaster risk is associated with unsustainable elements of development such as environmental degradation. Conversely, disaster risk reduction can contribute to sustainable development by reducing the process of development which meets the present day concerns without compromising with the fulfillment of the concerns of future generation.

Targeting: Identification and recruitment by local communities, the government, or external agencies of potential assistance recipients.

Tropical Cyclone: A tropical cyclone is a nonfrontal storm system that is characterized by a low pressure centre, spiral rain bands and strong winds. Usually it originates over tropical or sub-tropical waters and rotates clockwise in the southern hemisphere and counter-clockwise in the northern hemisphere. The system is fuelled by heat released when moist air rises and the water vapor it contains condenses ("warm core" storm system). Therefore, the water temperature must be >27°C. Depending on their location and strength, tropical cyclones are referred to as hurricane (western Atlantic/eastern Pacific), typhoon (western Pacific), cyclone (southern Pacific/Indian Ocean), tropical storm, and tropical depression (defined by wind speed; see Saffir-Simpson Scale). Cyclones in tropical areas are called hurricanes, typhoons, and tropical depressions (names depending on location).

Volcanic Eruption: All volcanic activity such as rock fall, ash fall, lava streams, gases etc. Volcanic activity describes both the transport of magma and/or gases to the Earth's surface, which can be accompanied by tremors and eruptions, and the interaction of magma and water (e.g., groundwater- and crater lakes) underneath the Earth's surface, which can result in phreatic eruptions. Depending on the composition of the magma, eruptions can be explosive and effusive and result in variations of rock fall, ash fall, lava streams, pyroclastic flows, emission of gases, etc.

Vulnerability: Characteristics and circumstances of a community, system, or asset that make it susceptible to the damaging effects of a hazard.

Vulnerable Groups: Groups or members of groups who are particularly exposed to the impacts of hazards. Examples are displaced persons, women, the elderly, the disabled, orphans, and any group subject to discrimination.

Watershed: An area of land from which all of the water under it or on it drains to the same place, which may be a river, lake, reservoir, estuary, wetland, sea, or ocean.

Wildfire: Wildfire describes an uncontrolled burning fire, usually in wild lands, which can cause damage to forestry, agriculture, infrastructure, and buildings.

Bibliography

Adelphi/EURAC. 2014. *The Vulnerability Sourcebook: Concept and Guidelines for Standardised Vulnerability Assessments* (p. 58). Bonn and Eschborn, Germany: Deutsche Gesellschaft für Internationale Zusammenarbeit (GIZ) GmbH.

Adinolfi, C., D.S. Bassiouni, H.F. Lauritzsen, and H.R. Williams. 2005. *Humanitarian Response Review* (p. 112). United Nations.

Ahrens, J., and P.M. Rudolph. 2006. "The Importance of Governance in Risk Reduction and Disaster Management." *Journal of Contingencies and Crisis Management,* 14 (4), 207–220.

Alcantara-Ayala, I., and A. Goudie., eds. 2010. *Geomorphological Hazards and Disaster Prevention.* Cambridge: Cambridge University Press, 291.

Alexander, D. 1993. *Natural Disasters.* London: UCL Press Ltd., University College.

Alexander, D. 2002. *Principles of Emergency Planning and Management* (p. 379). Terra Publishing, Dunedin Academic Press Ltd.

Alexander, D.E. 1993. *Natural Disasters* (p. 632). Dordrecht/Boston/London: Springer Science & Business Media; Kluwer Academic Publishers.

Apte, J. 2009. "Facilitating Transformative Learning: A Framework for Practice." *Australian Journal of Adult Learning,* 49 (1), 169–89.

Aryal, T.R. 2007a. "Age at First Marriage in Nepal: Differentials and Determinants." *Journal of Biosocial Science,* 39 (5), 693–706.

———. 2007b. "Differentials of Post-partum Amenorrhea: A Survival Analysis." *Journal of the Nepal Medical Association,* 46 (166), 66–73.

———. 2007c. "Post-partum Amenorrhea among Nepalese Mothers." *Journal of Population and Social Studies,* 16 (1), 33–64.

Balcik, B., B.M. Beamon, C.C. Krejci, K.M. Muramatsu, & M. Ramirez. 2010. "Coordination in Humanitarian Relief Chains: Practices, Challenges and Opportunities." International Journal of Production Economics, 126 (1), 22–34.

Bartels, S.A., and M.J. vanRooyen. 2012. "Medical Complications Associated with Earthquakes." Lancet,379 (9817), 48-57.

Beema Samiti. 2012. Micro Capital by Charlotte Newman–Microinsurance, South Asia, East Asia and Pacific. https://www.microcapital.org/microcapital-brief-insurance-regulatory-authority-of-nepal-to-launch-microinsurance-guidelines-in-march/

Beer, Tom, and Frank Ziolkowski. 1995. *Environmental Risk Assessment: An Australian Perspective.* Barton: Supervising Scientist, Department of the Environment and Energy.

Benson, Charlotte, and John Twigg 2007. *Tools for Mainstreaming Disaster Risk Reduction: Guidance Notes for Development Organisations.* ProVention Consortium, 2007.

Benz, A., & B. Eberlein. 1999. "The Europeanization of Regional Policies: Patterns of Multi-level Governance." *Journal of European Public Policy,* 6, 329-348.

Bhandary, N.P., R. Yatabe, K. Yamamoto, and Y.R. Paudyal. 2014. "Use of a Sparse Geo-Info Database and Ambient Ground Vibration Survey in Earthquake Disaster Risk Study–A Case of Kathmandu Valley." Journal of Civil Engineering Research, 4 (3A), 20-30. Scientific & Academic Publishing.

Blaikie, P., T. Cannon, I. Davis, and B. Wisner. 1994. *At Risk: Natural Hazards, People's Vulnerability, and Disasters.* London: Routledge.

Bronkhorst, Johannes. 2012. "Levels of Cognition: Did Indian Philosophers Know Something We Do Not?" *Asiatische Studien – Études Asiatiques*, 66 (2), 227–37.

Bui, T., S. Cho, S. Sankaran, and M. Sovereign. 2000. "A Framework for Designing a Global Information Network for Multinational Humanitarian Assistance/Disaster Relief." Information Systems Frontiers, 1 (4), 427–442.

Burrough, P.A. 1986. "Principles of Geographic Information Systems for Land Resource Assessment." Monographs on Soil and Resources Survey No. 12. New York, NY: Oxford Science Publications.

Burroughs, P.P., and R.A. McDonnel. 1998. *Principles of GIS.* London: Oxford University Press, 299.

Burton, I., R. Kates, and G. White. 1978. *The Environment as Hazard.* New York, NY: Oxford University.

Cannon, T., J. Twigg, and J. Rowell. 2003. *Social Vulnerability, Livelihoods and Disasters.* Report to DFID. Kent: University of Greenwich, 63.

Carter, W.N. 1999. Disaster Management: A Disaster Management Handbook (p. 416). Manila: Asian Development Bank.

Castanos, H., and C. Lomnitz. 2009. "Ortwin Renn, Risk Governance: Coping with Uncertainty in a Complex World." Natural Hazards, 48 (2), 313–314.

CEOS/IGOS Disaster Management Support Project. 1999. Available at: http://www.ceos.noaa.org/

Barrett E.C. 1996. "The Storm Project: Using Remote Sensing for Improved Monitoring and Prediction of Heavy Rainfall and Related events." *Remote Sensing Reviews*, 14, 282.

Center for Excellence, Disaster Management & Humanitarian Assistance (DMHA). 2012. *Nepal Disaster Management Reference Handbook.* Kathmandu, Nepal.

Chakraborti A.K. 1999. *Satellite Remote Sensing for Near-Real-Time Flood and Drought Impact Assessment—Indian Experience.* Workshop on Natural Disasters and their Mitigation - A Remote Sensing & GIS perspective, 11–15 October 1999, Dehradun, India.

Chakraborti D., C.R. Chanda, G. Samanta, U.K. Chowdhury, S.C. Mukherjel, A.B. Pal, B. Sharma, K.J. Mahanta, H.A. Ahmed, and B. Sing. 2000. "Fluorosis in Assam, India." *Current Science*, 78, 1421–23.

Champati Ray, P.K. 1996. "Landslide Hazard Zonation Using Fuzzy Logic and Probability Analysis in Parts of Western Himalayas." M.S. Thesis, ITC, Enschede, the Netherlands.

Champati Ray, P.K., Kenny Forest, and P.S. Roy. 2001. *Bhuj Earthquake*, GIS @ Development, V (8), 28–31.

Chen, Y., and D.C. Booth. 2011. *The Wenchuan Earthquake of 2008: Anatomy of a Disaster* (p. 272). Berlin: Springer.

Chen, R., R. Sharman, H.R. Rao, and S. Upadhyaya. 2008. "Coordination in Emergency Response Management." *Communications of the ACM,* 51 (5), 66–73.

Chen, A.Y., F. Peña Mora, and Y. Ouyang. 2011. "A Collaborative GIS Framework to Support Equipment Distribution for Civil Engineering Disaster Response Operations." *Automation in Construction,* 20 (5), 637–648.

Cohen, E. 2012. "Flooded: An Auto-Ethnography of the 2011 Bangkok Flood." *ASEAS— Austrian Journal of South-East Asian Studies,* 5(2), 316–34.

Colomina I., M. Blázquez, P. Molina, M.E. Parés, and M. Wis. 2008. "Towards A New Paradigm for High-Resolution Low-Cost Photogrammetry and Remote Sensing." In

The International Archives of the Photogrammetry, Remote Sensing and Spatial Information Sciences, ISPRS Congress, Beijing, China, XXXVII. Part B1, 1201–1206.

Corr D. 1983. "Production of DEM's from ERS-1 SAR Data." *Mapping Awareness*, 7, 18–22.

Das I., S. Sahoo, C. van Weston, A. Stein, and R. Hack. 2010. "Landslide Susceptibility Assessment Using Logistic Regression and Its Comparison with a Rock Mass Classification System, Along a Road Section in the Northern Himalayas (India). *Geomorphology*, 114, 627–37.

Davis, Mark H. 1996. *Empathy: A Social Psychological Approach*. New York: Westview Press, 260.

Davis. I. 2011. "What Have We Learned from 40 years' Experience of Disaster Shelter?" *Environmental Hazards*, 10 (3-4), 193-212.

Disaster Recovery Frameworks. 2015. *Guide to Developing Disaster Recovery Frameworks*, Sendai Conference Version, GFDRR, World Bank Group, EU, UNDP.

Doupnik, T.S., and S.B. Salter. 1995. *External Environment, Culture, and Accounting Practices: A Preliminary Test of a General Model of International Accounting Development*.

Dovers, Stephen R. 1995. "A Framework for Scaling and Framing Policy Problems in Sustainability." *Ecological Economics*, 12, 93–106.

Dovers, S., and J. Handmer. 2007. *The Handbook of Disaster and Emergency Policies and Institutions*. London: Earthscan Publications.

Doyle, H. 1995. *Seismology*. Chichester, London: Wiley and Sons.

DPNET. 2011. *Nepal: Disaster Preparedness Network Nepal (DPNET)*. Available at: http://www.dpnet.org.np/ (last accessed 13 December 2016).

Edrissi, A., H. Poorzahedy, H. Nassiri, and M. Nourinejad. 2013. "A Multi-agent Optimization Formulation of Earthquake Disaster Prevention and Management." *European Journal of Operations Research* 229 (1), 261–275.

Eisenbeiss H. 2008C. "UAV Photogrammetry in Plant Sciences and Geology." In 6th ARIDA Workshop on "Innovations in 3D Measurement, Modeling and Visualization", Povo (Trento), Italy.

Emergency Management Australia (EMA). 1995. *National Emergency Management Competency Standards*. Canberra: EMA.

ESCAP, and UNISDR. 2012. *Reducing Vulnerability and Exposure to Disasters: The Asia-Pacific Disaster Report 2012*. United Nations Economic and Social Committee for Asia and the Pacific (ESCAP) and United Nations Office for Disaster Risk Reduction(UNISDR), Bangkok.

EU 2001a. "European Governance: A White Paper." 52001DC0428. *Official Journal*, 287 (October 12, 2001), 0001- 0029.

Fiedrich, F., F. Gehbauer, and U. Rickers. 2000. "Optimized Resource Allocation for Emergency Response After Earthquake Disasters." *Safety Science*, 35 (1), 41–57.

Froberg, M. 2013. "Preparedness Saved Thousands of Lives During Cyclone Phailin: Now the Recovery Begins." International Federation of Red Cross and Red Crescent Societies. http://www.ifrc.org/en/news-and-media/news-stories/asia-pacific/india/recovering-from-cyclone-phailin-survivors-face-massive-hardships--63582/

GoO. 2013. *Memorandum on the Very Severe Cyclone Phailin and the Subsequent Flood*, October 12–15, 2013. The Revenue and Disaster Management Department Government of Odisha. http://www.osdma.org/userfiles/file/MEMORANDUMPhailin.pdf

Gee M. E., M.G. Anderson, & L. Baird. 1990. "Large-Scale Floodplain Modeling." *Earth Surface Processes and Landforms*, 15, 513–23.

Ghosh, T., and I. Pal. 2014. "Dust Storm and its Environmental Implications." *Journal of Engineering Computers & Applied Sciences (JECAS)*, 3 (4), 30–37.

Ghosh, T., G. Bhandari, and S. Hazra. 2004. "Application of a 'Bio-engineering' Technique to Protect Ghoramara Island (Bay of Bengal) From Severe Erosion." *Journal of Coastal Conservation*, 9, 171–78.

Ghosh, T., R. Hajra, and A. Mukhopadhyay. 2014. "Island Erosion and Afflicted Population: Crisis and Policies to Handle Climate Change." In *International Perspectives on Climate Change: Latin America and Beyond*, edited by Filho Leal, Fátima Alves, Sandra Caeiro, and Ulisses Azeiteiro, IX. Switerland: Springer, 217–26.

Giovanni, V. 1998. Mobile Agents and Security. Heidelberger, Berlin: Springer, 92–113.

GON. 2015. *National Strategy for Disaster Risk Management*. Government of Nepal.

Gopalakrishnan, C., and N. Okada. 2007. "Designing New Institutions for Implementing Integrated Disaster Risk Management: Key Elements and Future Directions." *Disasters,* 31 (4), 353–372.

Granot, H. 1997. "Emergency Inter-organizational Relationships." *Disaster Prevention and Management*, 6 (5), 305–310.

Guha-Sapir, D., Ph. Hoyois, and R. Below. 2012. Ann*ual Disaster Statistical Review 2012: The Numbers and Trends*. Brussels: CRED.

Guha-Sapir D., P. Hoyois, and R. Below. 2015. *Annual Disaster Statistical Review 2014*.

Hajra, R., S. Szabo, Z. Tessler, T. Ghosh, Z. Matthews, and E. Foufoula-Georgiou. 2017. "Unravelling the Association between the Impact of Natural Hazards and Household Poverty: Evidence from the Indian Sundarban Delta." *Sustainability Science*, 12, 453. doi:10.1007/s11625-016-0420-2

Handmer, J.W. 1995. "Managing Vulnerability in Sydney: Planning or Providence?" *Geo Journal*, 37, 355.

Harvey, D. 2010. *The Enigma of Capital and the Crises of Capitalism* (p. 296). New York: Oxford University Press.

Hohl, F. 1998. "Time Limited Blackbox Security: Protecting Mobile Agents from Malicious Hosts." *Proceedings: Mobile Agents and Security*, pp. 90–111.

Indian Standard (IS): 1893, Part 1. 2002. *Criteria for Earthquake Resistant Design of Structures*. New Delhi: Bureau of Indian Standards.

IFRC, Disaster relief emergency fund (DREF) India: Cyclone Phailin DREF operation No. MDRIN013 14 October 2013. International Federation of Red Cross (IFRC) and Red Crescent Societies.

International Federation of Red Cross and Red Crescent Societies (2002). The World Disasters Report 2002, edited by Jonathan Walter.

Inter-Agency Standing Committee (IASC). 2004. Civil Military Relationship in Complex Emergencies – An IASC Reference Paper, p. 17.

IPCC (Intergovernmental Panel on Climate Change). 2007. *Climate Change 2007 – Impacts, Adaptation and Vulnerability*. A contribution from Working Group II of the Intergovernmental Panel on Climate Change. Cambridge and New York: Cambridge University Press.

———— 2012. *Managing the Risks of Extreme Events and Disasters to Advance Climate Change Adaptation*. A Special Report of Working Groups I and II of the Intergovernmental Panel on Climate Change. Cambridge and New York: Cambridge University Press.

Inter-Agency Standing Committee (IASC). 2006. Guidance Note on Using the Cluster approach to Strengthen Humanitarian Response, p. 15.

IPCC. 2014. *Climate Change 2014: Impacts, Adaptation, and Vulnerability. Summaries, Frequently Asked Questions, and Cross-Chapter Boxes. A Contribution of Working Group II to the Fifth Assessment Report of the Intergovernmental Panel on Climate Change*, edited by C.B. Field, V.R. Barros, D.J. Dokken, K.J. Mach, M.D. Mastrandrea, T.E. Bilir, M. Chatterjee, K.L. Ebi, Y.O. Estrada, R.C. Genova, B. Girma, E.S. Kissel, A.N. Levy, S. MacCracken, P.R. Mastrandrea, and L.L. White, 190. Geneva, Switzerland: World Meteorological Organization.

IRIN (Integrated Regional Information Networks). 2005. Disaster reduction and the human cost of disaster. IRIN Web special, p. 41.

Jahre, M., and L. Jensen. 2010. "Coordination in Humanitarian Logistics through Clusters" *International Journal of Physical Distribution & Logistics Management*, 40 (8/9), 657–674.

Jain S., M. Sharma, and R. Thakur. 1996. "Seasonal Variations in Physico-chemical Parameters of Halai Reservoir of Vidisha District". *Indian Journal of Ecobiology*, 8 (3), 181–88.

Joshi, S.R. 2008. *Natural Disasters in North-East Region and its Management: An Essay.* Shillong: Centre for Science Education, North Eastern Hill University.

Kapucu, N. 2008. "Collaborative Emergency Management: Better Community Organising, Better Public Preparedness and Response." *Disasters,* 32 (2), 239–262.

Kilby, P. 2008. "The Strength of Networks: The Local NGO Response to the Tsunami in India." *Disasters,* 32 (1),120–130.

Kates, R.W., C. Hohenemser, and J. Kasperson. 1985. *Perilous Progress: Managing the Hazards of Technology* (p. 21). Boulder, CO: Westview Press.

Koirala, P.K. 2014. Institution for the management of natural disaster at national level, a sharing between Nepal and Japan, in reference to south Asian countries. Research Report, ADRC Visiting Researcher Program, November 2014.

Korte, Barbara. 1997. *Body Language in Literature.* Toronto: University of Toronto Press, ix, 329.

Kumar, A.V. 2013. "Cyclone Phailin Hits Odisha at 200 kmph, Over 7 lakh Evacuated." Z. *News.* http://zeenews.india.com/news/nation/live-cyclone-phailin-80-kms-from-gopalpur-three-dead-ahead-oflandfall_882698.html

Kusumasari, B., and Q. Alam. 2012. "Bridging the Gaps: The Role of Local Government Capability and the Management of a Natural Disaster in Bantul, Indonesia." Natural Hazards, 60 (2), 761–779.

Lindell, M.K. and C.S. Prater. 2003. "Assessing Community Impacts of Natural Disasters." *Natural Hazards Review,* 4 (4), 176–86.

Lu, Y., and J. Xu. 2015. "NGO Collaboration in Community Post-Disaster Reconstruction: Field Research Following the 2008 Wenchuan Earthquake in China." Disasters, 39 (2): 258-78.

Lyall, C., and J. Tait. 2004. "Shifting Policy Debates and the Implications for Governance." In New Modes of Governance: Developing an Integrated Approach to Science, Technology, Risk and the Environment , eds. C. Lyall and J. Tait (pp. 1-17). UK: Aldershot.

Madry, S. 2015. "Space Systems for disaster warning, Response and Recovery." Springer Briefs in Space Development, 131.

Marble, D.F. and D.J. Pequet. 1983. "Geographic Information Systems and Remote Sensing." In *Manual of Remote Sensing*, 2nd edition, 923–57, edited by R.N. Colwell. Falls Church, VA: American Society of Photogrammetry.

Marincioni, F. 2007. "Information Technologies and the Sharing of Disaster Knowledge: The Critical Role of Professional Culture." *Disasters,* 31 (4), 459–476.

McHarg, Ian L. 1969. Design with Nature. New York: American Museum of Natural History, 197.

Mitchell, J.K. 1989. "Hazards Research." In *Geography in America*, edited by G.L. Gaile and C.J. Willmott, 410–24. Columbus, OH: Merrill.

Mochalski, P., K. Unterkofler, G. Teschl, and A. Amann. 2015. "Potential of Volatile Organic Compounds as Markers of Entrapped Humans for Use in Urban Search and Rescue Operations." *Trends in Analytical Chemistry,*68 (2015),88–106.

Montague, P. 2004. "Reducing the Harms Associated with Risk Assessments." *Environmental Impact Assessment Review*, 24, 733–48.

OAS/DRDE. 1990. *Disaster, Planning and Development: Managing Natural Hazards to Reduce Loss.* Washington, DC: Department of Regional Development and Environment. Organization of American States, 80.

Mukherji, B., and V. Agarwal. 2013. "Early Warnings, Evacuations Saved Lives in Cyclone." *Wall Street Journal,* October 13.

National Disaster Management Guidelines—Incident Response System. 2010. National Disaster Management Authority, Government of India, July 2010.

National Disaster Response Framework (NDRF). 2013. Government of Nepal Ministry of Home Affairs, March, Nepal.

Nye, J.D., and J.D. Donahue (eds). 2000. *Governance in a Globalizing World* (p. 386). Washington, D.C.: Brookings Institution Press.

OASIS. 2010. *Advancing Open Standards for the Global Informational Society.* www.oasis-open. org.

O'Brien M. 2000. *Making Better Environmental Decisions: An Alternative to Risk Assessment.* Cambridge, MA: MIT Press.

OCHA. 2015. *World Humanitarian Data and Trends.* Geneva: United Nations, 96.

Ochoa, S.F., and R. Santos. 2015. "Human-centric Wireless Sensor Networks to Improve Information Availability During Urban Search and Rescue Activities." *Information Fusion,* 22, 71–84.

Okamura, Mitsu, Netra P Bhandary, Shinichiro Mori, Narayan Marasini, and Hemanta Hazarika. 2015. "Report on a Reconnaissance Survey of Damage in Kathmandu Caused by the 2015 Gorkha Nepal Earthquake." *Soils and Foundations,* 55 (5), 1015–1029.

Okuyama, Y., and S.E. Chang. 2004a. "Introduction." In *Modeling Spatial and Economic Impacts of Disasters,* edited by Y. Okuyama and S.E. Chang. New York, NY: Springer, 1–10.

Okuyama, Y., and S.E. Chang, eds. 2004b. *Modeling Spatial and Economic Impacts of Disasters.* New York, NY: Springer.

Özdamar, L., E. Ekinci, and B. Küçükyazici. 2004. "Emergency Logistics Planning in Natural Disasters." *Annals of Operations Research,* 129 (1), 217–245.

Pal, I. 2013, September. "Cloud Burst and Flash Floods at Uttarkashi, Uttarakhand: A Case Study of 2012." *Disaster—Response and Management,* 1 (1), 25–39. ISSN: 2347-2553.

———. 2015. "Land Use and Land Cover Change Analysis in Uttarakhand Himalaya and Its Impact on Environmental Risks." In *Mountain Hazards and Disaster Risk Reduction,* edited by Rajib Shaw and Hari Krishna Nibanupudi. Japan: Springer, 125–37. ISBN: 978-4-431-55241-3.

Pal, I., and T. Ghosh. 2014. "Fire Incident at AMRI Hospital, Kolkata (India): A Real Time Assessment for Urban Fire." *Journal of Business Management & Social Sciences Research,* 3 (1), 9–13.

Pal, I., T. Ghosh, and C. Ghosh. 2017. "Institutional Framework and Administrative Systems for Effective Disaster Risk Governance: Perspectives of 2013 Cyclone Phailin in India." *International Journal of Disaster Risk Reduction*, 21, 350–59.

Pal, I., T. Ghosh, A. Mukhopadhyay, and S. Ghosh. 2014. "A Case Study of Cloud Burst and Flash flood at Uttarkashi (INDIA): An Assessment Using Geoinformatics." In *Multiple Geographical Perspectives on Hazards and Disasters*, edited by Lina Maria Calandra, Giuseppe Forino and Andrea Porru. Italy: Valmar Publications.

Pal, I., S.K. Nath, K. Shukla, D.K. Pal, A. Raj, K.K. Thingbaijam, and B.K. Bansal. 2008. "Seismic Hazard Zonation of Sikkim Himalaya on GIS Platform." *Natural Hazards*, 45, 333–77.

Pal, I., S. Singh, and A. Walia. 2013, October. "Flood Management in Assam, INDIA: A Review of Brahmaputra Floods, 2012." *International Journal of Scientific and Research Publications*, 3 (10), 1–5. ISSN: 2250-3153.

Pelling, Mark., ed. 2003. *Natural Disasters in a Globalising World*. London: Routledge.

Pettit, S.J., and A.K.C. Beresford. 2005. "Emergency Relief Logistics: An Evaluation of Military, Non-military and Composite Response Models." HYPERLINK Pettit, S. J., and Beresford, A. K. C. 2005. Emergency relief logistics: an evaluation of military, non-military and composite response models. International Journal of Logistics Research and Applications, 8 (4), 313-331. http://www.tandfonline.com/toc/cjol20/8/4

Practical Action. 2010a. *Poor People's Energy Outlook 2010*. Rugby, UK.Annual Report 2009/10. Practical Action Regional Program (Sri Lanka, India & Pakistan). Colombo, Sri Lanka.

Practical Action. 2010b. *Understanding Disaster Management in Practice with reference to Nepal,* Practical Action, Kathmandu, Nepal.

Pusch, C. 2004. "Preventable Losses, Saving Lives and Property through Hazard Risk Assessment: A Comprehensive Risk Management Framework for Central Europe and Central Asia, Disaster Risk Management." Working Paper Series No. 9. Washington, DC: The World Bank, p. 88.

Quarantelli, E.L. 1993. "Organizational Response to the Mexico City Earthquake of 1985: Characteristics and Implications." *Natural Hazards*, 8 (1), 19–38.

Rao M.R., M.P. Singh, and R. Day. 2000. "Insect Pest Problems in Tropical Agroforestry Systems: Contributory Factors and Strategies for Management." *Agroforestry System*, 50 (3), 243–77.

Raschky, P., and S. Chantarat. 2013. "Natural Disaster Risk Financing and Transfer in ASEAN Countries". ERIA Discussion Paper 2013.

Rey, F. 2001. "The Complex Nature of Actors in Humanitarian Action and the Challenge of Coordination." In *Reflections on Humanitarian Action: Principles, Ethics and Contradictions,* ed. Humanitarian Studies Unit. London: TNI/Pluto Press with Humanitarian Studies Unit and ECHO (European Commission Humanitarian Office).

Robinson, Peter. 1988. "Root-N-Consistent Semiparametric Regression." *Econometrica*, 56 (4), 931–54.

Rosenau, J.N. 1992. "Normative Challenges in a Turbulent World." *Ethics & International Affairs*, 6 (1), 1–19.

Roy P.S., 2000. Assessment of Forest Fire in India through Remote Sensing. Century status report in project planning Workshop on the "Scientific dimension s of forest fires" organised by International Council for Science Committee on Science and Technology in Developing Countries. Chennai, 27–29 March 2000.

Saaty, T.L. 1980. *The Analytic Hierarchy Process*. New York, NY: McGraw-Hill.

————. 1990. "How to Make a Decision: The Analytic Hierarchy Process." *European Journal of Operational Research*, 48, 9–26.

Sawada, Y. and S. Oum. 2012. "Economic and Welfare Impacts of Disasters in East Asia and Policy Responses."In Y. Sawada and S. Oum (eds), *Economic and Welfare Impacts of Disasters in East Asia and Policy Responses* (pp. 1–25). ERIA Research Project Report 2011-8. Jakarta: ERIA.

Sawada, Y., and Zen, F. 2014. Disaster Management in ASEAN, edited by ERIA. ERIA Discussion Paper Series No: ERIA-DP-2014-03, p. 49.

Schooley, B.L., and T.A. Horan. 2007. "Towards End-to-End Government Performance Mnagement: Cse study of interorganizational information integration in emergency medical services (EMS). Gov Inf Q 24 (4), 755–784.

Senapati, A. 2013. Mass evacuation on in coastal districts, Earth, October 11, 2013.

Singh, J. 2013. How Phailin Was Different from Super Cyclone 1999." *Earth*, 15.

Smith, K. and D.N. Petley. 2008. *Environmental Hazards. Assessing Risk and Reducing Disaster.* London: Taylor & Francis.

———— 2009. *Environmental Hazards: Assessing Risk and Reducing Disaster.* 5th edition. New York: Routledge, 416.

The Sphere Project: Humanitarian Charter and Minimum Standards in Humanitarian Response. 2011. Practical Action Publishing, UK.

Stoddard, A., A. Harmer, K. Haver, D. Salomons, and V. Wheeler. 2007. *Cluster Approach Evaluation* (p. 111). Humanitarian Policy Group.

Tatham, P., R. Oloruntoba, and K. Spens. 2012. "Cyclone Preparedness and Response: An Analysis of Lessons Identified Using an Adapted Military Planning Framework." *Disasters,* 36 (1), 54–82.

Subramaniam, C., H. Ali, and F.M. Shamsudin. 2010. "Understanding the Antecedents of Emergency Response: A Proposed Framework." *Disaster Prevention and Management,* 19 (5), 571–581.

Suzuki, I., and Y. Kaneko. 2013. *Japan's Disaster Governance: How Was the 311 Crisis Managed?* (p. 116). Berlin: Springer.

Turnbull, S., and M. Pirson. 2011. "Corporate Governance, Risk Management, and the Financial Crisis: An Information Processing View." *Corporate Governance: An International Review,* 19 (5), 459-470.

Turner II, B.L., W.C. Clark, R.W. Kates, J.F. Richards, J.T. Mathews, and W.B. Meyer (eds). 1990. *The Earth as Transformed by Human Action: Global and Regional Changes in the Biosphere over the Past 300 Years* (p. 713). Cambridge: Cambridge University Press.

Swiss Re. 2016. Global insurance review 2016 and outlook 2017/18, p. 47.

Twigg, J. 2004. *Disaster Risk Reduction Mitigation and Preparedness in Development and Emergency Programming.* London: Overseas Development Institute.

UNDP (United Nations Development Programme). 2010. *A Guide to UNDP Democratic Governance Practice.* New York: Bureau for Development Policy, Democratic Governance Group.

UN-ISDR. 2004. *Terminology of Disaster Risk Reduction.* Geneva: United Nations International Strategy for Disaster Reduction. Available at: http://www.unisdr.org/eng/library/lib-ter-minology-eng%20home.htm (last accessed 27 October 2017).

————. 2009. UNISDR Terminology on Disaster Risk Reduction. Geneva, Switzerland.

————. 2011. Global Assessment Report on Disaster Risk Reduction: Revealing Risk, Redefining Development. Geneva, Switzerland: UNISDR.

UNISDR Terminology on Disaster Risk Reduction, United Nations International Strategy for Disaster Reduction (UNISDR), Geneva, Switzerland, 2009.

The United Nations. 2003. *Report on the World Social Situation. Social Vulnerability: Sources and Challenges.* Sales No. E. 03.IV.10, United Nations publication, New York. p. 95.

United Nations Office for Outer Space Affairs (UNOOSA). *Annual Report 2015* (p. 52). United Nations Office at Vienna.

Valcik, N.A., and P.E. Tracy. 2013. Case Studies in Disaster Response and Emergency Management (p. 220). ASPA Series in Public Administration and Public Policy. CRC Press.

Van Wassenhove, L.N. 2006. "Blackett Memorial Lecture. Humanitarian Aid Logistics: Supply Chain Management in High Gear." *Journal of the Operational Research Society*, 57 (5), 475–89.

Van Westen, C.J., ed. 2009. *Distance Education Course on the Use of Spatial Information in Multihazard Risk Assessment.* Available at: http://www.itc.nl/Pub/study/Courses/C11-AES-DE-0 (last accessed 15 August 2016).

Van Westen, C.J., E. Castellanos, and S.L. Kuriakose. 2008. "Spatial Data for Landslide Susceptibility, Hazard, and Vulnerability Assessment: An Overview." *Engineering Geology*, 102, 112–31.

Warfield, C. 2008. *The Disaster Management Cycle.* Available at: http://www.gdrc.org/uem/disasters/1-dm_cycle.html (last accessed 4 September 2016).

Waugh (Jr), W.L., and G. Streib. 2006. "Collaboration and Leadership for Effective Emergency Management." *Public Administration Review*, Special Issue, 131-140.

Wilder, A., and T. Morris. 2008. "Locals within Locals': Cultural Sensitivity in Disaster Aid." *Anthropology Today*, 24 (3), 1–3.

Wisner, B. 2004. At Risk: Natural Hazards, People's Vulnerability, and Disasters, 2nd ed, p. 471. London/ New York: Routledge.

Wolf, J., and W.G. Egelhoff. 2002. "A Reexamination and Extension of International Strategy–Structure Theory." *Strategic Management Journal*, 23 (2), 181–189.

World Bank. 2012. *Analysis of Disaster Risk Management in Colombia: A Contribution to the Creation of Public Policies.* Coordinators and Editors: Ana Campos G., Niels Holm-Nielsen, Carolina Díaz G., Diana M. Rubiano V., Carlos R. Costa P., Fernando Ramírez C. and Eric Dickson. Washington, D.C.: The World Bank and GFDRR.

World Bank. 2013. "Cyclone Devastation Averted: India Weathers Phailin." *The World Bank News*, October 17.

Web Resources

http://www.unisdr.org/
https://www.unisdr.org/we/inform/publications/31468
https://www.unisdr.org/we/coordinate/hfa
http://www.irdrinternational.org/
http://www.unesco.org/new/en/social-and-human-sciences/resources/reports/world-social-science-report-2010/
www.ndma.gov.in

http://pdf.usaid.gov/pdf_docs/Pnabj801.pdf

ftp://ftp.dmcii.com/pub/documents/International_Charter/UK%20NERC%20International %20Charter%20Presentation.pdf

https://unitar.org/unosat/

http://nidm.gov.in/PDF/pubs/HPC_Report.pdf

https://landslides.usgs.gov/

https://earthquake.usgs.gov/

http://www.unisdr.org/eng/library/lib-terminology-eng%20home.htm

https://www.ipcc.ch/report/ar5/syr/

http://www.undp.org/content/undp/en/home/librarypage/crisis-prevention-and-recovery/ reducing-disaster-risk–a-challenge-for-development.html

http://www.preventionweb.net/files/14452_genderincbdrm1.pdf

www.undp.org/cpr/we_do/integrating_risk.shtml

www.preventionweb.net

www.unisdr.org/

www.recoveryplatform.org

http://www.dpnet.org.np/

https://sites.google.com/site/drrtoolsinsoutheastasia/disaster-risk-reduction/school-based-risk-reduction/asean-safe-school-initiative-assi

http://go.worldbank.org/BCQUXRXOWO

http://earthtrends.wri.org/pdf_library/data_tables/cli1_2005.pdf

http://www.cred.be/sites/default/files/ADSR_2014.pdf

http://www.asean.org/

agreement.asean.org/media/download/20140119170000.pdf

www.apec-epwg.org

https://www.amcdrrindia.net/

www.ifrc.org/Global/Publications/disasters/WDR/32600-WDR2002.pdf

https://www.cfe-dmha.org/

www.odi.org.uk

www.erra.pk

Index

About the Authors

Indrajit Pal is Assistant Professor in Disaster Preparedness, Mitigation and Management at the Asian Institute of Technology, Thailand. Formerly, he was a Faculty Member at the Centre for Disaster Management in Lal Bahadur Shastri National Academy of Administration, Mussoorie. Dr Pal holds a PhD in seismotectonic and earthquake hazard assessment, and has several years of experience on teaching, research and consulting, focusing on disaster risk governance, risk management, hazard assessment, climate change adaptation, etc. He has published several books and has written articles for peer reviewed international journals. Dr Pal has been recognized as an IRDR Young Scientist by the Integrated Research on Disaster Risk, Beijing.

Tuhin Ghosh is a Faculty Member in the School of Oceanographic Studies, Jadavpur University, India. His research interests are coastal geomorphology, disaster management, climate change impacts, adaptation strategies, and human migration. He has several publications and books to his credit. He has collaborated in a project on the sustainability of deltaic systems funded by the Belmont Forum. Dr Ghosh is the Indian lead in an international project called DECCMA, a research on the impacts of climate change on deltas in Africa and Asia, and is working on the Ganges delta and Mahanadi delta in India.